HANDBOOK OF ADHESIVE RAW MATERIALS

HANDBOOK OF
ADHESIVE RAW MATERIALS

by

Ernest W. Flick

np | **NOYES PUBLICATIONS**
Park Ridge, New Jersey, U.S.A.

Published in the United States of America by
Noyes Publications
Mill Road, Park Ridge, New Jersey 07656

10 9 8 7 6 5 4 3 2 1

Library of Congress Cataloging in Publication Data

Flick, Ernest W.
    Handbook of adhesive raw materials.

    Includes index.
    1. Adhesives. I. Title.
TP968.F55   668.4'11   82-2251
ISBN 0-8155-0897-2   AACR2

To
Mario Zipilivan

Preface

This handbook contains descriptions of hundreds of raw materials which are available to the adhesives industry. It will be of value to technical and managerial personnel in adhesives manufacturing companies and companies which supply raw materials or services to these companies. This book will be useful to both those with extensive experience as well as those who are novices in the field.

The data consist of selections of manufacturers' raw materials made at no cost to, nor influence from, the makers or distributors of these materials. The coverage of raw materials is as complete as it could possibly be and any omissions are not intentional.

Only the most recent data have been compiled. Basically, only trademarked raw materials are included. Common chemicals are mostly excluded. Most solvent-based raw materials are omitted, with some exceptions which were considered of value to the book.

A detailed table of contents lists, in alphabetical order, the companies included in each section. The company names are listed in alphabetical order, and the raw materials from each company are listed in alphabetical and numerical order.

Acknowledgements

My fullest appreciation is expressed to the following companies and organizations who supplied the original raw material data included in this book:

Abbott Laboratories
Allied Chemical
Allied Colloids,Inc.
American Cyanamid Co.
The Ames Laboratories, Inc.
Amoco Chemicals Corp.
Argus Chemical
Arizona Chemical Co.
Ashland Chemicals
AZS Chemical Co.
Bareco Division
BASF Corp.
Burgess Pigment Co.
Cabot Corp.
Calcium Carbonate Co.
Carey Canada Inc.
Cargill, Inc.
Chemical Components, Inc.
Ciba-Geigy Resins
Cincinnati Milacron Chemicals
W.R. Cleary Chemical Corp.
Clinton Corn Processing Co.
Colloids, Inc.
Commercial Minerals Co.
Composition Materials Co., Inc.
Corn Products
Cosan Chemical Corp.
Cosden Oil & Chemical Co.
Crosby Chemicals, Inc.
Crowley Chemical Co.
Cyprus Industrial Minerals Co.
Davison Chemical Division
Degussa Corp.
Diamond Shamrock Corp.
Dover Chemical Corp.
Dow Chemical U.S.A.
DuPont Co.
Durez Division
Dynamit Nobel Chemicals
East Coast Chemicals Co.
Emery Industries, Inc.
Emkay Chemical Co.
The English Mica Co.
Essential Chemicals Corp.
Ethyl Corp.
Exxon Chemicals

Ferro Chemical Division
Firestone Plastics Co.
Firestone Synthetic Rubber & Latex Co.
Flintkote Stone Products Co.
Franklin Mineral Products Co.
FRP Co.
GAF Corp.
Georgia Kaolin Co.
Georgia Marble Co.
Goodyear Chemicals
The Goodyear Tire and Rubber Co.
W.R. Grace & Co.
Great Lakes Chemical Corp.
Grefco, Inc.
Gross Minerals Corp.
Harris Mining Co.
Henkel Corp.
Hercules, Inc.
Humphrey Chemical Corp.
ICI Americas, Inc.
Illinois Minerals Co.
O.G. Innes Corp.
International Dioxcide, Inc.
International Wax Refining Co., Inc.
Interstab Chemicals, Inc.
Isochem
ITT Rayonier, Inc.
Johns-Manville
Lawter Chemicals, Inc.
MacKenzie Chemical Works, Inc.
Martin Marietta Chemicals
Mayco Oil and Chemical Co., Inc.
M. Michel and Co., Inc.
Micro Powders, Inc.
Minerals and Chemicals Division
Mississippi Lime Co.
Mobay Chemical Products
Mobil Chemical Co.
Mona Industries
Mooney Chemicals, Inc.
Moore-Munger Marketing, Inc.
National Wax Co.
Natro Chem, Inc.
Neville Chemical Co.
Nyacol, Inc.
Ottawa Chemical Division

Pacific Anchor Chemical Co.
Pacific Smelting Co.
Penick & Ford, Ltd.
Pfizer Minerals, Pigments & Metals Division
Pioneer Division
Polymer Research Corp. of America
PPG Industries, Inc.
Procter and Gamble Distributing Co.
PVO International, Inc.
Raybo Chemical Co.
Reichhold Chemicals, Inc.
Rohm and Haas Co.
Frank B. Ross Co., Inc.
Schenectady Chemicals, Inc.
Sherwin-Williams Co.
Synthron, Inc.
Tammsco, Inc.
Chas. S. Tanner Co.

Texaco, Inc.
Thiele Kaolin Co.
Thiokol Corp.
Thiokol/Specialty Chemicals Division
Thiokol/Ventron Division
Thompson-Hayward Chemical Co.
Tri-Star Chemical Co.
Union Camp Corp.
Union Chemicals Division
Upjohn Polymer Chemicals
U.S. Industrial Chemicals Co.
U.S. Mica Co., Inc.
Uniroyal Chemical
R.T. Vanderbilt Co., Inc.
Velsicol Chemical Corp.
Virginia Chemicals, Inc.
S. Winterbourne & Co., Inc.
Witco Chemical

I also thank Mrs. Jackie Gianareles who did a superhuman job, through a low and high point of her life, of typing the excellent final manuscript draft. This is her eighth manuscript with me and she always does them well, in the good times and the bad. My thanks also go to her husband, Peter, and their children, Antigone and George.

Ernest W. Flick
February 1982

Contents and Subject Index

SECTION II: ALUMINUM SILICATES

SECTION III: ANTISKINNING AGENTS

SECTION IV: CALCIUM CARBONATES

SECTION V: CATALYSTS, CROSS-LINKING AND CURING AGENTS

SECTION VI: CLAYS

SECTION VII: CORN STARCH AND STARCH DERIVATIVES

SECTION VIII: DEFOAMERS AND ANTIFOAMS

SECTION IX: DISPERSING AND EMULSIFYING AGENTS

SECTION X: EPOXY RESIN DILUENTS

SECTION XI: FILLERS AND EXTENDER PIGMENTS

SECTION XII: FIRE AND FLAME RETARDANTS

SECTION XIII: LATICES

SECTION XIV: OILS

SECTION XV: PLASTICIZERS

SECTION XVI: POLYBUTENES

SECTION XVII: POLYVINYL ACETATES

SECTION XVIII: PRESERVATIVES AND FUNGICIDES

SECTION XIX: RESINS—ACRYLIC

SECTION XX: RESINS—EPOXY

SECTION XXI: RESINS–ESTER GUMS

SECTION XXII: RESINS–ETHYLENE/VINYL ACETATE

SECTION XXIII: RESINS–MALEIC

SECTION XXIV: RESINS–MISCELLANEOUS

SECTION XXV: RESINS–NATURAL

SECTION XXVI: RESINS–PHENOLIC

SECTION XXVII: RESINS–POLYAMIDE

SECTION XXXIII: RESINS–RADIATION COATING

SECTION XXXIV: RESINS–URETHANE

SECTION XXXV: RESINS–VINYL CHLORIDE

SECTION XXXVI: RESIN EMULSIONS AND DISPERSIONS

SECTION XXXVII: RESIN ESTERS

SECTION XXXVIII: ROSINS

SECTION XXXIX: SILICAS (SILICON DIOXIDE)

SECTION XL: STABILIZERS

SECTION XLI: SURFACTANTS/SURFACE ACTIVE AGENTS

SECTION XLII: TALCS

SECTION XLIII: THICKENERS AND THIXOTROPIC AGENTS

SECTION XLIV: WAXES

SECTION XLV: MULTIFUNCTIONAL AND MISCELLANEOUS COMPOUNDS

Introduction

This handbook is a compilation of the descriptions of hundreds of raw materials currently available to the chemical industry. It is the end result of information received from numerous industrial companies and other organizations. The data represent selections from manufacturers' descriptions made at no cost to, nor influence from, the makers or distributors of these materials. Only currently available raw materials are included. This handbook is the most complete and current of any of its type on the market today. Over 500 different companies and organizations were contacted. The coverage of raw materials is as complete as it could possibly be and any omissions are not intentional. Either the information was not received, or not requested, or did not apply.

Only the most recent data have been compiled. Basically, only trademarked raw materials are included. Common chemicals are mostly excluded. Most solvent-based raw materials are omitted, with some exceptions. Editorial judgement was used in selecting these raw materials. The raw materials in this book are divided into the following 45 sections:

I	Acids
II	Aluminum Silicates
III	Antiskinning Agents
IV	Calcium Carbonates
V	Catalysts, Crosslinking and Curing Agents
VI	Clays
VII	Cornstarch and Starch Derivatives
VIII	Defoamers and Antifoams
IX	Dispersing and Emulsifying Agents
X	Epoxy Resin Diluents
XI	Fillers and Extender Pigments
XII	Fire and Flame Retardants
XIII	Latices
XIV	Oils
XV	Plasticizers
XVI	Polybutenes
XVII	Polyvinyl Acetates
XVIII	Preservatives and Fungicides
XIX	Resins—Acrylic
XX	Resins—Epoxy
XXI	Resins—Ester Gums
XXII	Resins—Ethylene/Vinyl Acetate
XXIII	Resins—Maleic
XXIV	Resins—Miscellaneous
XXV	Resins—Natural
XXVI	Resins—Phenolic
XXVII	Resins—Polyamide
XXVIV	Resins—Polyethylene
XXIX	Resins—Polypropylene
XXX	Resins—Polyterpene
XXXI	Resins—Powder Coating
XXXII	Resins—Proprietary Composition
XXXIII	Resins—Radiation Coating
XXXIV	Resins—Urethane

XXXV	Resins—Vinyl Chloride
XXXVI	Resin Emulsions and Dispersions
XXXVII	Resin Esters
XXXVIII	Rosins
XXXIX	Silicas (Silicon Dioxide)
XL	Stabilizers
XLI	Surfactants/Surface Active Agents
XLII	Talcs
XLIII	Thickeners and Thixotropic Agents
XLIV	Waxes
XLV	Multifunctional and Miscellaneous Compounds

This many sections were selected, in detail, to enable the reader to easily find any raw material desired. A detailed Table of Contents lists, in alphabetical order, the companies included in each section. The company names are listed in alphabetical order and the raw materials in each section are also listed in chronological order. In addition to the above, there is one other section which will be helpful to the reader:

XLVI   Main office addresses (and telephone numbers), which are completely current, of the suppliers of information included, some of which are not available in the usual reference books.

Each raw material in this book lists as much data, as readily available, from the information given by each supplier. The descriptions of the raw materials are basically listed in the following order:

(1)   Chemical Name and Grade: Trade name or designation given by the supplier for the raw material. The grades available, if more than one.

(2)   Chemical Type: All the descriptive chemical information available from the literature supplied. Chemical name, formula and any other information is given. If the composition is proprietary, that is also stated.

(3)   Chemical Analysis: The complete chemical analysis of the raw material is given, if listed by the supplier.

(4)   Physical Properties: The complete physical properties of the raw material are listed. This information is sometimes combined with the Chemical Name and Grade, for the reader's convenience, at the beginning of the raw material description.

(5)   Key Properties: Herein is listed all the properties which the supplier considers to be of special importance to the raw material. Such things as unusual chemical characteristics, recommended end uses and important properties given the products containing it are given. This section is given in as great a detail, as possible, for the readers' value.

(6)   Specifications and Regulations: Any specifications and regulations which the raw material meets are given in this section. Anything listed by the supplier is given. If a regulation is not listed, the raw material may still meet it and the supplier should be contacted for further information.

(7)   Toxicity: Any toxicity information given by the supplier is listed. This may also be determined somewhat by the regulations met. The raw material is not necessarily toxic or nontoxic, if not listed, but was not given in the information supplied us. Special attention is given to any raw materials which the supplier considers to have toxic properties and any information furnished is listed.

Section I

Acids

EMERY INDUSTRIES, INC., 1300 Carew Tower, Cincinnati, OH 45202

Azelaic Acids

Product Name	Acid Value	Color (% trans) (440/550 nm) (min) (25% in methanol)	Melting Point (°C)	Azelaic Acid	Dibasic Acids Less Than C_9	Dibasic Acids Greater Than C_9	Mono-basic Acids
Emerox 1110	588	77	100	79.0	11.0	9.0	1.0
Emerox 1133	594	93	99	82.0	9.5	8.0	0.5
Emerox 1144 (Polymer Grade)	594	97	102	90.0	1.0	9.0	—

Chemical Type: Azelaic acid, a linear C_9 odd-carbon saturated acid. There is a polymer grade acid as well as acids for general applications.

Key Properties: Recommended for use in lacquers, alkyd resins, polyamides, polyester adhesives, low temperature plasticizers, urethane elastomers and other items.

Coconut Fatty Acids

Product Name	Titer (°C)	Iodine Value	Color (% trans) 440/550 nm (min)	Color (Gardner) (max)	Acid Value	Sap. Value
Emery 621 Coconut Fatty Acid	25	12	55	5	263	263
Emery 622 Coconut Fatty Acid	24	8	81	2	272	272
Emery 626 Low IV Coconut Fatty Acid	25	1 max	92	1	273	273
Emery 627 Low IV Stripped Coconut Fatty Acid	30	1 max	89	1	255	255
Emery 629 Stripped Coconut Fatty Acid	29	8	93	1	257	257

	Emery 621	Emery 622	Emery 626	Emery 627	Emery 629
Typical composition*					
Caprylic ($C_8H_{16}O_2$)	4	7	7	—	—
Capric ($C_{10}H_{20}O_2$)	5	6	6	1	1
Lauric ($C_{12}H_{24}O_2$)	48	48	51	55	55
Myristic ($C_{14}H_{28}O_2$)	20	19	18	22	23
Palmitic ($C_{16}H_{32}O_2$)	10	9	10	11	12
Stearic ($C_{18}H_{36}O_2$)	2	2	7	10	3
Oleic ($C_{18}H_{34}O_2$)	10	8	1	1	5
Linoleic ($C_{18}H_{32}O_2$)	1	1	—	—	1

*By GLC analysis (ASTM D1983-64T)

Food Grade Fatty Acids

Product Name	Titer (°C)	Iodine Value	Color (% trans) 440/550 nm (min)	Acid Value	Sap. Value	Unsap. (% max)
Emersol 6320 DP Stearic Acid	54.3	4.3	94	208	209	0.5
Emersol 6332 USP TP Stearic Acid	55.2	0.5	96	208	210	0.5
Emersol 6349 Stearic Acid	59.8	0.5	94	205	206	0.5
Emersol 6353 USP Stearic Acid	68	1.0	89	197	199	0.5
Emersol 6313 Low-Titer Oleic Acid	6 max	91	87	203	209	0.5
Emersol 6321 Low-Titer White Oleic	6 max	90	92	203	204	0.5
Emersol 6333 USP LL Oleic Acid	9	89	92	202	204	0.5

	Emersol 6320	Emersol 6332	Emersol 6349	Emersol 6353	Emersol 6313	Emersol 6321	Emersol 6333
Typical Composition							
Myristic ($C_{14}H_{28}O_2$)	2.5	1.5	3	—	3	3	3
Pentadecanoic ($C_{15}H_{30}O_2$)	0.5	0.5	0.5	—	Trace	Trace	Trace
Palmitic ($C_{16}H_{32}O_2$)	50	50	26.5	5	5	5	6.5
Margaric ($C_{17}H_{34}O_2$)	1	1	1	—	1	1	1
Stearic ($C_{18}H_{36}O_2$)	40	47	69	95	Trace	Trace	1.5
Myristoleic ($C_{14}H_{26}O_2$)	—	—	—	—	3	3	3
Palmitoleic ($C_{16}H_{30}O_2$)	—	—	—	—	6	6	5.5
Oleic ($C_{18}H_{34}O_2$)	6	—	Trace	—	75	75	73.5
Linoleic ($C_{18}H_{32}O_2$)	—	—	—	—	6	6	5.5
Linolenic ($C_{18}H_{30}O_2$)	—	—	—	—	1	1	0.5

Isostearic Acids

Product Name	Titer (°C)	Iodine Value	Color (% trans) 440/550 nm (min)	Acid Value	Sap. Value	Unsap. Value (% max)
Emersol 871 Isostearic Acid	10	12 max	30/85	175 min	180 min	6.0
Emersol 875 Isostearic Acid	10	3 max	85/98	196	201	3.0
Emery 894 Fatty Acid	46	34	80	195 min	203 min	2.0
Emery 896 Fatty Acid	50	6 max	88	201 min	201 min	2.5

Oleic and Linoleic Acids

Product Name	Titer (°C)	Iodine Value	Color (% trans) 440/550 nm (min)	Acid Value	Sap. Value	Unsap. Value (% max)
Emersol 210 Oleic Acid	10	91	2/30	202	204	1.5
Emersol 213 Low-Titer Oleic Acid (USP)	5 max	92	30/85	202	204	1.5
Emersol 220 White Oleic Acid	10	91	71/99	202	204	1
Emersol 221 Low-Titer White Oleic Acid (USP)	5 max	92	71/99	202	204	1
Emersol 233 LL Oleic Acid	4	88	78/99	202	204	0.5
Emersol 315 Linoleic Acid	5 max	153	60/98	198	200	1

Typical Composition

	Emersol 210	Emersol 213	Emersol 220	Emersol 221	Emersol 233	Emersol 315
Saturated Acids						
Lauric ($C_{12}H_{24}O_2$)	Trace	Trace	Trace	Trace	Trace	—
Myristic ($C_{14}H_{28}O_2$)	3	3	3	3	3	0.5
Pentadecanoic ($C_{15}H_{30}O_2$)	Trace	Trace	Trace	Trace	Trace	Trace
Palmitic ($C_{16}H_{32}O_2$)	6	5	6	4	5	4
Margaric ($C_{17}H_{34}O_2$)	1	1	1	1	1	Trace
Stearic ($C_{18}H_{36}O_2$)	1	Trace	Trace	Trace	Trace	0.5
Unsaturated Acids						
Myristoleic ($C_{14}H_{26}O_2$)	2	2	1	2	3	0.5
Palmitoleic ($C_{16}H_{30}O_2$)	6	8	6	6	11	1
Oleic ($C_{18}H_{34}O_2$)	73	74	75	76	74	24
Linoleic ($C_{18}H_{32}O_2$)	8	6	7	7	3	62
Linolenic ($C_{18}H_{30}O_2$)	1	1	1	1	Trace	7.5

Short-Chain Acids

Product Name	Titer (°C)	Iodine Value (max)	Color (% trans) 440/550 nm (min)	Acid Value	Sap. Value
Emery 657 Caprylic Acid	15	0.2	88/99	388	389
Emery 658 Caprylic-Capric Acid	4	0.3	88/99	359	361
Emery 659 Capric Acid	30	0.2	88/99	324	326
Emery 650 Lauric Acid	34	0.4	85/97	270	270
Emery 651 Lauric Acid	42	0.2	90/99	278	279
Emery 652 Lauric Acid	43 min	0.2	90/99	279	280
Emery 655 Myristic Acid	53	0.5	90/99	245	246
Emery 1202 Pelargonic Acid	10	0.5	90/99	350	—
Emery 1210 LMW Acid	—	1.5	80/96	415	—
Emery 1207 Special Acid	40	2	47/93	255	270
Emery 1205 Special Acid	—	1	70/95	305	—

Typical Composition

	Emery 657	Emery 658	Emery 659	Emery 650	Emery 651	Emery 652
Caproic ($C_6H_{12}O_2$)	Trace	1	—	—	—	—
Enanthic ($C_7H_{14}O_2$)	—	—	—	—	—	—
Caprylic ($C_8H_{16}O_2$)	99	58	1	—	Trace	Trace
Pelargonic ($C_9H_{18}O_2$)	—	—	—	—	—	—
Capric ($C_{10}H_{20}O_2$)	1	40	97	—	1	0.3
Lauric ($C_{12}H_{24}O_2$)	—	1	2	71	96	99
Myristic ($C_{14}H_{28}O_2$)	—	—	—	28	3	0.7
Palmitic ($C_{16}H_{32}O_2$)	—	—	—	—	—	—

Typical Composition

	Emery 655	Emery 1202	Emery 1210	Emery 1207	Emery 1205
Caproic ($C_6H_{12}O_2$)	—	—	26	80%	Analysis
Enanthic ($C_7H_{14}O_2$)	—	—	33	Mono	not
Caprylic ($C_8H_{16}O_2$)	—	4	9	basic	given
Pelargonic ($C_9H_{18}O_2$)	—	94	27	acids,	by
Capric ($C_{10}H_{20}O_2$)	—	2	—	20%	supplier
Lauric ($C_{12}H_{24}O_2$)	1	—	—	dibasic	
Myristic ($C_{14}H_{28}O_2$)	97	—	—	acids	
Palmitic ($C_{16}H_{32}O_2$)	2	—	—		

Stearic and Palmitic Acids

Product Name	Titer (°C)	Iodine Value (max)	Color (% trans) 440/550 nm (min)	Acid Value	Sap. Value
Emersol 110 Stearic Acid	53.2	12	60/94	208	209
Emersol 120 Stearic Acid	54.3	6	88/99	208	209
Emersol 132 USP Lily Stearic Acid	55.0	0.5	93/99	208	209
Emersol 140 Palmitic Acid	54.5	2 max	93/99	212	213
Emersol 142 Palmitic Acid	56.5	1 max	93/99	216	216
Emersol 143 Palmitic Acid	59.5	1 max	93/99	219	219
Emersol 150 Stearic Acid	64.5	1 max	93/99	200	201
Emersol 153 Stearic Acid	68.0	1 max	80/97	198	199
Emersol 400 Stearic Acid	52.0 min	10 max	1/29	201	203
Emersol 404 Stearic Acid	54.0	8	1/50	203	204
Emersol 410 Stearic Acid	58.1	7 max	40/86	202	204
Emersol 420 Stearic Acid	58.6	2 max	71/97	204	205
Emersol 422 Stearic Acid	56.5	1 max	90/99	206	207

	Emersol 110	Emersol 120	Emersol 132	Emersol 140	Emersol 142	Emersol 143
Typical Composition						
Saturated Acids						
Myristic ($C_{14}H_{28}O_2$)	2.5	2.5	2.5	1.5	7	Trace
Pentadecanoic ($C_{15}H_{30}O_2$)	0.5	0.5	0.5	0.5	1	0.5
Palmitic ($C_{16}H_{32}O_2$)	50	50	50	74.5	80	91
Margaric ($C_{17}H_{34}O_2$)	2	2.5	1.5	0.5	1	4.5
Stearic ($C_{18}H_{36}O_2$)	37	39.5	45.5	23	11	4
Unsaturated Acids						
Palmitoleic ($C_{16}H_{30}O_2$)	—	Trace	—	—	—	—
Oleic ($C_{18}H_{34}O_2$)	8	5	—	Trace	—	—
Linoleic ($C_{18}H_{32}O_2$)	Trace	Trace	—	—	—	—

	Emersol 150	Emersol 153	Emersol 400	Emersol 404	Emersol 410	Emersol 420	Emersol 422
Typical Composition							
Saturated Acids							
Myristic ($C_{14}H_{28}O_2$)	3	—	. . Analysis . .		3	4	3
Pentadecanoic ($C_{15}H_{30}O_2$)	—	—	. . not given . .		0.5	0.5	Trace
Palmitic ($C_{16}H_{32}O_2$)	13	5	. . . . by		30	29	41
Margaric ($C_{17}H_{34}O_2$)	0.5	—	. . supplier . .		2	1.5	1
Stearic ($C_{18}H_{36}O_2$)	83.5	95	—	—	58	65	55
Unsaturated Acids							
Palmitoleic ($C_{16}H_{30}O_2$)	—	—	—	—	2.5	Trace	—
Oleic ($C_{18}H_{34}O_2$)	1	—	—	—	4	Trace	—
Linoleic ($C_{18}H_{32}O_2$)	—	—	—	—	—	—	—

Tall Oil and Tall Oil Fatty Acids

Product Name	Acid Value	Unsap. (%)	Moisture (max %)	Color (Gardner)	Rosin Acids
Emtall 786 Tall Oil Pitch	48	30	—	Black	37
Emtall 745 Fatty Acid Heads	93	30	—	Dark	10
Emtall 743 Fatty Acid Heads	143	30	0.5	15	2.0 max
Emtall 753 Special Crude Tall	162	8 max	0.3	—	—
Emtall 729 Distilled Tall Oil	185	4 max	—	8 max	37
Emtall 731 Distilled Tall Oil	185	4 max	—	8 max	33

Tallow Fatty Acids

Product Name	Titer (°C)	Iodine Value	Color (% trans) 440/550 nm (min)	Acid Value	Sap. Value	Unsap. (% max)
Emery 500 Fatty Acid	37	57.5	2/20	185	186	10
Emery 531 Fatty Acid	41.8	50.5	19/81	205.5	206	1.5
Emery 540 Fatty Acid	47	39	80/95	203.5	205	–
Emery 880 Fatty Acid	15	70	17/64	229	232.5	4

Linoleic and Vegetable Fatty Acid

Product Name	Titer (°C)	Iodine Value	Color (% trans) 440/550 nm (min)	Acid Value	Sap. Value	Unsap. (% max)
Emery 315 Linoleic Acid	5 max	152.5	60/98	198	200	1
Emery 610 Soya Fatty Acid	24.5	125 min	50/95	200	201.5	1

	Emery 315	Emery 610
Typical Composition		
Saturated Acids		
Lauric ($C_{12}H_{24}O_2$)	–	Trace
Myristic ($C_{14}H_{20}O_2$)	0.5	0.5
Pentadecanoic ($C_{15}H_{30}O_2$)	Trace	–
Palmitic ($C_{16}H_{32}O_2$)	4	16
Margaric ($C_{17}H_{34}O_2$)	Trace	–
Stearic ($C_{18}H_{36}O_2$)	0.5	4
Unsaturated Acids		
Myristoleic ($C_{14}H_{26}O_2$)	0.5	–
Palmitoleic ($C_{16}H_{30}O_2$)	1	1
Oleic ($C_{18}H_{34}O_2$)	24	25.5
Linoleic ($C_{18}H_{32}O_2$)	62	48
Linolenic ($C_{18}H_{30}O_2$)	7.5	5

Key Properties: All Emery fatty acids are characterized by excellent color and oxidation stability, low content of impurities and unmatched uniformity from shipment to shipment.

Recommended for use in paints and lacquers, as an emulsifier and many other uses.

Empol Dimer and Trimer Acids

Product Name	Acid Value	Saponification Value	Color (Gardner) (max)	Unsaponifiable (% max)	Dimer Acid (C_{36} Dibasic Acids, MW ~565)	Trimer Acid (C_{54} Tribasic Acids, MW ~845)	Monobasic Acid (C_{18} Fatty Acids, MW ~282)
1010 Dimer	194	197	1	Nil	97	3	0
1012 Dimer	194	–	4	2	87	3	10
1014 Dimer	196	199	5	0.5	95	4	1

Product Name	Acid Value	Saponification Value	Color (Gardner) (max)	Unsaponifiable (% max)	Dimer Acid (C$_{36}$ Dibasic Acids, MW ~565)	Trimer Acid (C$_{54}$ Tribasic Acids, MW ~845)	Monobasic Acid (C$_{18}$ Fatty Acids, MW ~282)
1016 Dimer	194	197	6	0.5	87	13	Trace
1018 Dimer	192	195	8	1	83	17	Trace
1022 Dimer	193	195	9	1	75	22	3
1024 Dimer	193	195	9	1	75	25	Trace
1028 Dimer	190	—	6	—	75	19	7
1031 Dimer	203	—	6	—	68	18	15
1040 Trimer	184	196	—	—	20	80	—
1041 Trimer	171	188	11	—	10	90	—

Chemical Type: Dimer acids, C$_{36}$ liquid aliphatic acid.
Trimer acids, C$_{54}$ tribasic acid.

PROCTER AND GAMBLE DISTRIBUTING CO., P.O. Box 599, Cincinnati, OH 45201

Fatty Acids

Product Name	Titer (°C)	Acid Value	Iodine Value
Stearic Acid (T.P. Type)	54.3	208	1.8

Chemical Type: Palmitic—C$_{16}$ (50%) and Stearic—C$_{18}$ (48%).

Stearic Acid (S.P. Type)	53	208	5.0

Chemical Type: Myristic—C$_{14}$ (1%), Palmitic—C$_{16}$ (51%), Stearic—C$_{18}$ (43%) and Oleic-C$_{18}$ (6%).

High Palmitic	54.5	211	4.0

Chemical Type: Palmitic—C$_{16}$ (75%), Stearic—C$_{18}$ (20%) and Oleic—C$_{18}$ (5%).

Solid Fatty Acid (Hydrogenated Tallow Type)	57.5	202	5.5

Chemical Type: Myristic—C$_{14}$ (2%), Palmitic—C$_{16}$ (35%), Stearic—C$_{18}$ (59%) and Oleic—C$_{18}$ (4%).

Solid Fatty Acid (Hydrogenated Vegetable Type)	60.5	202	5.5

Chemical Type: Myristic—C$_{14}$ (1%), Palmitic—C$_{16}$ (29%), Stearic—C$_{18}$ (68%) and Oleic—C$_{18}$ (2%).

Solid Fatty Acid (Hydrogenated Fish Type)	52	196	8.0

Chemical Type: Myristic—C$_{14}$ (10%), Palmitic—C$_{16}$ (35%), Stearic—C$_{18}$ (28%), Arachidic—C$_{20}$ (15%), Behenic—C$_{22}$ (10%) and Oleic-C$_{18}$ (2%).

High Lauric Acid (Distilled Coconut)	24.5	271	10

Chemical Type: Caprylic—C$_8$ (1%), Capric—C$_{10}$ (7%), Lauric—C$_{12}$ (49%), Myristic-C$_{14}$ (17%), Palmitic—C$_{16}$ (9%), Stearic—C$_{18}$ (2%), Oleic-C$_{18}$ (6%) and Linoleic—C$_{18}$ (2%).

Oleic Acid (Red Oil)	7.5	197	90

Chemical Type: Myristic-C$_{14}$ (2%), Palmitic-C$_{16}$ (7%), Stearic—C$_{18}$ (2%), Oleic—C$_{18}$ (79%) and Linoleic—C$_{18}$ (10%).

Oleic Acid (Multiple Distilled)	6	199	92.5

Chemical Type: Myristic-C$_{14}$ (3%), Palmitic-C$_{16}$ (5%), Stearic—C$_{18}$ (1%), Oleic—C$_{18}$ (85%) and Linoleic—C$_{18}$ (6%).

Animal Fatty Acid (Distilled)	41	200	55.5

Chemical Type: Myristic—C$_{14}$ (4%), Palmitic—C$_{16}$ (25%), Stearic—C$_{18}$ (15%), Oleic—C$_{18}$ (50%), Linoleic—C$_{18}$ (4%) and Linolenic—C$_{18}$ (2%).

Vegetable Fatty Acid (Distilled Cottonseed)	35.5	199.5	100

Chemical Type: Myristic—C$_{14}$ (1%), Palmitic—C$_{16}$ (24%), Stearic—C$_{18}$ (2%), Oleic—C$_{18}$ (33%), Linoleic—C$_{18}$ (39%) and Linolenic—C$_{18}$ (1%).

Product Name	Titer (°C)	Acid Value	Iodine Value
Vegetable Fatty Acid (Distilled Soybean)	27.5	197.5	125

Chemical Type: Palmitic–C_{16} (11%), Stearic–C_{18} (4%), Oleic–C_{18} (29%), Linoleic–C_{18} (51%) and Linolenic–C_{18} (5%).

Vegetable Fatty Acid (Fractionated Soybean)	23.5	199.5	137.5

Chemical Type: Oleic–C_{18} (39%), Linoleic–C_{18} (60%), Linolenic–C_{18} (1%).

Tall Oil Fatty Acid	7.5	194	134

Chemical Type: Palmitic–C_{16} (2%), Oleic–C_{18} (51%), and Linoleic–C_{18} (47%).

Distilled Linseed Fatty Acid	21	199.5	172.5

Chemical Type: Palmitic–C_{16} (6%), Stearic–C_{18} (3%), Oleic–C_{18} (19%), Linoleic–C_{18} (24%) and Linolenic–C_{18} (47%).

Key Properties: All of the above are recommended for both paints and adhesives.

REICHHOLD CHEMICALS, INC., RCI Bldg., White Plains, NY 10603

Distilled Tall Oil Fatty Acids

	Aconew Extra (35-801)	Aconew 500 (35-802)	Acofor (35-803)	Acofor 8 (35-804)
Color (Gardner-1963) (max)	3	4	4	6
Acid Number	196	196	192	192
% Unsaponifiables	1.8 max	1.8 max	2.5	3.4
Iodine Number	130	130	132	129
% Rosin Acids (max)	1.2	1.2	2.5	3.5
Saponification Number	197	197	195	192
Iodine Number	—	—	132	129
Fatty Acids	97.5	97.5	94.5	93
% Ash	Trace	Trace	Trace	Trace
% Moisture	Trace	Trace	Trace	Trace
Refractive Index @ 20°C	1.464	1.464	1.469	1.472
Viscosity (SUS) @ 100°F	95	95	100	110
Viscosity (Gardner-Holdt) @ 25°C	A-2	A-2	A-2	A-1
Flash Point (Pensky Martens Closed Cup) (°F)	398	423	408	412
Fire Point (Cleveland Open Cup)	455	455	435	435
Specific Gravity @ 25°/25°C	0.900	0.900	0.903	0.906
Weight per Gallon (lb) @ 25°C	7.48	7.48	7.52	7.54
Titer (°C)	5	5	4	4

Chemical Type: Five pale distilled tall oil fatty acids are offered, as follows:

(1) Aconew 450: Extremely pale tall oil fatty acids containing, typically, 0.7% of rosin acids (other data not furnished by supplier).

(2) Aconew Extra: Very pale tall oil fatty acids containing, typically, 0.8% of rosin acids.

(3) Aconew 500: Pale tall oil fatty acids, almost identical to Aconew Extra, except for having a slightly darker color and a typical rosin acids content of 0.9%.

(4) Acofor: Pale tall oil fatty acids, containing a somewhat higher rosin acids content than Aconew 450, Aconew Extra and Aconew 500. It is, typically, darker in color than either Aconew 450 or Aconew Extra.

(5) Acofor 8: Moderately pale tall oil fatty acids containing a slightly higher rosin acids content than Acofor and being slightly darker than it in color.

As a group, these products are characterized by their pale colors and mild odors. There is a wide industry acceptance of distinction of the definition of tall oil fatty acids, as they refer to products generally containing 90% or more of fatty acids. The remainder consists of rosin and neutral materials. (Defined by ASTM Standard D804-63).

Key Properties: The growth of the Tall Oil Industry was very rapid and this led to highly diversified uses for these oils. The extent of their usefulness may be seen from the following list:

Alkyd resins

Asphalt additives

Caulking compounds

Driers

Emulsion polymerization

Epoxy resins

Latex emulsifiers

Oleoresinous varnishes

Paint removers

Pigment wetting agents

Tar removers

Urethane resins

Vinyl plasticizers

Many other uses

Section II

Aluminum Silicates

BURGESS PIGMENT CO., P.O. Box 349, Sandersville, GA 31082

Hydrous Aluminum Silicates

Product Name	Particle Size (μ)	pH	Oil Absorption (Gardner Coleman)	Color (G.E.) (%)	Plus 325 Mesh (% Max)
No. 10	0.5	4.6	43.0	87.0	0.02
No. 80	0.5	4.6	42.0	81.0	0.03
No. 20	0.75	7.0	35.0	86.0	0.015
No. 60	0.75	4.6	37.0	86.0	0.015
No. 40	4.50	4.6	30.0	82.0	0.15
Thermo Glace H	*	7.0	39.0	89.0	0.10

Note: Recommended for aqueous systems only.

Product Name	Particle Size (μ)	pH	Oil Absorption (Gardner Coleman)	Color (G.E.) (%)	Plus 325 Mesh (% Max)
Thermo Glace HB	*	7.0	39.0	89.0	0.10

Note: Recommended for aqueous or solvent systems.

*0.1 to 0.4 micron thickness. (The standard particle size measurement does not apply.)

Chemical Analysis (all of the above)

Ignition Loss	13.7-14.1%	Iron Oxide (Fe_2O_3)	Trace
Silica (SiO_2)	44.8-45.3%	Titanium Dioxide	1.35-2.27%
Alumina (Al_2O_3)	37.5-39.7%	(TiO_2)	
		Manganese (Mn)	None

Physical Properties (all of the above)

Specific Gravity	2.63	Moisture Content	1.0%
Bulking Value	0.0456	Refractive Index	1.56

Thermo-Optic Aluminum Silicates

Product Name	Particle Size (μ)	pH	Oil Absorption (Gardner Coleman)	Color G.E. (%)	Plus 325 Mesh (% Max)	Particle Shape
Optiwhite	1.4	5.9	58	91.5	0.5	Amorphous

Key Properties: Represents the most versatile pigment of the Thermo-Optic Series. Not only capable of substantial titanium dioxide reduction, but also acts as a flatting agent. Properly formulated Optiwhite approaches have resulted in significant property improvements while saving substantially. Typical approaches result in equal or better hiding, 25% better scrub resistance with raw material cost savings.

Product Name	Particle Size (μ)	pH	Oil Absorption (Gardner Coleman)	Color G.E. (%)	Plus 325 Mesh (% Max)	Particle Shape
Tisyn	0.8	4.3	70	84.5	0.3	Thin flat plate

Key Properties: Exhibits the greatest degree of opacity in this pigment series. Film properties and sheen properties more approach what is expected of conventional calcined clay. Coatings may exhibit a buff-tone, depending on the level of Tisyn and other pigments in the formula.

Product Name	Particle Size (μ)	pH	Oil Absorption (Gardner Coleman)	Color G.E. (%)	Plus 325 Mesh (% Max)	Particle Shape
Optiwhite P	1.3	4.3	62	84.0	0.3	Amorphous

Key Properties: Exhibits more opacity than Optiwhite. Film property advantages associated with Optiwhite are not as good, but can still be demonstrated in many formulation approaches. Does contribute a buff-tone to coatings, due to the 84 brightness. Angular sheen may be equal to or slightly higher than Optiwhite. Overall, Optiwhite P offers more hiding and greater savings than Optiwhite without the exceptional advantages in film properties associated with Optiwhite.

30P	1.0	4.3	62	83.0	0.3	Amorphous

Key Properties: Patented product especially designed for polyvinyl chloride compounds. Combines ease of incorporation with unique uniformity of compound color, excellent electrical properties and low specific gravity.

Physical Properties (all of the above)

Specific Gravity	2.2	Moisture Content	0.5% max
Bulking Value	0.0545	Refractive Index	1.62

Anhydrous Aluminum Silicates

Product Name	Particle Size (μ)	pH	Oil Absorption (Gardner Coleman)	Color G.E. (%)	Plus 325 Mesh (% Max)	Particle Shape
Icecap K	1.0	5.5	52.5	91	0.03	Thin flat plate

Key Properties: Excellent calcined extender for titanium dioxide in coatings where a high Hegman grind is required. The low screen residue, due to the removal of agglomerators, makes it a leader in the field of calcined kaolins. Applications include wire and cable, molded and extruded rubber and plastic products, paper coatings and paints. Icecap K has good electricals and low compression set and low water absorption.

Iceberg	1.4	5.5	52.5	91	0.35	Thin flat plate

Key Properties: Iceberg is the original workhorse in the Burgess Pigment calcined clay group. It is used in paints, paper, board, rubber and plastics and various other applications similar to Icecap K.

Physical Properties (all of the above)

Specific Gravity	2.63	Moisture Content	0.5% max
Bulking Value	0.0456	Refractive Index	1.62

Both the Thermo-Optic and Anhydrous Aluminum Silicates have the following Chemical Analysis:

Ignition Loss	0-1.0%	Iron Oxide (Fe_2O_3)	Trace
Silica (SiO_2)	51.0-52.4%	Titanium Dioxide (TiO_2)	1.56-2.50%
Alumina (Al_2O_3)	42.1-44.3%	Manganese (Mn)	None

U.S. Patents: 3,309,214, Optiwhite; 3,021,195, Tisyn, Optiwhite P, 30P.

GEORGIA KAOLIN CO., 433 North Broad St., Elizabeth, NJ 07207

Product Name	Particle Size (μ)	pH (20% Solids Slurry)	Oil Absorption (Gardner- Coleman)	Brightness (G.E.) (%)	Plus 325 Mesh (% Max)	Surface Area (Sq.M./ Gram)
Glomax JDF	0.9	4.7	70	91	0.01	9.0
Glomax LL	1.8	4.7	60	91	0.02	5.1

Chemical Type: Extender pigments produced by calcining fractionated hydrated aluminum silicates. Further processing assures accurate control of particle size distribution and minimizes 325 mesh residue (grit).

Chemical Analysis (of both)

	Percent by Weight
Aluminum Oxide (Al_2O_3)(Combined)	44.40
Silicon Dioxide (SiO_2)(Combined)	53.80
Iron Oxide (Fe_2O_3)	0.44
Titanium Dioxide (TiO_2)	1.55
Calcium Oxide (CaO)	0.05
Magnesium Oxide (MgO)	0.23
Sodium Oxide (Na_2O)	0.25
Potassium Oxide (K_2O)	0.09

Physical Properties (of both)

Refractive Index	1.56	Bulking Value (Gallons/Pound)	0.0456
Weight per Solid Gallon	21.91 lb	Specific Gravity	2.62

Key Properties: Extreme low grit content reduces abrasion and improves dispersibility in paint and other polymeric systems. Calcination markedly changes the properties and crystal structure of kaolins. Brightness and oil adsorption are increased along with overall color improvement. Important advantages offered in various paint systems: Contrast ratios are increased with the same prime white pigment volume. Increasing the amount of Glomax extender pigments improves reflectance and decreases yellowness.

Increasing the Glomax content reduces gloss.

Imparts excellent salt spray resistance and fast grinding characteristics to a wide range of industrial primers.

Significant advantages provided in rubber and plastics applications: Imparts excellent physical and electrical properties to rubber and other polymer systems.

Ease of dispersion improves processability in rubber and plastic compounds.

Improves performance of a wider range of ceramic bodies.

Kaopaque Delaminated Aluminum Silicates

Product Name	Particle Size (μ)	pH (20% Solids Slurry)	Oil Absorption (Gardner- Coleman)	Color (G.E.) (%)	Plus 325 Mesh (% Max)	Moisture (% Max)
10	1	4.3	45	89	0.02	1
10S	1	7.0	45	89	0.01	1
10 Slurry	1	7.0	45	89	0.01	32
20	2	4.7	51	87	0.1	1
20 Slurry	2	7.0	51	87	0.01	32
30	3	4.7	55	86	0.01	1

Chemical Type: Kaopaque delaminated aluminum silicates are produced from high purity kaolins by an exclusive Georgia Kaolin process. These delaminated clay grades have an average diameter to thickness ratio of about 11:1 compared to 6:1 for natural kaolin. This characteristic offers unique properties for a variety of process industries.

Chemical Analysis (of all grades)

	Weight %
Silica (SiO_2)	45.20
Alumina (Al_2O_3)	39.30
Ferric Oxide (Fe_2O_3)	0.32
Titanium Dioxide (TiO_2)	0.66
Calcium Oxide (CaO)	0.21
Magnesium Oxide (MgO)	0.03
Sodium Oxide (Na_2O)	0.03
Potassium Oxide (K_2O)	0.09
Loss on Ignition (at 1050°F)	13.92

Physical Properties (of all grades)

Refractive Index	1.56
Specific Gravity	2.61
Bulking Value (Pounds/Gallon)	0.04617
Weight per Gallon	21.66 pounds
Hardness Index (Mohs Scale)	No. 2
Weight per Gallon of Slurries	14.38 pounds

Key Properties: Paint and Coating Systems—These large platy particles are readily dispersed in aqueous systems. They impart the following:

(1) Excellent whiteness, brightness and durability.
(2) Outstanding hiding power.
(3) Excellent color and cleaner tints.
(4) Improved scrubbability and stain removal.
(5) Excellent holdout.
(6) Reduced "mud cracking."
(7) Control of gloss and sheen.

For Ceramic Systems: Kaopaque aluminum silicates improve color in a variety of ceramic bodies.

GROSS MINERALS CORP., P.O. Box 116, Aspers, PA 17304

Mesh-Ratio Factors (in %) for Ser-X

Grades	...Ser-X #200...		...Ser-X #325...		...Ser-X #400...	
	Pass	Retain	Pass	Retain	Pass	Retain
325 Mesh Screen	88	12	96	4	99+	Trace
200 Mesh Screen	93	7	99+	Trace	99+	Trace
100 Mesh Screen	97	3	99+	Trace	99+	Trace
60 Mesh Screen	99+	Trace	100	0	100	0

Chemical Type: Best described as a hydrous alumina silicate, produced from the basic mineral, sericite. Sericite is similar to pyrophyllite, or talc, in its physical properties.

Chemical Analysis (calcined basis)

	Weight %
Silica (SiO_2)	70.40
Alumina (Al_2O_3)	17.11
Potassium Oxide (K_2O)	4.40
Ferric Oxide (Fe_2O_3)	3.72
Magnesium Oxide (MgO)	2.85
Sodium Oxide (Na_2O)	1.01
Various Other Oxides	Small Degree

Physical Properties

Specific Gravity	2.7
Weight per Gallon	21.41
Pounds per Cubic Foot	43–46
Bulking Value	0.04671
Oil Absorption per 100 pounds	38.1

Average Particle Size	3.66 μ
Surface Area (Sq.Cm./Gram)	6,000
Refractive Index	±1.51
Hardness (Mohs Scale)	4.00–4.50
pH (20% Aqueous Solution)	6.8
Ion Exchange Capacity per 100 Grams	1.8 M.E.
Loss on Ignition	2.51
Asbestos	Free

Key Properties: Extreme fine, dry product of flakelike particle structure, with excellent free-flow characteristics—chemically inert—which will not pick up moisture in handling or shipment. This makes it an excellent filler for a variety of products. Ser-X wets almost instantly, when mixed with water, and disperses readily to a moderately stable suspension.

Paints: Used extensively as a base filler in both water and oil-paints, Ser-X provides high shear mixing and imparts toughness, scrubbability, flatting and suspension in the final paint product. Because of its chemical inertness, Ser-X is in particular demand in paint specifications calling for silicates, or insoluble siliceous material. It may be used with any color.

Joint Cements: One of the prime uses for Ser-X is as a basic ingredient in producing high quality pipe and ceramic joint cements. To preserve uniform color characteristics, material is taken only from a set section of the mine.

Other Uses: Many other uses are being discovered, such as extenders for automotive undercoating, nonmoisture absorbing cement-product fillers, plastic body fillers, and protective coatings.

Toxicity: Ser-X is free of asbestos and other fibrous materials.

R.T. VANDERBILT CO., INC., 30 Winfield St., Norwalk, CT 06855

Pyrax WA

Chemical Type: Hydrated aluminum silicate produced from the North Carolina deposits of the mineral pyrophyllite.

Chemical Analysis (expressed as oxides)

	% by Weight
Aluminum Oxide (Al_2O_3)	15–19
Silicon Dioxide (SiO_2)	75–81
Iron Oxide (Fe_2O_3)	0.20–0.30
Titanium Dioxide (TiO_2)	0–0.15
Calcium Oxide (CaO)	0–0.10
Magnesium Oxide (MgO)	0–0.04
Sodium Oxide (Na_2O)	0.10–0.40
Potassium Oxide (K_2O)	0.30–2.50
Ignition Loss	2.8–3.5

Physical Properties

Dry Brightness	80
Specific Gravity	2.8
Oil Absorption (Gardner-Coleman)	36
Residue on 200 Mesh	3%
Specific Resistance (Ohms)	30,000

Key Properties: Recommended for use in crack fillers, joint cements and the like.

Section III
Antiskinning Agents

THE AMES LABORATORIES, INC., 200 Rock Lane, Milford, CT 06460

Antiskinning Agents

	Molecular Weight	Flash Point (°F)	Boiling Point or (Melting Point) (°C)	Refractive Index n_D^{25}	Specific Gravity
Acetaldehyde Oxime	59	140	115	1.4240	0.969

Key Properties: Recommended for use as antiozonant.

Acetone Oxime	73	>200	135	—	0.901

Key Properties: Recommended for use as antioxidant and polymerization short stopper.

Acetophenone Oxime	135	>280	(60)	—	1.04

Key Properties: Recommended for use as perfume base, additive, antioxidant in water-oleoresinous emulsion systems and antiskinning agents in baking schedule paints.

N-Acetyl Ethanolamine (85% solution in de-ionized water)	105	—	—	1.4550	1.12

Key Properties: Recommended for use as humectant, protein glue extender and dissolver.

Butyraldehyde Oxime	87	110	151	1.4353	0.921

Key Properties: Recommended for use as antiskinning agent for special printing inks, paints and varnishes.

Isobutyraldehyde Oxime	87	95	140	1.4289	0.920

Key Properties: Recommended for use as antiskinning agent for polyurethane finishes, etc.

Isobutyraldehyde Oxime/ Mineral Spirits Mixture (50/50)	—	120	155	—	0.914

Key Properties: Recommended for use as antiskinning agent for polyurethane finishes.

Butanone Oxime (Methyl Ethyl Ketoxime)	87	125	152	1.4406	0.923

Key Properties: Recommended for use as paint and varnish antiskinning agent.

FERRO CHEMICAL DIVISION, 7050 Krick Road, Bedford, OH 44146

Product Name	Active Content (%)	Specific Gravity @ 80°F	Weight per Gallon @ 80°F (lb)	Flash Point, C.O.C. (°F)
Skinfoil KE	100	0.92	7.70	140

Chemical Type: Methyl ethyl ketoxime of high purity.

Key Properties: Antiskinning agent with solubility complete in commonly used solvents.

Section IV
Calcium Carbonates

CALCIUM CARBONATE CO., P. O. Box 4005, 3150 Gardner Expressway, Quincy, IL 62301

Calcium Carbonates

Product Name	Particle Size (μ)	Specific Surface Diameter (μ)	Dry Brightness (MgO = 100)	Oil Absorption (lb/100 lb)	Water Requirements (lb/100 lb)	Bulk Density (lb/ft³)
H-White	2.5	1.5	91	16	65	45
		Meets ASTM D1199, Type GC, Grade I				
G-White	5.5	2.2	90	13.5	44	55
		Meets ASTM D1199, Type GC, Grade II				

Chemical Type: Fine, double classified, ball-milled natural calcium carbonate produced from high purity, white, calcitic limestone quarried from underground mines. These functional filler extenders are characterized by easy dispersion and unusually low binder demand.

Chemical Analysis:

Calcium Carbonate ($CaCO_3$)	97.7 wt %	Sulfur (S)	0.03 wt %
Magnesium Carbonate ($MgCO_3$)	1.3 wt %	Manganese Oxide (MnO)	0.02 wt %
Silica and Silicates (SiO_2)	0.6 wt %	Copper Oxide (CuO)	5 ppm
Aluminum Oxide (Al_2O_3)	0.08 wt %	Lead Oxide (PbO)	0.01 ppm
Iron Oxide (Fe_2O_3)	0.05 wt %	Mercury (Hg)	0.08 ppb

Physical Properties:

	H-White	G-White
Screen Analysis (% on 325 mesh)	0.001	0.01
Screen Analysis (% through 500 mesh)	99.99	99
Hegman Fineness of Grind	6-6.5	4-4.5

Physical Properties (typical of both):

Color	White
Solubility	0.003 g/100 ml H_2O at 100°C
Alkalinity (as NaOH)	0.2 mg/g (pH 9.3 saturated solution)
Refractive Index	1.6
Hardness	3 (Mohs Scale), nonabrasive
Linear Expansion Coefficient	$4.3 \times 10^{-6}/°C$
Particle Shape (Microphotographs)	Irregular, uniaxial
Specific Gravity	2.71
Weight per Solid Gallon	22.6 lb/solid gal (0.0443 solid gal/lb)
Moisture	0.1% max

Recommended Applications:

Coatings—Latex and oil base paints; textile coatings; wallboard and ceiling tile coatings; dry color mixes.

Plastic Filler—Thermoplastics and thermosets. Molded, extruded, cast, calendered, coated products and plastisols.

Building Products—Caulks, glazes and sealants; tile grouts, spackling compounds and patching plaster.

Miscellaneous Products—Adhesives and many others.

Product Name	Particle Size (μ)	Dry Brightness (MgO = 100) (% min)	Oil Absorption (lb/100 lb)	Water Requirement (lb/100 lb)	Bulk Density Compacted (lb/ft³)	Bulk Density Loose (lb/ft³)
Q-White (spray dried)	0.9	95	28	38	62	52

(The spray dried beads range from 40 to 325 mesh and easily disperse to their ultimate particle size.)

Product Name	Viscosity (max cp)	% Solids	pH	Weight per Gallon (lb)	Residue on 325 Mesh
Q-White (70% slurry)	200	70	8.8	14.9	0.01

Chemical Type: A water-ground, classified, high purity and brightness natural calcium carbonate produced from white, calcitic limestone. It is suitable as a replacement for precipitated calcium carbonates in many applications. Q-White is available in two forms: spray dried free flowing beads and 70% solids slurry.

Chemical Analysis:

Calcium Carbonate ($CaCO_3$)	97.7 wt %	Sulfur (S)	0.03 wt %
Magnesium Carbonate ($MgCO_3$)	1.5 wt %	Manganese Oxide (MnO)	0.02 wt %
Silica and Silicates (SiO_2)	0.6 wt %	Copper Oxide (CuO)	5 ppm
Aluminum Oxide (Al_2O_3)	0.08 wt %	Lead Oxide (PbO)	0.01 ppm
Iron Oxide (Fe_2O_3)	0.05 wt %	Mercury (Hg)	0.08 ppb

Physical Properties: Dispersed Particle Size Distribution (% Finer by Weight, Sedigraph Method):

5 μ	100%	Moisture	0.5% maximum
2 μ	90%	Refractive Index	1.6
1 μ	57%	Hardness (Mohs Scale)	3 (Nonabrasive)
0.5 μ	26%	Solubility, at 100°C	0.0035 g/100 ml H_2O
0.2 μ	8%	Particle Shape	Irregular, uniaxial
Residue on 325 Mesh	0.01%	Specific Gravity	2.71
Alkalinity (as NaOH)	0.2 mg/g (pH 9.7 saturated soln)	Weight per Solid Gallon	22.6 lb (0.0443 solid gal/lb)
		Linear Expansion Coefficient	4.3×10^{-6}/°C

Key Properties: A new superfine natural calcium carbonate for water base coatings, sealants and adhesives, described as follows:
1. Lower cost replacement for precipitated calcium carbonate.
2. Free-flowing, spray dried beads or predispersed 70% slurry.
3. 5 micron top size.

In Latex Coatings:
Saves titanium dioxide with improved reflectance and hide.
Yields a 7 Hegman Fineness of Grind in semigloss, eggshell or flats.
Disperses more readily than powders.
Improves stain resistance.
Reduces settling, sag, flooding and porous surface penetration.
Exhibits moderate binder demand for high PVC and viscosity control.

In Latex Sealants and Adhesives:
Improves adhesion, slump and stain resistance.
Permits a smoother finish.
Results in better "length".
Exhibits moderate binder demand for improved "gunnability".
Suitable for many other applications.

Product Name	Particle Size (μ)	Specific Surface Diameter (μ)	Dry Brightness (MgO = 100)	Oil Absorption (lb/100 lb)	Water Requirements (lb/100 lb)	Bulk Density (lb/ft³)
Q-1	10	3.9	87	13.5	42	60

Meets ASTM D1199, Type GC, Grade II

Typical Screen Analysis:

>Cumulative through U.S. Screen 200 Mesh (76 μ)—99.99%
>Cumulative through U.S. Screen 325 Mesh (44 μ)—99.5%
>Typical fine "325 Mesh" grade.

Product Name	Particle Size (μ)	Specific Surface Diameter (μ)	Dry Brightness (MgO = 100)	Oil Absorption (lb/100 lb)	Water Requirements (lb/100 lb)	Bulk Density (lb/ft³)
Q-2	11	4.1	86.5	13	41	60
		Meets ASTM D1199, Type GC, Grade III				

Typical Screen Analysis:

>Cumulative through U.S. Screen 200 Mesh (76 μ)—99.99%
>Cumulative through U.S. Screen 325 Mesh (44 μ)—98.5%

Q-30	12	4.4	86	12.5	40	65
		Meets ASTM D1199, Type GC, Grade III				

Typical Screen Analysis:

>Cumulative through U.S. Screen 100 Mesh (150 μ)—99.99%
>Cumulative through U.S. Screen 200 Mesh (76 μ)—99.8%
>Cumulative through U.S. Screen 325 Mesh (44 μ)—95.0%

Q-31	13	4.8	85.5	12	39	65
		Meets ASTM D1199, Type GC, Grade III				

Typical Screen Analysis:

>Cumulative through U.S. Screen 100 Mesh (150 μ)—99.99%
>Cumulative through U.S. Screen 200 Mesh (76 μ)—99%
>Cumulative through U.S. Screen 325 Mesh (44 μ)—90%

Q-3	15	5.6	85	11.5	37	70
		Meets ASTM D1199, Type GC, Grade III				

Typical Screen Analysis:

>Cumulative through U.S. Screen 100 Mesh (150 μ)—99.95%
>Cumulative through U.S. Screen 200 Mesh (76 μ)—95%
>Cumulative through U.S. Screen 325 Mesh (44 μ)—83%
>Typical "200 Mesh" Grade.

Q-3SPP	20	6.1	84.5	11	35	75
		Meets ASTM D1199, Type GC, Grade IV				

Typical Screen Analysis:

>Cumulative through U.S. Screen 60 Mesh (250 μ)—99.99%
>Cumulative through U.S. Screen 100 Mesh (150 μ)—98.5%
>Cumulative through U.S. Screen 200 Mesh (76 μ)—85%
>Cumulative through U.S. Screen 325 Mesh (44 μ)—70%

Q-4	19.5	6.05	84	11.0	35	75
		Meets ASTM D1199, Type GC, Grade IV				

Typical Screen Analysis:

>Cumulative through U.S. Screen 60 Mesh (250 μ)—99.95%
>Cumulative through U.S. Screen 100 Mesh (150 μ)—97.5%
>Cumulative through U.S. Screen 200 Mesh (76 μ)—85%
>Cumulative through U.S. Screen 325 Mesh (44 μ)—70%
>Typical "100 Mesh" product.

Q-32	26	6.3	83	10.5	33	80

Typical Screen Analysis:

>Cumulative through U.S. Screen 60 Mesh (250 μ)—99.5%
>Cumulative through U.S. Screen 100 Mesh (150 μ)—90%
>Cumulative through U.S. Screen 200 Mesh (76 μ)—75%
>Cumulative through U.S. Screen 325 Mesh (44 μ)—60%
>Typical "80 Mesh" product.

Chemical Type: Dry ground and air fractionated high calcium, white, calcitic limestone, quarried from underground mines. "Whiting", "calcite", "marble dust", "putty powder" and ground limestone are synonymous with ground natural "calcium carbonate".

Chemical Analysis:

Calcium Carbonate ($CaCO_3$)	97.7 wt %	Sulfur (S)	0.03 wt %
Magnesium Carbonate ($MgCO_3$)	1.3 wt %	Manganese Oxide (MnO)	0.02 wt %
Silica and Silicates (SiO_2)	0.6 wt %	Copper Oxide (CuO)	5 ppm
Aluminum Oxide (Al_2O_3)	0.08 wt %	Lead Oxide (PbO)	0.01 ppm
Iron Oxide (Fe_2O_3)	0.05 wt %	Mercury (Hg)	0.08 ppb

Physical Properties:

Color	White
Alkalinity (as NaOH)	0.2 mg/cm (pH 9.3 saturated solution)
Hardness (Mohs Scale)	3 (Nonabrasive)
Solubility	0.0035 g/100 ml H_2O at 100°C
Particle Shape	Irregular, uniaxial
Specific Gravity	2.71
Refractive Index	1.6
Weight per Solid Gallon	22.6 lb (0.0443 solid gal/lb)
Linear Expansion Coefficient	4.3×10^{-6}/°C
Moisture	0.1% max

Recommended Applications:

Building Products—Dry wall cements and texture paints; putties, glazes, caulks and sealants; tile grouts, spackling compounds and patching plaster; floor tile and seamless flooring.

Rubber Filler—Elastomers, hard rubber, cellular and latex goods.

Plastic Filler—Thermoplastics and thermosets: Molded, extruded, cast, coated products; plastisols.

Coatings—Latex and oil-base paints, textile coatings, wallboard and ceiling tile coatings, dry color mixes.

Asphalt Filler—Paring, roofing and undercoating filler.

Miscellaneous—Industrial and foundry sealants, adhesives and mastics and many other uses.

COMMERCIAL MINERALS CO., 6899 Smith Ave., P.O. Box 363, Newark, CA 94560

Chemical Type: Finely ground calcium carbonate. Pigment grade white extender produced from natural limestone (dry ground and air classified).

	Specification	
Product	Mesh No.	Percent Pass
L220 (Calcium Carbonate 130)	200	80
L306	325	94
L303	325	97
L3002 (RM)	325	99.8
Tube Mill Whitings		
L215T (1300)	200	85
L212T (TMO)	200	88
L3002T	325	99.8

Chemical Analysis:

	Percent by Weight
Calcium Carbonate ($CaCO_3$)	98.0 min
Magnesium Carbonte ($MgCO_3$)	0.5 max
Aluminum Iron Oxide (R_2O_3)	0.5 max
Iron Oxide (Fe_2O_3)	0.005
Silica (SiO_2)	0.5 max

FLINTKOTE STONE PRODUCTS CO., Executive Plaza IV, Hunt Valley, MD 21031

Product Name	Particle Diameter (μ)	Particle Range (μ)	Dry Brightness (Hunter Reflectometer) (min)	Oil Absorption (lb/100 lb)	Bulk Density (Loose) (lb/ft³)	Solubility in Water (%)
Camel-WITE	3.0	0.3–12.0	95	15	40	0.08
Camel-TEX	5.0	0.3–25.0	93	14	50	0.04
Camel-KOTE	11.0	0.4–25.0	93	10	50	0.04
Camel-CARB	7.0	0.3–44.0	93	13	58	0.03

Physical Characteristics (all of the above):

Specific Gravity	2.70–2.71	pH (saturated solution)	9.5
One Pound Bulk Gallons	0.0443	Index of Refraction	1.6
Weight per Solid Gallon	22.57 lb	Moh Hardness	3.0

Particle Size Distribution:

	Camel-WITE	Camel-TEX	Camel-KOTE	Camel-CARB
	. . . . *(percent by weight)*			
Finer than 44 μ	—	—	—	99.5
Finer than 30 μ	—	—	—	99
Finer than 25 μ	—	99.7	99.7	96
Finer than 20 μ	—	96	86	83
Finer than 12 μ	99.9	90	60	68
Finer than 10 μ	95	78	48	62
Finer than 8 μ	85	70	40	59
Finer than 7 μ	78	65	36	50
Finer than 5 μ	67	50	28	42
Finer than 4 μ	59	42	20	36
Finer than 3 μ	50	33	15	24
Finer than 2 μ	36	20	10	22

Camel-WITE Calcium Carbonate

Chemical Type: Distinctive type of fine-ground limestone (calcium carbonate).

Key Properties: Readily wettable, but does not pick up moisture from the atmosphere. Agglomerate-free, which is a property essential to rapid dispersion and minimum process time. Low oil absorption and rigid control of particle size distribution result in lower binder requirements compared to other pigments. Permits formation of more nearly continuous paint films giving better resistance to stain, less ink penetration, better wearing qualities and better adhesion to substrates.

Paints—Used extensively in manufacturing gloss, semigloss, flats, primers and sealers. Ease of dispersion and excellent suspension properties permit use of high speed modern equipment. The product is essentially a "stir-in" material. Can be used in vehicle systems ranging from oils and alkyds to the latex systems. Can be used to prevent pH drift in some latex systems and contributes to shelf stability of water-based paints.

Paper Coatings—Used as an inexpensive pigment in coating formulations. In addition to brightness and whiteness, it offers distinct advantages of easy dispersibility and high solids coating color.

Other Uses—Internal filler for paper, PVC and rubber, polyester-fiber glass, caulk and glazing compounds, linoleum and floor tile, ceramics, adhesives, food processing, and in thermosetting plastic moldings containing polyesters and epoxy resins.

Camel-TEX Calcium Carbonate

Chemical Type: General purpose grade of limestone (calcium carbonate) produced from the same white calcite as Camel-WITE.

Key Properties: High brightness, high whiteness, freedom from agglomerates, low vehicle demand, rapid dispersibility and chemical inertness combine to make it a highly valuable product in its field of application.

The product is readily dispersed in high-speed mixing equipment, permitting maximum production scheduling. It is used in interior flat paints and in primers and sealers where good color, low angle sheen, scrubbability and tinting qualities are important. Hegman fineness of grind values range from 4 to 5.

Other Uses—Plastics, rubber, putties, caulks, bath tub sealers, body deadeners and adhesives.

Camel-KOTE Calcium Carbonate

Chemical Type: Fine-ground limestone.

Key Properties: Designed for use where maximum filler loading is desired and a minimum oil and water absorption are required. 50% of the particles are less than 11 microns. It is agglomerate-free and easily dispersed.

Paint—It is of particular value in the manufacture of paints. Low water and oil demand and ease of dispersion combine to permit maximum filler use without adversely affecting finish properties. Its use provides good tint retention, excellent color and good sheen properties. Hegman values of 4 to 5 are obtainable with high speed dispersion.

Caulks and Sealants—Properties readily suggest their use in caulks and sealants where vehicle demand and working properties are critical.

Camel-CARB Calcium Carbonate

Chemical Type: Quality extender in the low-priced field, manufactured from white calcite.

Key Properties: Larger in particle size than the other products in the line, but made to the same rigid quality control standards. This assures the user of uniform low vehicle demand, good color and high brightness.

Used extensively in the production of interior flat and exterior house paints. In flats, uniform sheen is obtainable together with good nonpenetrating properties. The low water demand of Camel-CARB is a very useful characteristic. Latex formulations exhibit superior film strength, good scrubbability and resistance to staining. Unsurpassed in the formulation of exterior tint bases-tint retention after exposure for years is superior to that of other products tested. Durability on exposure is also excellent. Panels exposed for years show no cracking or peeling. Camel-CARB also finds use as an inexpensive filler in rubber compounding. It is widely used in putties and caulks, ceramics, adhesives, linoleum, floor tile and textile coatings.

Product Name	Viscosity Brookfield (cp)	TSC (%)	pH	Weight per Gallon (lb)	Refractive Index	Oil Absorption (lb/100 lb)
Camel-WITE Slurry	150	72.0	9.5	15.0	1.6	15

Chemical Type: The Slurry has the same physical characteristics as Camel-WITE. Contains an anionic dispersant, nonreactive and fugitive biocide. Surfactants and biocides have been chosen which are acceptable in almost all emulsion formulas.

Key Properties: Solids concentration is as high as possible to minimize the amount of water added to the user's costs. Settling will occur on standing, as in any practical slurry. Minimum periodic agitation in storage will keep the solids evenly dispersed and at constant density.

GEORGIA MARBLE CO., 2575 Cumberland Parkway, N.W., Atlanta, GA 30339

Gama-Sperse Natural Ground Calcium Carbonates

Product Name	Particle Size (μ)	Dry Brightness (Hunter Reflectometer) (% min)	Plus 325 Mesh (% max)	Hegman Fineness of Grind	Oil Absorption (lb/100 lb)
80	1.30	96	0.005	6¾	19

Product Name	Particle Size (μ)	Dry Brightness (Hunter Reflectometer) (% min)	Plus 325 Mesh (% max)	Hegman Fineness of Grind	Oil Absorption (lb/100 lb)
140	1.60	96	0.005	6¼	16
6532	1.70	96	0.005	6	16
6451	2.25	96	0.008	5	13
255	4.00	94	0.020	3¾	11

Key Properties: Line of calcium carbonate products with the characteristics that give a better product at lower costs. Gama-Sperse is finely ground, but is free of the hard agglomerates common to other finely divided products, resulting in fast dispersion in formulations mixed with high-speed equipment.

Precise uniformity of particle size distribution is another valuable virtue. Improved classification methods provide a clean-cut top size with no over-size particles to cause problems in paints, paper, plastics or rubber. Gama-Sperse will give a very clean grind.

Gama-Sperse can be used in many products that have previously required expensive precipitated calcium carbonates. The savings can be substantial. Formulations can provide for high loading of Gama-Sperse because of its extremely low binder demand.

Screened Products

Product Name	U.S. #16	U.S. #20	U.S. #30	U.S. #40	U.S. #50	U.S. #200
#XO	1% Retained			15% Passing		
3050			1% Retained		20% Passing	
#20*	1% Retained					
40-200				1% Retained		20% Passing

*This product is made by using only one screen. Therefore, there is no specification on the fine end.

Maximum % Retained or Passing U.S. Screens or Openings in Inches (Dry Rotap-5 minutes).
Retained On: Designates the amount of material that stays on a particular size screen.
Passing: Designates the amount of material that passes a given screen.

Chemical Type: Crushed and screened white marble containing a very high pecentage of calcium carbonate is the source of these products, which differ only in their particle size distribution.

Recommended Uses: Neutralization of acids; synthetic marble; aggregate for cement finishes; vinyl asbestos tile; welding rods; and polyester and epoxy floor tiles.

Surface Modified Calcium Carbonates

Physical Properties:

	Gama-Sperse CS-11	Gama-Sperse CR-12
Wet Screen Analysis (retained on U.S. 325 Mesh, % max)	0.005	0.005
Oil Absorption per 100 pounds (pounds) (ASTM D281)	10–12	12–14
Specific Gravity	2.7	2.7
Hardness (Mohs Scale)	3.0	3.0
Bulking Factor (gallons per pound)	0.0443	0.0443
Finer than 44 μ (passing through 325 Mesh)	99.995	99.995
Finer than 10 μ	98.00	99.00
Finer than 5 μ	70.00	75.00
Finer than 1 μ	18.00	18.00
Finer than 0.5 μ	8.00	10.00
Mean Average Particle Size (μ)	3	<3
Color	White	White
Surface Modification	Calcium Stearate	Calcium Resinate

Gama-Sperse CS-11 Calcium Carbonate

Chemical Type: Finely ground, exceptionally white surface treated calcium carbonate. Has a chemically bonded calcium stearate surface which renders it extemely hydrophobic. This surface is produced by a chemical reaction. Therefore, the treatment is firmly bonded to and actually a part of the surface of the particles.

Key Properties: Free-flowing powder developed for use in plastics and other compounds which require rapid and complete dispersion, enhanced processing characteristics and low abrasion at high loadings. The surface will neither wash off in processing nor segregate in shipping and handling.

Toxicity: Has National Sanitation Foundation acceptance of ingredients for chemical extraction, taste and odor when tested in specific compounds. The basic coating material has been approved by the FDA for use in the following food contact applications: adhesives, components of paper and paperboard in contact with aqueous and fatty foods, components of paper and paperboard in contact with dry food.

Gama-Sperse CR-12 Calcium Carbonate

Chemical Type: Calcium resinate surface modified calcium carbonate. White free-flowing filler that is relatively hydrophobic. Produced by the reaction of an oxidation resistant resin with calcium carbonate.

The calcium resinate coating is chemically bonded to the surface of the calcium carbonate particles. The coating becomes a part of the surface of the particle. It is not a physical coating or a mixture.

Key Properties: Has good compatibility with most noncrystalline polymers and organic-based additives commonly used in solvent-based paints, plastisols, printing inks and non-water sealants, caulks and mastics. Its rapid and complete dispersion makes it of great interest. Its ability to reduce the "strike-in" and improve the hiding power is of great interest to solvent based paint formulators.

Toxicity: The basic coating material has been approved by the FDA for use in the following food applications: adhesives, components of paper and paperboard in contact with aqueous and fatty foods and components of paper and paperboard in contact with dry food.

MISSISSIPPI LIME CO., Alton, IL 62002

Precipitated Calcium Carbonates (Technical Grade)

Chemical Analysis:

Calcium Carbonate ($CaCO_3$)	98.60 wt %	Silica (SiO_2)	0.45 wt %
Magnesium Carbonate ($MgCO_3$)	0.60 wt %	Alumina (Al_2O_3)	0.20 wt %
Calcium Sulfate ($CaSO_4$)	0.10 wt %	Ferric Oxide (Fe_2O_3)	0.05 wt %

Physical Properties:

	Heavy M-24	Light M-48	Light M-60	Extra Light M-60	Filler
Particle Shape	Cubical	Acicular	Acicular	Acicular	Acicular
Particle Size (By Electron Microscope)					
Length (μ)	0.3–1.2	0.8–2.0	0.7–1.5	0.7–1.5	0.5–2.5
Thickness (μ)	0.3–1.2	0.2–0.7	0.1–0.5	0.1–0.5	0.1–0.8
Particle Size (By Air Permeability)					
Average (μ)	1.00	0.55	0.43	0.43	0.58
Surface Area (cm^2/g)					
(Blaine Air Permeability)	13,350	21,000	31,000	31,000	20,000
Oil Absorption (ml/100 g)					
(Gardner-Coleman Method)	36–39	42–46	56–60	90–100	80–100
G.E. Brightness (%)	97.0	98.0	98.0	98.0	98.0
Specific Gravity	2.60	2.72	2.85	2.85	2.80
Plus 325 Mesh (%)	0.02	0.01	0.01	0.02	0.02
Apparent Density (lb/ft^3)					
Loose	20	17	16	11	13
Packed	40	32	30	21	27

PFIZER MINERALS, PIGMENTS & METALS DIVISION,640 North 13th St., Easton, PA 18042

Surface Treated, Precipitated Calcium Carbonate

Super-Pflex 200

Chemical Analysis (before treatment):

Calcium Carbonate ($CaCO_3$)	98.4 wt %	Alumina (Al_2O_3)	0.25 wt %
Magnesium Carbonate ($MgCO_3$)	1.2 wt %	Ferric Oxide (Fe_2O_3)	0.05 wt %
Silica (SiO_2)	0.07 wt %		

Physical Properties (after surface treatment):

Bulk Density		Average Particle Size (μ)	0.5
Loose-Scott	23 lb/ft^3	Surface Area (N_2 Absorption)	6.0 m^2/g
Tapped	38 lb/ft^3	Dry Brightness (IDL Color Eye)	94
Specific Gravity	2.7	Wet Out in Water	0%
Oil Absorption (lb/100 lb)	26		

Key Properties: Developed specifically as a reinforcing agent for plastic and rubber compounds. The latest production technology has been employed to chemically bond the surface coating to the particles. This new coating process ensures an extremely uniform surface treatment as well as high purity and brightness.

Especially designed for significant impact strength and flexural modulus improvement in various polymer systems, ease of dispersion and low abrasion. In addition, it improves processability and can be added at high loading levels. Also, lowers overall material and production costs and extends resin supply.

The surface treatment renders the product completely hydrophobic.

Ultrafine Precipitated Calcium Carbonates

Product Name	Specific Gravity	Oil Absorption (lb/100 lb)	Dry Brightness	Loose Bulk Density (lb/ft^3)	Packed Bulk Density (lb/ft^3)
Multifex MM	2.71	55	98	15	24

Chemical Type: Uncoated grade.

Key Properties: Disperses more easily in aqueous systems.

Multifex SC	2.67	40	96	16	22

Chemical Type: Surface treated grade.

Ultra-Pflex	2.65	35	95	17	22

Chemical Type: Surface treated grade.

Key Properties (both Multifex SC and Ultra-Pflex): Specifically surface treated to improve dispersion and physical properties in polymeric systems. Where the utmost in physical properties is desired, Ultra-Pflex should be chosen. Both disperse more easily in nonaqueous media. Both are ideal choices for adhesives and sealants, because of their superior dispersion properties.

Chemical Type (all three grades): Pure synthetic calcite, especially developed as reinforcing agents in plastic and rubber compounds.

Chemical Analysis (all three grades before surface treatment):

	Percent by Weight
Calcium Carbonate ($CaCO_3$)	98
Magnesium Carbonate ($MgCO_3$)	<1
Ferric Oxide (Fe_2O_3)	<0.05
Silica (SiO_2)	<0.1
Alumina (Al_2O_3)	<0.02
Water Loss at 110°C	<0.5

Vicron Ground Limestones (Eastern)

Product Name	Dry Brightness (Hunter)	Hegman Fineness of Grind	Oil Absorption (lb/100 lb)	Surface Area (m²/g)	Bulk Density (lb/ft³)	Specific Gravity
15-15	96	6	21	3.7	37.5	0.60
25-11	96	5	19	2.8	40	0.64
31-6	94	3.5	18	1.6	45	0.72
41-8	94	3.0	18	2.2	49	0.78

Chemical Type: Vicron Ground Limestone is available in a full line of high brightness, high purity extender pigments of carefully controlled particle size distribution.

Chemical Analysis:

Calcium Carbonate ($CaCO_3$)	96.0 wt %	Alumina (Al_2O_3)	0.3 wt %
Magnesium Carbonate ($MgCO_3$)	1.5 wt %	Ferric Oxide (Fe_2O_3)	0.08 wt %
Silicon Dioxide (SiO_2)	1.2 wt %	Moisture (H_2O)	0.25 wt %

Physical Properties (all grades):

Weight per Gallon-Pound	22.57	pH Value	9.4
One Pound Bulks-Gallon	0.04431	Specific Resistance-Ohms	23,000
Specific Gravity	2.71	Particle Shape	Rhombic

Key Properties (all grades): Manufactured from selected natural calcite which is processed in modern plant facilities at Adams, MA, which has been designed to assure uniform and consistent control of product quality.

Each grade is controlled for top-size, size distribution and other important physical properties.

Chemical composition is the same for all grades.

The uses include: adhesives and glues, paint; paper; plastics; printing inks; putty and caulking; rubber; and many other applications.

THOMPSON-HAYWARD CHEMICAL CO., P.O. Box 2383, Kansas City, KS 66110

T-Carb Ultra

Chemical Type: Finest grade. A high purity calcium carbonate made from extremely white natural marble. Water ground and water classified allows extremely tight tolerances to uniform particle size distribution and color specifications.

Key Properties: Excellence in manufacturing technique and quality control enables case in dispersion and low binder demand for highest quality formulations in coatings, paper, plastics, rubber and adhesives.

Typical Uses: Paints—Gloss, semigloss and flat enamels, interior and exterior latex primers. Adhesives—Hot melt, rubber based, rubber sealants, organic caulking compounds, ceramic mastic compounds.

Also used in paper, plastics, rubber and ceramics.

T-Carb Centra

Chemical Type: Intermediate range water ground calcium carbonate, produced from the same high quality marble as the top of the line. Developed to offer a narrow particle size distribution.

Key Properties: Mean particle size (6.5 μ) to enable a wider formulation capability. Color characteristics are still the finest in the industry with a minimum of 96% brightness. Flotation technique allows the lowest possible impurity contamination, providing excellent characteristics in flat interior finishes, semigloss enamels, various plastics systems and for reinforcement to many rubber and vinyl recipes.

Typical Uses: Paints—Interior latex, semigloss, eggshell enamels, flat wall, ceiling and exterior house paints.

Adhesives—Latex and oil based putties and caulks, rubber, synthetic sealants and gaskets.

Also used in plastics and rubber.

T-Carb Endura

Chemical Type: The durable one in the family of water ground, water classified calcium carbonates displays the same high quality of brightness and purity found in T-Carb Ultra and Centra.

Key Properties: Specifically produced to allow a more durable particle size (7.5 μ), yet a much narrower particle distribution range found in most dry ground, air classified grades. Dependable characteristics found in T-Carb Endura are much lower binder demand as well as absence of unwanted high and low particle sizes in distribution. T-Carb Endura is an excellent recommendation for use in exterior paints, putties, joint cements and adhesives.

Typical Uses: Paints—Flat interior, exterior flats, alkyd and latex.
Adhesives—Joint cements, spackling compounds, caulks and putties.
Also used in plastics and rubber.

T-Carb Magna

Chemical Type: Magna is produced from the highest quality calcite available, but is dry ground and air classified.

Key Properties: Most economical member of the T-Carb family. Produced to avail a lower cost extender for lower ranged Hegmans. Provides exceptional characteristics in that "stir-in" quality is introduced wth a much finer particle size distribution than many lower quality dry ground products. Principal applications are low binder demand, high pigment load coatings, urethane foams and thermoset compounds.

Typical Uses: Paints—Low cost flat wall, latex and oil based exterior house paints, economical barn coatings, curing compounds.
Adhesives—Putty, caulking compounds, glazing compounds, polyester gunks, epoxy cold solders.
Also used in rubber.

T-Carb Ultra-D

Chemical Type: Dry ground, air classified equivalent to T-Carb Ultra, is produced from highest purity marble and dry ground to very tight tolerances. The particular distribution displays 62% between 10 and 2 microns in diameter, with a mean particle diameter of 2.5 microns.

Key Properties: Displays many of the same characteristics as T-Carb Ultra, and some of its own. Because of a very closely packed particle size distribution, it is considered a "stir-in" grade with extremely low binder demand. Ultra-white in color. Recommended in many formulations where paint, plastics, rubber and adhesives call for extremely fine particles producing a 6½ Hegman.

Typical Uses: Paints—Gloss, semigloss enamels, interior and exterior latex primers.
Adhesives—Hot melt, rubber-based, rubber sealants, organic caulks, ceramic mastic compounds.
Also used in paper, plastics, rubber and ceramics.

T-Carb Centra-D

Chemical Type: Companion and all purpose dry ground counterpart to T-Carb Centra. The purest of all Alabama marbles is used to produce a univeral calcium carbonate white enough and fine enough to enhance a wide range of formulations.

Key Properties: It is noted for rapid dispersion in high speed equipment and is packed narrowly with a top particle size of less than 30 micons and a mean particle size of 5.4 microns. Multistep classsification allows extremely clean cut particle distribution with no agglomerates. It provides excellent characteristics in enamel, interior coatings, plastics and rubber formulations.

Typical Uses: Paints—Interior latex, semigloss, eggshell enamels, flat wall, ceiling and exterior house paints.
Adhesives—Latex, oil-based putties and caulks, rubber, synthetic sealants and gaskets.
Also used in plastics and rubber.

T-Carb Endura D

Chemical Type: Dry ground companion to T-Carb Endura. Endura D is also a durable calcium carbonate which displays the same high quality found in all dry ground T-Carbs.

Key Properties: When there is a need for larger particle size yet narrower distribution found in most dry ground products, it is often specified with a mean size of 7.5. It provides no agglomerates, no oversized particles, high loading and no binder demand. Endura-D is an excellent recommendation for use in exterior house paints, putty formulations, joint cements and adhesives.

Typical Uses: Paints—Interior and exterior flats.
Adhesives—Joint cements, spackling compounds, caulks and putty.
Also used in plastics and rubber.

T-Carb 2000–2001

Chemical Type: Produced from high quality Alabama marble, but more economical methods of manufacture are used.

Key Properties: Recommended when economy is of prime importance. The top and bottom ends of the particle distribution are controlled to a lesser degree. Excellent results are achieved in formulations for traffic paints, carpet underlay, joint cements, rubber products and flooring products.

Chemical Analysis (Ultra, Centra, Endura, Magna, Ultra-D, Centra-D, Endura-D):

Calcium Carbonate ($CaCO_3$)	98.20 wt %	Alumina (Al_2O_3)	0.36 wt %
Magnesia (MgO)	0.60 wt %	Ferric Oxide (Fe_2O_3)	0.06 wt %
Lime (CaO)	54.97 wt %	Carbon Dioxide (CO_2)	43.85 wt %
Silica (SiO_2)	0.05 wt %		

Chemical Analysis (2000–2001):

Calcium Carbonate ($CaCO_3$)	95.00 wt %	Ferric Oxide (Fe_2O_3)	0.06 wt %
Magnesia (MgO)	3.00 wt %	Alumina (Al_2O_3)	0.36 wt %
Silica (SiO_2)	0.05 wt %	Moisture (H_2O)	0.15 wt %

T-Carb Calcium Carbonates

Physical Properties:

	Ultra	Centra	Endura	Magna	Ultra-D	Centra-D	Endura-D	2000	2001
% Brightness (Luminous Green Filter)	96.0+	96.0+	96.0+	94.0+	94.0+	94.0	95.0	94.0	85.0
Oil Absorption (Spatula Rubout)	15.6	14.2	12.4	12.5	17.0	15.5	13.5	—	—
Oil Absorption (Gardner Coleman)	16.0	14.25	13.75	11.0	16.0	14.0	12.0	—	—
Loose Bulk Density (lb/ft^3)	49.5	59.50	49.50	65.5	48.00	58.0	62.0	64.0	69.5
Weight/One Pound									
Solid-Gallon	22.57	22.57	22.57	22.57	22.57	22.57	22.57	22.57	22.57
Bulk-Gallon	0.0443	0.0443	0.0443	0.0443	0.0443	0.0443	0.0443	0.0443	0.0443
Specific Gravity	2.71	2.71	2.71	2.71	2.71	2.71	2.71	2.71	2.71
pH	9.4	9.4	9.4	9.4	9.4	9.4	9.4	9.4	9.4
Hardness	3.00	3.00	3.00	3.00	3.00	3.0	3.0	3.0	3.0
Refractive Index	1.59	1.59	1.59	1.59	1.59	1.59	1.59	1.59	1.59
Specific Resistance (Ohms)	20,000	20,000	20,000	20,000	20,000	20,000	20,000	20,000	20,000
Grade ASTM D1199	1	1	1	1	1	1	1	11	111

Particle Size Analysis:

	Ultra	Centra	Endura	Magna	Ultra-D	Centra-D	Endura-D	2000	2001
% Retained on									
200 Mesh (74 μ)						—	—	Nil	1.0
325 Mesh (44 μ)	Nil	Nil	Nil	0.3	Nil	0	0.0	99.0	91.8
% Passed									
20 μ	—	—	—	81.0	—	—	—	78.0	62.0
15 μ	99.6	88.0	76.20	70.2	99.60	98.0	86.0	64.3	46.0
10 μ	97.0	76.0	53.00	52.0	96.00	91.0	71.0	41.5	28.5

	Ultra	Centra	Endura	Magna	Ultra-D	Centra-D	Endura-D	2000	2001
% Passed									
5 μ	71.0	38.0	22.10	26.1	66.30	54.0	41.0	15.1	10.0
2 μ	32.0	15.0	7.90	13.0	34.20	26.0	20.0	5.8	2.8
1 μ	20.0	9.0	3.80	4.0	22.00	18.4	14.0	5.0	1.0
Mean Particle Value									
(μ)	3.2	6.5	7.5	12.5	2.5	5.4	7.5	13.0	17.5
Hegman Value	6+	5.0	4.5	3.0	6.5	5.0	4.0	–	–

Section V
Catalysts, Cross-Linking and Curing Agents

BASF Corp., 491 Columbia Ave., Holland, MI 49423

Laromin Epoxy Resin Hardeners

Product Name	Equivalent Weight	Specific Gravity (20°C)	Refractive Index	Boiling Point (°C)
A327	27	0.928	1.482	110

Key Properties: Suitable for both solvent-type and solvent-free epoxy resin formulations. The mixture of reactants should be allowed to stand for some time before application to ensure tack-free coatings.

C252	52	0.916	1.482	108

Key Properties: Used for epoxy resin two-can finishes that can be applied immediately after they have been made up. High gloss coatings with no haze are obtained.

M252	52	0.938	1.493	126
C260	52	0.945	1.495	200

Key Properties: Suitable for solvent-free coatings that yield high gloss coatings that are resistant to chemicals and do not discolor. It is used together with low-viscosity epoxy resins with a high epoxy value.

Chemical Type (of all): Polyamine hardeners suitable for curing epoxy resins.

Key Properties (all grades): Colorless to yellowish, alkaline liquids with a faint ammoniacal odor. The first of the three digits in the Laromin-type numeral indicates the number of nitrogen atoms in the molecule of the hardener. The other two digits represent the equivalent weight with respect to active hydrogen. This figure must be multiplied by the epoxy value of the epoxy resin concerned in order to obtain the stoichiometrical proportion of Laromin per 100 grams of epoxy resin.

CIBA-GEIGY RESINS, Saw Mill River Road, Ardsley, NY 10502

Amine Hardeners

Product Name	Melting Point (°C)	Percent Assay (min)	Water Content (% max)
Eporal	175	98	1

Chemical Type: 4,4'-diaminodiphenyl sulfone (technical), a solid aromatic amine hardener, in a tan, nearly odorless powder form.

Key Properties: It is recommended for use with Araldite 6005 in casting and laminating applications. Some characteristics of the cured systems are high ASTM deflection temperature, excellent thermal stability and outstanding chemical resistance. Long pot life, so accelerators may be used. Recommended uses are adhesives, castings, electrical, laminating, pre-preg and tooling.

Product Name	Viscosity (Brookfield) (cp)	Color (Gardner) (max)	Weight per Gallon (lb)	Flash Point (°F)
Lancast A	718	11	8.0	450 min

Chemical Type: Low viscosity proprietary composition liquid hardener for epoxy resin systems.

Key Properties: Particularly recommended for use with Araldite resins in applications where flexibility or resilience is desired. Resin/hardener mixing ratios are convenient. Mixed systems have the following characteristics: low viscosity, high impact strength, low shrinkage and low exotherm. Typical uses are adhesives, casting, flooring, electrical, laminating-wet lay-up and tooling.

Product Name	Viscosity (Brookfield) (cp)	Color (Gardner) (max)	Weight per Gallon (lb)	Flash Point (°F)	Active Hydrogen Equivalent Weight
Experimental Hardener XU205	3,500	—	9.0	200	56

Chemical Type: Low viscosity clear, dark red liquid aromatic diamine hardener curing agent designed for various structural applications.

Key Properties: Its advantages are: noncrystallizing, nonstaining, long pot life, good mechanical and electrical properties, excellent chemical resistance, good adhesion, low exotherm and low shrinkage. Suggested applications are: adhesives, filament winding, structural laminates and tooling compounds.

Developmental Hardener XU224	75	1	7.9	—	73

Chemical Type: Low viscosity, very lightly colored liquid modified alipahtic amine hardener.

Key Properties: Moisture-insensitive, room temperature curing agent. Can be used in thicknesses ranging from several mils to one inch or more and provides coatings that are free of blush and exudation. Its advantages are long pot life, low temperature cure, resistance to blushing and exudation, excellent adhesion, high gloss, good color retention, good impact resistance, good chemical and water resistance.

Developmental Hardener XU225	5,500	1	8.1	—	143

Chemical Type: Medium viscosity, very light colored, liquid modified aliphatic amine hardener, designed for extremely fast cure in thin sections at room temperature. Compositions using this hardener cure at low temperature and high humidity.

Key Properties: Its advantages are high gloss, good adhesion, low temperature cure, resistance to blushing and exudation and good water and alkali resistance. Suggested applications are adhesives—for rapid cure, casting, coating—for tough, clear and high gloss coatings and floorings—for seamless floors and patchings, even at low temperatures.

Hardeners

Product Name	Viscosity (Brookfield) (cp)	Color (Gardner) (max)	Percent Amino Nitrogen	Amine Value	Weight per Gallon (lb)	Flash Point (°F)
830	4,500	—	6.3	252	9.3	365
850	23,000	—	6.3	252	9.4	360

Chemical Type: Proprietary composition hardener systems designed to cure solvent-free epoxy resin coatings, using 100% solids Araldite epoxy resins.

Key Properties: Different pot lives and cure times can be obtained by varying the ratio between the hardeners. Suggested for applications where hard, tough extremely corrosion resistant coatings are required. Films ranging from 5 mils to more than 120 mils in thickness may be applied. Their advantages are: chemical resistance, complete cure (even in the presence of moisture), low temperature cure, solvent-free, with no solvent hazards, and provides high-build, 100% solids, air-dry coatings. Applications include: tank repair and lining, linings and coatings for concrete and metal pipe, chemical resistant flooring, concrete structure finishes, marine coatings and splash zone coatings.

Product Name	Viscosity (Brookfield) (cp)	Color (Gardner) (max)	Percent Amino Nitrogen	Amine Value	Weight per Gallon (lb)	Flash Point (°F)
Hardener HT939	–	–	–	–	–	–

Chemical Type: Softening point of 221°F with equivalent weight of 93, active hydrogen. Latent modified polyamide hardener in finely ground yellow powder form.

Key Properties: Features latent curing action with epichlorohydrin-bisphenol A resins. Combinations of this latent hardener and epoxy resins are stable at room temperature for at least six months.

Product Name	Viscosity (Brookfield) (cp)	Color (Gardner) (max)	Percent Amino Nitrogen	Amine Value	Weight per Gallon (lb)	Flash Point (°F)
Hardener HV837	3,000	1	10	–	–	158

Chemical Type: Clear, amber, medium viscosity, long chain aliphatic amine adduct hardener for solvent-free, fast curing, chemical and water resistant coatings.

Key Properties: Its advantages are: light color, excellent adhesion, high gloss, resistance to blushing and exudation, high reactivity, low temperature and high relative humidity curing. The viscosity can be reduced and pot life extended by modification with some liquid resins. Its applications are maintenance paints, flooring, tank, pipe and drum linings, in high humidity areas, chemical resistant metal finishes, architectural coatings and marine coatings.

Product Name	Viscosity (Brookfield) (cp)	Color (Gardner) (max)	Percent Amino Nitrogen	Amine Value	Weight per Gallon (lb)	Flash Point (°F)
Hardener HY940	600 M	–	–	–	–	–

Chemical Type: Moderately high viscosity dispersion of a solid polyamide with 41% active hardener content in a liquid epichlorohydrin-bisphenol epoxy resin.

Key Properties: Its advantages are outstanding stability of mixed systems, rapid curing at temperatures as low as 100°C, easy combination with liquid resins, nonsensitive mixing ratio, good adhesion to many substrates and good mechanical properties. Typical applications are adhesives, sealants, plastisols and toolings.

Product Name	Viscosity (Brookfield) (cp)	Color (Gardner) (max)	Percent Amino Nitrogen	Amine Value	Weight per Gallon (lb)	Flash Point (°F)
Hardener HY956	395	3	25.5	–	8.5	200

Chemical Type: Low viscosity, modified proprietary composition amine hardener for use with Araldite epoxy resins.

Key Properties: Exhibits low irritation and skin sensitizing effects on contact, as compared to unmodified aliphatic polyamines, such as diethylenetriamine. Its advantages are: low viscosity and excellent wetting characteristics, low shrinkage and excellent dimensional stability, good electrical properties, good chemical resistance properties, low irritation and skin sensitizing effects. Typical applications are: adhesives, casting, coatings, encapsulation and potting, laminating and tooling.

Product Name	Viscosity (Brookfield) (cp)	Color (Gardner) (max)	Percent Amino Nitrogen	Amine Value	Weight per Gallon (lb)	Flash Point (°F)
Hardener HY2964	55	2	–	–	8.30	226

Chemical Type: Very low viscosity liquid modified aliphatic amine solvent-free curing agent, based on aliphatic and cyclo-aliphatic polyamines. Has 93.5% active hydrogen equivalent weight.

Key Properties: Its advantages are good all-round resistance to chemicals, extremely low viscosity, readily cures with short dust-dry time, even at low temperatures and high relative humidity, resistant to blushing and exudation, excellent surface appearance and high gloss, very pale color, good flexibility and easily applied by conventional spray equipment, by roller and brush. Particularly suitable for high performance, solvent-free coatings, chemically resistant linings for pipe, tanks and containers, sewage treatment facilities, ship hull coatings and mortar and floor repair compounds.

Product Name	Viscosity (Brookfield) (cp)	Color (Gardner) (max)	Percent Amino Nitrogen	Amine Value	Weight per Gallon (lb)	Flash Point (°F)
Hardener HY2969	750	12	6.65	—	9.32	212

Chemical Type: Low viscosity modified aromatic amine adduct (with 115 active hydrogen equivalent) room temperature curing agent for liquid Araldite epoxy resins.

Key Properties: Produces solvent-free systems which are easily processed and possess high chemical resistance. Its advantages are: long pot life, films with good hardness and flexibility, can be blended with polyamine hardeners for faster cure and improved flexibility, low ammonia resistance and is suitable for spray application. Its applications are: coatings for gasoline, fuel oil tanks, metal, concrete and asbestos cement pipes; chemical resistant floorings; sealants; protective coatings for chemical plants, locks, canal installations, ships and effluent treatment plants.

Hardener HY9130	370	12	—	375	7.90	275

Chemical Type: Low viscosity liquid modified aliphatic polyaminoamide room temperature curing hardener for liquid Araldite epoxy resins.

Key Properties: Suitable for solvent-free systems with a relatively long pot life. Has good mechanical properties. Its applications are adhesives, sealants, floorings and coatings.

Araldite HZ949U	F-H (Gardner-Holdt)	10	—	—	8.2	115

Chemical Type: Coreacting etherified resol-type resin (50% solids in butanol) based on bisphenol A for heat-cured coatings.

Key Properties: Designed for use in high performance baked industrial finishes. Its advantages are: fast curing, outstanding hardness and distensibility with excellent resistance to solvents, chemicals and acids, and excellent resistance to sterilization and pasteurization processing. Typical uses include linings for cans and collapsible tubes, protective coatings for pipes and fittings and coatings applications where anticorrosive properties are required.

Hardener 955	700	—	—	—	8.2	300

Chemical Type: Modified liquid amido-amine hardener (equivalent weight of 65) especially developed for use with liquid epoxy resins.

Key Properties: Combines the advantages of aliphatic amines and amino polyamides and is particularly useful for applications on concrete. It has the following advantages: cures well in high humidity atmospheric conditions, bonds well to concrete, has a simple and noncritical mixing ratio and cures well at normal temperature.

Toxicity (all grades): Some of the above are slightly toxic and are skin irritants. Refer to manufacturer's Material Safety Data Sheets.

Araldite Accelerators

Product Name	Molecular Weight	Refractive Index n_D^{25}	Specific Gravity @ 25°C	Weight per Gallon (lb)
062	135	1.500	0.90	7.51

Chemical Composition: Benzyl dimethylamine (BDMA), in water white liquid form.

064	265	1.514	0.97	8.10

Chemical Composition: Tri(dimethylaminomethyl)phenol (DMP-30), in dark red liquid form.

066	151	1.530	1.02	8.54

Chemical Composition: Dimethylaminomethylphenol (DMP-10), in dark red liquid form.

Key Properties: Tertiary amines, generally used in combination with anhydride hardeners for curing epoxy resins. These accelerators are also used with liquid polysulfide polymers and polyamides. They are advantageous for two package systems and for reducing the gel time and cure schedule.

Curing Agent/Accelerator XU213

Physical Properties

Melting Point	28°C
Viscosity @ 30°C	70 cp

Chemical Type: Boron trichloride-amine complex in yellowish brown solid/liquid form.

Key Properties: May be used either as a latent catalytic curing agent for selected epoxy resins or as a latent accelerator for anhydride-cured epoxy resins. Its advantages are exceptional latency at temperatures up to 80°C, both as curing agent and accelerator, high reactivity at temperatures above 120°C, does not degrade electrical properties of the cured systems, readily dissolves in liquid epoxy resins and hardeners—forms stable preaccelerated resins and hardeners.

Toxicity (all of the above): Slightly toxic. Refer to manufacturer's Material Safety Data Sheets.

Araldite Reactive Modifiers

Product Name	Weight per Epoxide	Epoxy Value	Color (Gardner) (max)	Weight per Gallon (lb)	Flash Point (°F)
RD-1	140	0.72	2	7.6	120

Chemical Type: Butyl glycidyl ether, a monoepoxy reactive modifier.

Key Properties: Used primarily as an additive to reduce viscosity of Araldite epoxy resins. It is chemically combined in the cured system. Resins modified with RD-1 are specifically recommended for casting, electrical, laminating, tooling and other applications where very low viscosity is required for good impregnation or maximum filler loading. Mixes easily with liquid epoxy resins.

RD-2	134	0.75	1	9.2	280

Chemical Type: 1,4-Butanediol diglycidyl ether, (technical grade), a diepoxy reactive modifier.

Key Properties: Used primarily as an additive to reduce viscosity of Araldite epoxy resins. It is chemically combined in the system. Resins modified with RD-2 are specifically recommended for casting, electrical, laminating, tooling and other applications where very low viscosity is required for good impregnation, maximum filler loading, etc. and low vapor pressures. RD-2, as compared with RD-1, exhibits the following: higher flash point; higher boiling point; faint, pleasant scent; little effect on pot life and less efficient viscosity reduction.

Epoxide No. 7	227	0.4	1	7.6	200°C

Chemical Type: Essentially a mixture of aliphatic glycidyl ethers where the alkyl groups are predominantly C_8 and C_{10}, a reactive modifier.

Key Properties: Exhibits a low volatility and very low order of toxicity. Used primarily as an additive to reduce the viscosity of epoxy resins. It is chemically combined in the system. Resins modified with Epoxide No. 7 are specifically recommended for casting, electrical, laminating, tooling and flooring where low viscosity is required for good impregnation or maximum filler loading. (Product of Procter & Gamble Co.)

Epoxide No. 8	286	—	15 (APHA)	—	310

Chemical Type: Consists predominantly of C_{12} and C_{14} alkyl groups, a clear water-water aliphatic glycidyl ether.

Key Properties: Particularly suitable for efficient viscosity reduction of liquid epoxy resins. Provides low surface tension, good wetting character and maintains a high level of mechanical and chemical properties. Its advantages are: FDA sanction for coatings in contact with dry bulk food, low toxicity, good flowability, good wetting, low surface tension and good leveling. Its applications are: potting, casting, tooling, laminating, filament winding, coating and floorings. (Product of Procter & Gamble Co.)

Product Name	Weight per Epoxide	Epoxy Value	Color (Gardner) (max)	Weight per Gallon (lb)	Flash Point (°F)
Araldite DY 023	184	0.55	4	9.0	240

Chemical Type: Cresyl glycidyl ether, a reactive diluent for epoxy resins.

Key Properties: Used primarily as an additive to reduce viscosity of Araldite epoxy resins. It is chemically combined in the system. DY 023 produces smaller reductions in viscosity than RD-1, but is less volatile and shows good resistance to water. Can be added to epoxy resins in quantities of 2 to 20% by weight. Recommended for coatings (especially for thick films), flooring, adhesives, casting, impregnating and tooling.

The reduction of viscosity of epoxy systems by addition of any of the above reactive diluents will offer the following advantages:

(1) Easier handling and mixing.
(2) Higher filler loading.
(3) Improved wetting characteristics.
(4) Lower cost of formulation.

Toxicity (all of the above): Practically nontoxic. Refer to manufacturer's Material Safety Data Sheets.

CINCINNATI MILACRON CHEMICALS, West St., Reading, OH 45215

Mercaptate Q-43 Ester

Chemical Type: A polyfunctional mercaptan ester [pentaerythritol tetrakis(mercaptopropionate)]–$C_{17}H_{28}O_8S_4$. Amber colored liquid.

Physical Properties

Molecular Weight	488	Fire Point (C.O.C.)	545°F
Refractive Index @ 25°C	1.530	Purity (SH Iodine Assay)	92%
Specific Gravity	1.280	Volatility	0.1% max
Viscosity @ 25°C	390–400 cs	Color (Gardner)	3
Flash Point (C.O.C.)	490°F	Specific Gravity	1.28

Key Properties: An effective crosslinking agent in radiation curing; Effective chain transfer agent; Relatively nonvolatile material; and Possesses a high flash and fire point.

Toxicology: Considered toxic but not highly toxic by oral ingestion. Exposure to the eyes and skin of rabbits indicates this product is nonirritating.

CROSBY CHEMICALS, INC., 600 Whitney Bldg., New Orleans, LA 70130

Crosby Epoxy Curing Agents

Chemical Type: Complete line of polyamide epoxy curing agents. These resins are made from Crosby dimer acids (Crodym) and linear polyamines.

Product Name	Amine Value	Viscosity (poises)
Cropolamid L-100	90	10 @ 150°C
Cropolamid LM-250	240	35 @ 75°C
Cropolamid LM-350	340	8 @ 75°C
Cropolamid LM-400	380	4 @ 75°C
Cropolamid LVM-450	455	8 @ 25°C
Cropolamid L-520	540	7 @ 25°C

Key Properties: These liquid polyamides range in amine value from 90 through 600. These polyamides are available in common solutions. There is a polyamide epoxy curing agent for most paint and adhesive systems.

HENKEL CORP., 425 Broad Hollow Road, Melville, Long Island, NY 11746

Genamid Epoxy Resin Coreactants

Product Name	Amine Value	Color (Gardner) (max)	Viscosity (Brookfield) (cp) @ 25°C	% Resinous	Specific Gravity @ 25°C	Flash Point (°C)
250	437.5	10	750	100	0.95	231
747	467.5	11	350	100	0.94	210
2000	600	10	1750	100	0.98	217
5701-H65	237.5	9	1100 @ 55°C	65	1.04	212

Chemical Type: Water-based epoxy resin coreactants.

Usages

	250	747	2000	5701-H65
Coatings (General)	X	X	X	X
Maintenance	−	−	−	X
Primers	−	−	−	X
Enamels	−	−	−	X
High Build/High Solids	X	X	X	X
Vinyl Modified	−	−	−	−
Coal Tar Modified	−	−	−	−
Masonry	−	−	−	X
Metallic	−	−	−	X
Paper	−	−	−	−
Flexible	−	−	−	−
Grouts	X	X	X	X
Toppings	X	X	X	X
Water Systems	−	−	−	X

Note: X is recommended usage

PACIFIC ANCHOR CHEMICAL CO., 1145 Harbour Way South, Richmond, CA 94804

Aliphatic Amine Curing Agents

Product Name	Viscosity (Brookfield) (cp)	Color (Gardner)	Weight per Gallon (lb)	Equiv. Weight (per active H)	Amine Value (mg KOH/g)	Gel Time (150 g @ 77°F) (min)
Ancamine AD	1,700	10	9.00	110	490	5

Key Properties: Room temperature cure. An extremely rapid curing agent. Cures under damp conditions, down to 32°F. Recommended for adhesives and fast setting civil engineering applications.

| Sur-Wet R | 5,300 | 6 | 8.15 | 210 | 185 | 75 |

Key Properties: Room temperature cure. Adheres and cures well when applied under water. Recommended for underwater coatings and adhesives.

| Ancamine S-4 | 300 | 11 | 7.90 | 190 | 179 | 55 |

Key Properties: Room temperature cure. Economical hardener with good chemical resistance. Recommended for binder for industrial flooring mortars. Certain electrical potting and foundry applications.

Product Name	Viscosity (Brookfield) (cp)	Color (Gardner)	Weight per Gallon (lb)	Equiv. Weight (per active H)	Amine Value (mg KOH/g)	Gel Time (150 g @ 77°F) (min)
Ancamine S-5	900	16	8.20	380	112	800

Key Properties: Room temperature cure. Economical hardener with very low working life and tolerance to moisture. Primer for flooring mortars and bonding agent for new concrete to old.

| Ancamine S-6 | 100 | 4 | 7.6 | 190 | 149 | 40 |

Key Properties: Room temperature cure. Economical hardener with light color. Binder for epoxy terrazzo.

| Ancamine T | 250 | 1 | 8.6 | 37 | 1145 | 18 |

Key Properties: Room temperature cure. Technical 2-hydroxy ethyl diethylene triamine. Exhibits excellent color and good MEK resistance. Recommended for auto and boat body patch kits, laminating and tooling.

| Ancamine T-1 | 2000 | 4 | 8.9 | 47 | 860 | 12 |

Key Properties: Room temperature cure. Similar to Ancamine T, but shorter pot life and faster cure. Recommended for auto and boat patch kits, laminating and tooling.

| Ancamine XT | 100 | 8 | 8.7 | 42 | 820 | 9 |

Key Properties: Room temperature cure. Combines fast cure rate with good chemical resistance. Recommended for adhesives, coatings, accelerator for polyamides.

| EDA Adduct 870 | Solid | 3 | 8.4 | 200 | 190 | — |

Chemical Type: Elevated cure temperature. Pure isolated adducts of EDA and DETA with an epoxy resin. Has softening point of 230°F.

| Ancamine 1062 | Solid | 2 | 8.4 | 155 | 290 | — |

Chemical Type: Elevated cure temperature. Pure isolated adducts of EDA and DETA with an epoxy resin. Has softening point of 206°F.

Key Properties (both of the above): Approved under Section 175.300 for use in coatings in contact with food. Products can be used to accelerate DICY or novolacs in epoxy powder coatings. Coatings show better heat resistance than DICY cures. Recommended for epoxy powder can coatings and sanitary food liners. Very rapid, elevated temperature curing, one component adhesives molding powders and gunk molding compounds with both liquid epoxy resin as a dispersion and solid epoxy resin as a powder blend.

| Ancamine 1062-BTC-42 | 300 | 3 | — | 375 | — | — |

Chemical Type: Room temperature cure. Isolated DETA/epoxy adduct solution in n-butanol, Cellosolve and toluene.

Key Properties: In comparison to conventional adduct solutions, provides lower toxicity, no fuming and lower film blushing. Recommended for solvent based coatings, mastic coatings and resistance to boiling water. Conforms to Rule 66 and FDA 175.300.

| Ancamine 1483 | 7,700 | 4 | 9.1 | 45 | 820 | 22 |

Key Properties: Room temperature cure. Used in castings, adhesives and laminates where rigidity and high physical strength are required. Recommended for laminates, laminate gel coats, patch repair kits, small castings.

| Ancamine 1510 | 50 | 3 | 7.6 | 60 | 515 | 25 |

Key Properties: Room temperature cure. Coatings exhibit reduced yellowing on exposure to U.V. Gives very tough cure with excellent electrical properties. Recommended for solventless coatings when applied in thick films. Also, for electrical potting compounds.

Product Name	Viscosity (Brookfield) (cp)	Color (Gardner)	Weight per Gallon (lb)	Equiv. Weight (per active H)	Amine Value (mg KOH/g)	Gel Time (150 g @ 77°F) (min)
Ancamine 1608	3,000	6	9.0	38	795	15

Key Properties: Rapid cure at room temperature. Good chemical resistance. Recommended for laminates, gel-coats, adhesives and small castings.

Ancamine 1636	20	2	8.2	38	950	80

Key Properties: Room temperature cure. Designed for use with CTBN-modified epoxy resins to provide high peel strength after room temperature cure. Recommended for coatings, adhesives, laminates. Combination of long pot life, room temperature cure and excellent adhesion to glass and carbon fiber for filament winding applications.

Ancamine 1637	4,500	12	9.1	50	690	15

Key Properties: Room temperature fast cure. Good chemical resistance. Cures epoxy novolacs at room temperature. Recommended for laminates, adhesives and solventless coatings.

Ancamine 1638	150	4	8.6	31	1070	13

Key Properties: Room temperature cure. Combines fast cure rate with good chemical resistance and high heat distortion temperature. Recommended for adhesives, coatings and accelerators for other hardeners.

Ancamine 1644	3,000	2	8.1	141	390	6

Key Properties: Room temperature cure. Resistant to blushing and formation of oily surface film during resin cure. Recommended for floor coatings, fast-setting adhesives, cold weather patching compounds, 100% solid coatings and decoupage.

Ancamine 1767	6,000	2	8.1	188	320	6

Key Properties: Similar to Ancamine 1644, but more flexible.

Ancamine 1768	300	2	8.1	94	–	2

Key Properties: Similar to Ancamine 1644, but more rigid.

Ancamine 1769	550	4	8.4	41	1060	25

Key Properties: Room temperature cure. Standard hydroxy alkyl amine hardener, similar to Ancamine T, but lower toxicity. Recommended for auto and boat body patch kits, laminating and tooling.

Amido-Amine Curing Agents

Product Name	Viscosity (Brookfield) (cp)	Color (Gardner)	Weight per Gallon (lb)	Equiv. Weight (per active H)	Amine Value (mg KOH/g)	Gel Time (150 g @ 77°F) (min)
Ancamide 500	300	10	7.9	90	445	130

Key Properties: Room temperature cure. Combination of low viscosity and good adhesion makes it suitable for concrete coatings. Long pot life.

Ancamide 501	600	10	8.2	65	550	50

Key Properties: Room temperature cure. Cures in the presence of moisture and shows excellent adhesion to concrete.

Ancamide 502	350	10	8.0	90	460	80

Key Properties: Similar to Ancamide 500, but higher reactivity.

Ancamide 503	700	10	7.9	95	435	60

Key Properties: Similar to Ancamide 500 and 502, but more reactive than both.

Key Properties (all of the above): This series of curing agents is used mainly in civil engineering applications. Examples include surface coatings for concrete, bonding new concrete to old, grouts, adhesives, tooling, potting compounds and flooring mortars and high solids coatings.

Product Name	Viscosity (Brookfield) (cp)	Color (Gardner)	Weight per Gallon (lb)	Equiv. Weight (per active H)	Amine Value (mg KOH/g)	Gel Time (150 g @ 77°F) (min)
Ancamide 506	400	11	7.9	115	400	350

Key Properties: Room temperature cure. Similar to Ancamide 500, but slow curing. Useful where long pot life is required or large masses are involved. Combination of long pot life and low viscosity means this product is suitable for high solids epoxy coatings to meet new CARB regulations.

| Ancamide 507 | 1,500 | 9 | 8.2 | 65 | 600 | 35 |

Key Properties: Room temperature fast cure. Cures in the presence of moisture. Improved chemical resistance. Intended to meet MIL-P-24441. Recommended for high solids coatings, concrete floor toppings, adhesives and laminates.

Product Name	Viscosity (Brookfield) (cp)	Color (Gardner)	Weight per Gallon (lb)	Equiv. Weight (per active H)
Ancamine S-475	30	8	9.1	—

Chemical Type: 50% solution of morpholinium-p-tosylate in ethylene glycol monoethyl ether.

Key Properties: Catalyst permitting reduction of cure schedule required for epoxy/PF resin and epoxy/UF resin coatings systems. Recommended for industrial epoxy/PF and epoxy/UF coatings.

Aromatic Amine Curing Agents

Product Name	Viscosity (Brookfield) (cp)	Color (Gardner)	Weight per Gallon (lb)	Equiv. Weight (per active H)	Amine Value (mg KOH/g)	Gel Time (150 g @ 77°F) (min)
Ancamine LO	1,800	16	9.4	99	280	35

Key Properties: Cures at room temperature. Similar to Ancamine LT, but has low odor and presents low dermatitic hazard. Recommended for floorings and coatings.

| Ancamine LOS | 300 | 16 | 9.13 | 99 | 280 | 17 hr |

Key Properties: Cures at room temperature. Used in combination with Ancamine LO to provide a wide range of cure times from 35 minutes to 17 hours at room temperature. Recommended for flooring and coatings. Can be used for large room temperature coatings where low exotherm is required.

| Ancamine LT | 2,000 | 18 | 9.65 | 99 | 280 | 28 |

Key Properties: Low temperature curing agent. Cures down to 25°F and also under water. Offers the ultimate in chemical resistance to organic acids. Recommended for flooring, coatings for chemical and oil tanks.

| Ancamine TL | 3,000 | 18 | 9.26 | 120 | 245 | 35 |

Key Properties: Room temperature cure. Coatings based on TL show low extractability when immersed in distilled water, 5% acetic acid, 25% ethyl alcohol and heptane. Recommended for tank linings.

| Ancamine TLS | 3,000 | 18 | 9.26 | 115 | 255 | 16 hr |

Key Properties: Room temperature cure. Used in combination with Ancamine TL to provide a wide range of cure times from 35 minutes to 16 hours. Recommended for tank linings.

| Ancamine 1482 | 800 | 18 | 9.6 | 37 | 784 | 4 hr @ 122°F |

Key Properties: Elevated temperature cure. Liquid eutectic designed for use in place of MDA because of easier handling. Recommended for preimpregnated laminates and filament-wound pipe.

Product Name	Viscosity (Brookfield) (cp)	Color (Gardner)	Weight per Gallon (lb)	Equiv. Weight (per active H)	Amine Value (mg KOH/g)	Gel Time (150 g @ 77°F) (min)
Ancamine 1648	2,000	18	9.6	104	255	4

Key Properties: Room temperature cure. Very fast version of Ancamine LT. Recommended for adhesives.

Ancamine A58/6	Solid: Soft. Pt of 176°F	11	—	—	290	—

Chemical Type: Isolated adduct of aromatic amine and epoxy resin.

Key Properties: Gives high flexibility and good chemical resistance in powder coatings. Recommended for powder coatings for pipe.

Boron Trifluoride Amine Complex Curing Agents

Product Name	Viscosity (Brookfield) (cp)	Color (Gardner)	Weight per Gallon (lb)	Equiv. Weight (per active H)	Amine Value (mg KOH/g)	Gel Time (150 g @ 77°F) (min)
Ancaflex 70	1,100	18	8.9	—	—	8 hr
Ancaflex 150	1,100	13	7.8	—	—	8 wk

Chemical Type (both of the above): These amine complexes of boron trifluoride are flexibilized integrally.

Key Properties (both of the above): Flexibilized integrally and impart a marked degree of resilience and extensibility to epoxy cures. Recommended for adhesives and oil-resistant sealants.

Anchor 1040	8,400	15	9.4	—	—	6–10 wk

Key Properties: Recommended for powder coatings, epoxy molding powders and laminates.

Anchor 1115	1,700	17	9.6	—	—	6–10 wk

Key Properties: Recommended for laminates, encapsulation and insulating varnishes.

Anchor 1170	2,400	14	10.4	—	—	2–3 hr

Key Properties: Recommended for laminates, adhesives and stopping compounds.

Anchor 1171	12,000	14	10.2	—	—	6–8 hr

Key Properties: Recommended for hot dip coatings, capacitors and coil impregnation.

Anchor 1222	2,200	9	9.2	—	—	6 mo

Key Properties: Recommended for one-pack coatings, laminates, molding powders and vacuum impregnation of electrical motors.

Chemical Type (all of the above): Chemically modified amine complexes of boron trifluoride.

Key Properties (all of the above): All nonhygroscopic and freely compatible with epoxy resins. One component epoxy/amine complex mixes may be prepared having an extended shelf life (according to complex choice). High HDT's are obtained upon heat cure, the HDT attained being dependent upon complex choice, loading choice and cure conditions employed.

Ancaflex 1532	2,200	18	10.8	—	—	—

Key Properties: Room temperature cure. This complex has a very rapid cure rate and exhibits good adhesion and flexibility. Recommended as rapid cure metal-to-metal adhesive.

Cyclo-Aliphatic Amine Curing Agents

Product Name	Viscosity (Brookfield) (cp)	Color (Gardner)	Weight per Gallon (lb)	Equiv. Weight (per active H)	Amine Value (mg KOH/g)	Gel Time (150 g @ 77°F) (min)
Ancamine MCA	250	13	8.6	102	305	20

Key Properties: Room temperature cure. Exhibits excellent adhesion to cold wet concrete. Good color stability for coating applications. Recommended for flooring and coatings.

Ancamine 1561	90	2	—	85	330	20

Key Properties: Room temperature cure. Good color stability in solventless coatings. Recommended for solventless coatings, flooring, concrete repair compounds.

Ancamine 1618	420	1	8.6	115	275	40

Key Properties: Room temperature cure. Excellent color, high gloss, nonblush makes 1618 ideal for clear coatings. Recommended for solventless coatings, self-levelling flooring and mortars.

Ancamine 1680	70	13	8.7	85	345	22

Key Properties: Room temperature cure. More flexible modification of Ancamine 1721. Recommended for flooring, patching compounds and coatings.

Ancamine 1693	100	4	8.7	95	310	43

Key Properties: Room temperature cure. Light colored curing agent providing good gloss and gloss retention. Recommended for solventless coatings and self-leveling flooring.

Ancamine 1704	100	15	8.7	78	450	18

Key Properties: Room temperature cure. Harder version of Ancamine 1721. Recommended for flooring and coatings.

Ancamine 1721	100	15	8.7	76	450	18

Key Properties: Curing agent designed as lower cost alternative for Ancamine MCA. Recommended for floorings and coatings.

Ancamine 1770	10	1	7.8	29	982	100

Chemical Type: Unmodified cycloaliphatic diamine.

Key Properties: Lower cost and similar performance to isophorone diamine (IPD) and Iaromin C260. Recommended for pottings, castings and laminating.

Polyamide Curing Agents

Product Name	Viscosity (Brookfield) (cp)	Color (Gardner)	Weight per Gallon (lb)	Equiv. Weight (per active H)	Amine Value (mg KOH/g)	Gel Time (150 g @ 77°F) (min)
Ancamide 100	2,500 (60% solution in xylol)	9	—	400	120	—

Key Properties: Room temperature cure. Very flexible polyamide. Normally sold as a solvent cut because of high viscosity. Recommended for solvent-based coatings (with solid epoxy resins), as an additive for nonreactive polyamide-based inks and for hot-melt adhesives.

Ancamide 220, 220x70	330,000	9	8.0	190	240	260

Key Properties: Room temperature cure. Standard polyamide offering good flexibility. Cure can be accelerated with 3 phr of K-54. 70% solids cut in xylol is also available. Can be used with solid resins, in the formulation of solvent-based surface coatings.

Product Name	Viscosity (Brookfield) (cp)	Color (Gardner)	Weight per Gallon (lb)	Equiv. Weight (per active H)	Amine Value (mg KOH/g)	Gel Time (150 g @ 77°F) (min)
Ancamide 260A	37,500	8	8.0	120	360	150

Key Properties: Room temperature cure. Standard polyamide. Lower viscosity than Ancamide 220. Recommended for surface coatings, adhesives and sealants, small castings, encapsulation, laminates, cables, jointing compounds.

| Ancamide 350A | 13,000 | 10 | 8.0 | 110 | 375 | 110 |

Key Properties: Room temperature cure. Standard polyamide. Lower viscosity than Ancamide 220 or 260A. Recommended for surface coatings, flooring adhesives, small castings, laminates and encapsulation.

| Ancamide 400 | 1,500 | 11 | 8.1 | 95 | 400 | 80 |

Key Properties: Room temperature cure. Unlike most polyamides, Ancamide 400 is compatible with bisphenol F resins. Low viscosity makes it useful for solventless coatings. Recommended for floorings, coatings and adhesives.

| Anquamine 100 | 7,000 | 12 | 8.3 | 200 | 370 | 60 |

Key Properties: Room temperature cure. Modified liquid polyamide designed for water dispersible coatings. Recommended for coatings for walls, floors, etc. Reinforcing latex additive for epoxy cement screeds.

| Ancamide 700-B-75 | 6,000 | 10 | 8.0 | — | 248 | — |

Chemical Type: Ancamide 350A/epoxy adduct (75% solids in n-butanol).

Key Properties: Provides good epoxy resin compatibility and better cure under adverse conditions than standard polyamides. Intended to meet MIL-P-24441. Recommended for solvent-based coatings.

Tertiary Amine Curing Agents

Product Name	Viscosity (Brookfield) (cp)	Color (Gardner)	Weight per Gallon (lb)	Equiv. Weight (per active H)	Amine Value (mg KOH/g)	Gel Time (150 g @ 77°F) (min)
Ancamine K54	300	6	8.2	—	615	—

Chemical Type: Tris(dimethylaminomethyl)phenol.

Key Properties: Accelerator for polyamide, anhydrides, amine, and polysulfide cures. Catalyst for isocyanate/polyol reaction. Recommended as epoxy cure accelerator (various applications). Rigid and semirigid polyurethane foam catalyst.

| Ancamine K61-B | 600 | 8 | 7.95 | — | 230 | 8 hr |

Chemical Type: Tri-2-ethyl hexoate salt of K-54.

Key Properties: Exhibits longer pot life, lower toxicity and lower exotherm than parent amine K-54. Recommended for small and medium-size castings, wet lay-up laminates, potting compounds.

| Ancamine 1110 | 300 | 13 | 7.4 | — | 370 | — |

Chemical Type: Dimethylaminomethylphenol. Slower version of K-54.

Key Properties: Recommended for adhesives, flooring, coatings and castings.

THE SHERWIN-WILLIAMS CO., Chemicals Division, 1500 Higgins Road, Park Ridge, IL 60068

Benzyldimethylamine

Chemical Type: N,N-dimethylbenzylamine or N,N-dimethylbenzenemethanamine ($C_9H_{13}N$), with a molecular weight of 135.2.

Chemical Analysis and Physical Properties

Appearance	Clear, nearly colorless liquid	Ionization Constant (K)	8.5×10^{-6}
		Assay	98.8%
ASTM Distillation (below 170°C)	10%	Moisture (Karl Fischer)	<0.1%
ASTM Distillation (below 182°C)	95%	Chloride	<0.1%
Flash Point (Tag Open Cup)	142°F (61°C)	Specific Gravity @20°C	0.900
Solubility in Water @ 25°C	1.1%	Refractive Index @ 25°C	1.498

Key Properties: Functions as a relatively slow acting tertiary amine type catalyst in most epoxy systems, affording good pot life, low color, medium high heat distortion temperatures and good electrical properties. It is a satisfactory replacement for o-methylbenzyldimethylamine.

Toxicity: Benzyldimethylamine is corrosive to the skin, can be absorbed through the skin and is an eye irritant. Oral LD_{50} (rat), 239 mg/kg and dermal LD_{50} (rabbit), 2,510 mg/kg. Inhalation should be avoided. Approved hand and eye protection should be used.

Epoxy Resin Curing Agents

Chemical Type **(MXDA)**: m-Xylylenediamine, with molecular weight of 136.13.

Chemical Type **(1,3-BAC)**: 1,3-Bis(aminomethyl)cyclohexane, with molecular weight of 142.18.

Chemical Analysis and Physical Properties

	MXDA	1,3-BAC
Molecular Weight	136.13	142.18
Appearance	Colorless liquid	Colorless liquid
Solidification Point (°C)	14.1	<-70°
Viscosity @20°C	6.8 cp	9.06 cp
Flash Point (Open Cup) (°C)	134	107
Vapor Pressure	15 mm Hg @ 145°C	14 mm Hg @ 20°C
Vapor Pressure	760 mm Hg @ 247°C	760 mm Hg @ 220°C
Specific Gravity (d_4^{20})	1.050	0.943
Color (APHA)	5	5
Assay	99.5	99.7
p-Xylylenediamine	0.3%	—
3-Methylbenzylamine	0.005%	0.05%
3-Cyanobenzylamine	0.02%	—
1-Methyl-3-Aminomethylcyclohexamine	—	0.10%

Key Properties (both): Versatile chemicals which offer a variety of applications, including use as epoxy curing agents. Studies show that curing is possible at room temperature and handling is easier and safer because of low vapor pressure. The films have much lower tendency to blush and better thermal degradations than films made with conventional curing agents. The Polymer Industry at the Univeristy of Detroit has conducted studies which confirm the excellent properties which both offer as curing agents, including: (1) Good chemical resistance; (2) Excellent film transparency; (3) Room temperature curing; (4) Longer pot life; (5) Little or no blushing; (6) High adhesion; (7) Low heat evolution; and (8) Low vapor pressure.

Toxicology: The compounds are regarded as corrosive to skin and may cause sensitization. Aerosol inhalation should be avoided. Prolonged overexposure to vapor may produce asthmalike symptoms. Adequate ventilation or other engineering controls should be used to reduce employee exposure below OSHA permissible limits. If controls are not adequate or available, use a respirator approved by NIOSH/MESA under schedule TC-23C for protection against *not more* than 1,000 ppm organic vapors, dusts, fumes and mists with a permissible exposure limit of not less than 0.05 mg/m³ or 2 mppcf based on an eight-hour time-weighted average.

SYNTHRON, INC., 44 East Ave., Pawtucket, RI 02860

Actiron

Chemical Type: Actiron NX products are a series of dimethylaminomethylphenol compounds that have found widespread use as accelerators and hardening agents for epoxy resins. Actiron NX-1: Dimethylaminomethylphenol and Actiron NX-3: 2,4,6-Tris(dimethylaminomethyl)phenol.

Chemical Analysis and Physical Properties

	NX-1	NX-3
Appearance	Amber Liquid	Amber Liquid
Specific Gravity @ 25°C	1.023	0.973
Refractive Index	1.530	1.514
Distillation Range	78% at 80°–130°C under 2 mm Hg	96% at 130°–160°C under 1 mm Hg
Odor	Phenolic	Amine
Flash Point	100°C min	150°C min

Key Properties: Actiron NX-1 and NX-3 are used widely as accelerators and hardening agents for epoxy resins. In epoxy systems, they react similarly to other tertiary amines except that their activity is enhanced by the phenolic structure. Actiron NX accelerators are particularly suited for low temperature (25°C) cures. NX-3, at 8 to 12 parts by weight per 100 parts of epoxy resin, gives a short pot life of 30 minutes. NX-1, used at the same concentration, gives a more extended pot life of several hours. Mixtures of NX-1 and NX-3 give intermediate pot lives to allow maximum flexibility in formulating epoxy systems. Both are especially recommended for use:

(1) In flexibilized epoxy-polysulfide formulations.
(2) With polyamine and anhydride epoxy hardeners.
(3) Can be used to increase the amount of amine used in an epoxy resin formulation and have a relatively long pot life.
(4) As an effective additive in naphthenate printing inks.
(5) As chemical intermediates for many types of compounds with applications in many areas.

Toxicity: Both are moderately toxic and care should be taken to prevent skin or eye contact and inhalation should be avoided.

TEXACO, INC., 2000 Westchester Ave., White Plains, NY 10650

Jeffamine Polyoxypropyleneamines

	D-230	D-400	D-2000	T-403
Type of Amine	Diamine	Diamine	Diamine	Triamine
Total Acetylables, meq/g	8.75	5.17	1.01	6.75
Total Amines, meq/g	8.45	4.99	0.96	6.45
Prim. Amines, meq/g	8.30	4.93	0.95	6.16
Water, Weight %	0.10	0.13	0.10	0.08
Color, Pt-Co	30	50	100	10
Specific Gravity @ 20°/20°C	0.9480	0.9702	0.9964	0.9812
Viscosity @ 20°C (cs)	14.4	30	344	97
Flash Point (C.O.C.) (°F)	256	347	460	380
Flash Point (P.M.C.C.) (°F)	255	330	395	340
pH (5% Aqueous Solution)	11.7	11.6	10.5	11.6

Chemical Type: Low viscosity, light colored liquids exhibiting low vapor pressure and high primary amine content.

Key Properties: Recommended for the following uses:

(1) Curing agent for epoxy resin systems.
(2) In preparing polyamide, polyurea and polyurethane systems.

(3) In adhesive, elastomer and foam formulations.

(4) As intermediates for textile and paper treating chemicals.

(5) Polymeric derivatives may be useful as dispersants in paints, fuels and lubricants and as viscosity improvers for lube oils.

Thancat Amine Catalysts

Product Name	Boiling Range (°C)	Color (Pt-Co Scale) (max)	Flash Point (T.C.C.) (°F)	Freezing Point (°F)	Specific Gravity (20°/20°C)	Water Content (% by wt) (max)
Thancat NEM**	131–141	15	90	−63	0.914	0.2
Thancat NMM**	112–118	15	61	−66	0.920	0.3
Thancat DME**	130–137	20	105	−59	0.888	0.5
Thancat DMP**	125–135	75	72	−1.8	0.852	0.5
Thancat DM-70**	150+	4*	102	−32	0.992	0.5
Thancat DD**	202–217	30	170	−49	0.848	0.5
Thancat TD-33	—	100	205***	−25	1.044	0.5

 *Gardner Scale

 **Contains tertiary nitrogen

 ***Pensky-Martens Closed Tester

UNION CAMP CORP., 1600 Valley Road, Wayne, NJ 07470

Uni-Rez Epoxy Curing Agents

Product Name	Major Uses	Solvent	% Solids Content	Amine No.	Viscosity (Brookfield) @ 25°C (cp)	Color (Gardner)
2100	Adhesives, Coatings	—	100	90	1,000 @ 150°C	7
2100-C80	Coatings	Ethylene Glycol Ethyl Ether	80	72	38,000	7
2115	Coatings	—	100	240	3,300 @ 75°C	8
2115-C70	Coatings	Ethylene Glycol Ethyl Ether	70	165	1,800	7
2115-I75	Coatings	Isopropanol	75	180	3,000	7
2115-P80	Coatings	n-Propanol	80	192	5,300	7
2125	Adhesives, Coatings, Grouts, Castings	—	100	350	800 @ 75°C	8
2140	Adhesives, Coatings, Grouts, Coatings	—	100	380	350 @ 75°C	8
2180-B75	Coatings	n-Butanol	75	248	2,500 @ 40°C	7
2188	Coatings	n-Butanol	75	255	2,100	8
2341	Adhesives, Coatings, Grouts, Castings	—	100	410	2,400	17
2355	Adhesives, Coatings, Floor Toppings	—	100	540	800	8

Product Name	Major Uses	Solvent	% Solids Content	Amine No.	Viscosity (Brookfield) @ 25°C (cp)	Color (Gardner)
2392	Coatings, Grouts, Castings	—	100	500	600	Dark
2400	Coatings	(Toluene:Iso-propanol—1:1)	60	55	1,250	6
2401	Coatings	(Xylene:Ethyl-ene Glycol Ethyl Ether—9:1	60	55	2,700	6
2415	Coatings	Xylene	70	168	900	7
2800	Adhesives, Coatings, Grouts, Castings, Floor Toppings	—	100	445	350	9
2810	Adhesives, Coatings	—	100	600	1,400	7
2811	Adhesives, Grouts, Castings, Floor Toppings	—	100	480	520	8
2850	Adhesives, Coatings, Grouts, Castings, Floor Toppings	—	100	440	600	8

Uni-Rez Epoxy Curing Agents for Water-Based Coatings

Product Name	Major Uses	Solvent	% Solids Content	Amine No.	Viscosity (Brookfield) @ 25°C (cp)	Color (Gardner)
2510	Coatings	—	100	275	1,400 @ 75°C	8
2511	Coatings	Ethylene Glycol Butyl Ether	80	215	6,300	7

R.T. VANDERBILT CO., INC., 30 Winfield St., Norwalk, CT 06855

Vanoxy Ketimine Curing Agents

Product Name	Viscosity (Brookfield) (cp)	Color (Gardner) (max)	Amine Value	Equiv. Weight	Weight per Gallon (lb)	Flash Point (O.C.) (°F)
H-1	8.5	5	16.5	52	7.3	172
H-2	3.5	8	13.5	55	7.1	172
H-3	350	8	10.5	101	8.1	230

Key Properties (all grades): Gives latent room temperature cures with Vanoxy resins. Ideally suited for use in high solids coating with sufficient pot life to permit use of spray equipment.

Section VI

Clays

COMMERCIAL MINERALS CO., 6899 Smith Ave., Newark, CA 94560

Lincoln Clay C301

Chemical Type: Inexpensive, finely particled filler for adhesives, stains, foundry facings, etc. Produced by kiln drying and roll-mill pulverizing the best quality fire clay from the Lincoln area of the Sierra Nevada foothills. The base material is similar to Ione clays of the highest grades.

Physical Properties

Specific Surface Diameter	1.4 μ
Through 325 Mesh	99%
Color	Beige
Relatively free of grit	

Chemical Analysis (dry basis)

	Percent by Weight
Silica (SiO_2)	53.5
Alumina (Al_2O_3)	31.6
Ferric Oxide (Fe_2O_3)	2.4
Calcium Oxide (CaO)	0.01
Magnesium Oxide (MgO)	0.01
Moisture (H_2O)	3.0
Loss on Ignition	9.55

GEORGIA KAOLIN CO., 433 North Broad St., Elizabeth, NJ 07207

Chemical Type: Hydrite Kaolinites are carefully processed to meet rigid specifications. By using centrifugal fractionation techniques, Georgia Kaolin can offer a wide range of products with precisely-controlled particle size distribution.

Product Name	Particle Size (μ)	pH (20% Solids Solution)	Oil Absorption (Gardner-Coleman)	Brightness (G.E.) (%)	Plus 325 Mesh (% Max)	Surface Area (Sq.M./ Gram)
Flat D	4.5	4.7	34	81.6	0.25	7

Chemical Type: Large particle size, lower water and oil demand.

Key Properties: Good enamel holdout, suspension and hiding power. Low chalking rate. Improves physical properties, nonreactive, permits high loadings. Reduces shrinkage. Very low modulus characteristics. Nonreactive, thixotropic. Controls penetration. Recommended for paints: interior emulsions, primer systems, exterior emulsions, exterior oleoresinous; plastics: bulk molding compounds and preform epoxy molding, phenolic molding; mechanical rubber goods and epoxy and water type adhesives.

Product Name	Particle Size (μ)	pH (20% Solids Solution)	Oil Absorption (Gardner-Coleman)	Brightness (G.E.) (%)	Plus 325 Mesh (% Max)	Surface Area (Sq.M./ Gram)
MP	9.5	4.7	30	81.0	0.5	6

Chemical Type: Largest particle size available, lowest water and oil demand.

Key Properties: Highest flatting efficiency, low burnishing, improves color uniformity and lapping characteristics, minimum chalking. Permits highest loading and gives best physical properties. Least penetration and least thixotropic. Recommended for paints: flatting agent for oleoresinous and emulsion paints, and exterior paints; plastics: phenolic molding compounds; adhesives: epoxy type and water type.

R	0.77	4.7	41	85.8	0.03	10

Chemical Type: Good brightness, free of coarse particles, medium oil demand.

Key Properties: Excellent color, stain removal, hiding power and suspension. Rheological control and strike-in control. Reduces cost with good finish. Recommended for interior emulsion paints, primer systems, universal tints; bulk molding compound plastics; water base adhesives and letterpress inks.

RS	0.77	7.0	38	85.8	0.03	10

Chemical Type: Predispersed form of Hydrite R.

Key Properties: Good physical, electrical and flow characteristics. Recommended for mechanical and electrical rubber goods; bulk molding plastic compounds and thermoplastic calendered goods.

121	1.5	4.7	39	83.8	0.15	8

Chemical Type: Intermediate particle size and intermediate water and oil demand.

Key Properties: Good stain removal, hiding power and suspension. Flow control. Recommended for interior emulsion paints, primer systems and polyester premix plastics.

UF	0.20	4.7	47	83.5	0.03	21

Chemical Type: Finest particle size kaolinite commercially available with highest water and oil demand.

Key Properties: Highest gloss characteristics, best suspension characteristics, highest thixotropic characteristics, least abrasion. Recommended for gloss and semigloss paint emulsions, primer systems, thixotropic aid in adhesives and caulks, in letterpress inks and in engineering plastics.

PX	0.68	4.7	43	88.5	0.03	12

Chemical Type: Good brightness, fine particle size, high water and oil demand.

Key Properties: Excellent gloss, color stain removal, hiding power and suspension. Excellent rheological properties and excellent flow control. Recommended for interior emulsions, primer systems, emulsion floor paints; bulk molding compounds and gel coats, plastics and letterpress inks.

PXS	0.68	7.0	42	88.5	0.03	12

Chemical Type: Predispersed form of Hydrite PX.

Key Properties: High modulus characteristics. Recommended for mechanical rubber goods.

Chemical Analysis (of all grades):

	Percent by Weight
Aluminum Oxide (Al_2O_3)	38.38
Silicon Dioxide (SiO_2)(Combined)	45.30
Ignition Loss at 950°C (Combined Water)	13.97
Iron Oxide (Fe_2O_3)	0.30
Titanium Dioxide (TiO_2)	1.44
Calcium Oxide (CaO)	0.05
Magnesium Oxide (MgO)	0.25
Sodium Oxide (Na_2O)	0.27
Potassium Oxide (K_2O)	0.04

Chemical Reactivity: Kaolinites are chemically inert and react with acids and bases only under extreme conditions. Hydrite Kaolinites are water processed to reduce soluble salt contents to extremely low levels.

Physical Properties (of all grades):

Refractive Index	1.56
Weight per Solid Gallon	21.66 pounds
Bulking Value	0.04617 gallon/pound
Moisture	1% maximum
Specific Gravity	2.58
Abrasion Index	Very low
Moh's Hardness Scale	No. 2

MINERALS AND CHEMICALS DIVISION, Engelhard Corp., Menlo Park, Edison, NJ 08817

Attapulgus Clay Products

Chemical Type: Attapulgus clay, commonly called attapulgite after the principal mineral it contains, is a crystalline hydrated magnesium aluminum silicate with a unique chain structure that gives it unusual colloidal and sorptive properties. It is the principal member of a group of sorptive clays known collectively as fuller's earth. Attapulgite is the principal mineral constituent (70–80%), but it also contains 10–15% montmorillonite, sepiolite and other clays, 4–8% quartz and 1–5% calcite or dolomite. Attapulgite derives two unusual characteristics from its unique hydrated magnesium aluminum silicate structure. First, because the structure consists of three dimensional chains, it cannot swell like clays such as montmorillonite which has three layer sheets. Second, specific clearage of the crystal structure yields a porous attapulgite clay, that is high in surface area, sorptivity and decolorizing power with its slips or slurries which are viscous and thixotropic.

Chemical Analysis (volatile-free basis):

	Percent by Weight
Silica (SiO_2)	68.0
Aluminum Oxide (Al_2O_3)	12.0
Magnesium Oxide (MgO)	10.5
Ferric Oxide (Fe_2O_3)	5.0
Calcium Oxide (CaO)	1.7
Phosphoric Anhydride (P_2O_5)	1.0
Potassium Oxide (K_2O)	1.0
Titanium Dioxide (TiO_2)	0.7
Trace Elements	0.1
	100.0
Heavy Metals and Arsenic	10–100 ppm*

The major constituents shown in the analysis are combined as complex magnesium aluminum silicate and do not exist as free oxides.

*This applies to all Attapulgus Clay products with the exception of Pharmasorb which is con-controlled for pharmaceutical use at less than 20 ppm.

Physical Properties

	Granular Products 4/8,8/16,16/30,18/35, 25/50,30/60,50/80 (U.S. Sieve)	Mesh Products 100/up,100/200, 200/up
Free Moisture @ 220°F (Weight %–As Produced)		
*RVM Grades	6	6
*LVM Grades	1	1
Ignition Loss @ 1800°F (Weight %)		
RVM Grades	16	16
LVM Grades	6	6

	Granular Products 4/8,8/16,16/30,18/35, 25/50,30/60,50/80 (U.S. Sieve)	Mesh Products 100/up, 100/200, 200/up
Volatile Matter (As Produced-Weight %)		
RVM Grades	9	9
LVM Grades	5	5
Bulking Value		
Pounds per Gallon		
RVM Grades	20.41	20.41
LVM Grades	20.58	20.58
Gallons per Pound		
RVM Grades	0.0490	0.0490
LVM Grades	0.0486	0.0486
pH (ASTM D 1208)	7.5-9.5	7.5-9.5
Color		
RVM Grades	Light Cream	Light Cream
LVM Grades	Cream	Cream
Specific Gravity		
RVM Grades	2.45	2.45
LVM Grades	2.47	2.47
Bulk Density (lb/ft³)	28-36	28-36
BET Surface Area (m²/gram)	125	125

*RVM = Regular Volatile Matter
*LVM = Low Volatile Matter

Physical Properties

	Attaclay	Attaclay X-250	Attacote	LVM Attasorb	RVM Attasorb
Average Particle Size (μ)	18.0	20.0	5.3	2.9	2.9
% Finer than 10 μ	28	28	73	95	95
% Finer than 0.2 μ	−	−	−	−	−
Free Moisture @ 220°F (As Produced)	1	6	2	1	2
Ignition Loss @ 1800°F (Weight %)	6	16	6	6	12
Volatile Matter (As Produced-Weight %)	5	9	5	5	9
Bulking Value (lb/gal)	20.58	20.41	20.58	20.58	20.41
Bulking Value (gal/lb)	0.0486	0.0490	0.0486	0.0486	0.0490
pH (ASTM D1208)	7.5-9.5	7.5-9.5	7.5-9.5	7.5-9.5	7.5-9.5
Color	Cream	Light Cream	Cream	Cream	Light Cream
Specific Gravity	2.47	2.45	2.47	2.47	2.45
Residue, 325 Mesh (Wet Weight %)	10.0	15.0	1.0	0.10	0.01
Bulk Density—Tamped (lb/ft³)	25-29	26-30	15-18	16-18	13-15
B.E.T. Surface Area (m²/gram)	125	125	125	125	125

	Regular Pharmasorb	Colloidal Pharmasorb	Attagel 40	Attagel 50	Attagel 150 & 350
Average Particle Size (μ)	2.9	0.14	0.14	0.14	0.12
% Finer than 10 μ	95	−	−	−	−
% Finer than 0.2 μ	−	65	65	65	65
Free Moisture @ 220°F (As Produced)	1.0	12	12	12	15
Ignition Loss @ 1800°F (Weight %)	6	22	22	22	24.5
Volatile Matter (As Produced-Weight %)	5	11	11	11	11
Bulking Value (lb/gal)	20.58	19.70	19.70	19.70	19.70
Bulking Value (gal/lb)	0.0486	0.0507	0.0507	0.0507	0.0507
pH (ASTM D 1208)	7.5-9.5	7.5-9.5	7.5-9.5	7.5-9.5	7.5-9.5
Color	Cream	Light Cream	Light Cream	Light Cream	Light Cream
Specific Gravity	2.47	2.36	2.36	2.36	2.36
Residue, 325 Mesh (Wet Weight %)	0.10	0.30	0.30	0.01	8.0

	Regular Pharmasorb	Colloidal Pharmasorb	Attagel 40	Attagel 50	Attagel 150 & 350
Bulk Density-Tamped (lb/ft³)	15–18	18–21	19–22	19–22	38–45
B.E.T. Surface Area (m²/gram)	125	210	210	210	210

Key Properties: Has a wide variety of industrial applications. Colloidal grades are well known thickening, gelling, stabilizing and thixotropic agents in products as diverse as paints and fertilizer solutions. Sorptive grades find use as decolorizing and clarifying agents, floor absorbents, pesticide carriers, catalysts and refining aids.

These naturally sorbent attapulgite clays are improved by specific thermal activation, milling and screening and are available in sizes ranging from extremely fine powders (100 mesh to submicron particle sizes) to coarse granules (⅛ to ⁵⁰/₈₀ mesh). High heat calcination results in LVM (low volatile matter) grades, intermediate heat calcination results in RVM (regular volatile matter) grades, while low heat calcination produces colloidal or gelling grades. Granular clays are further produced in two qualities as related to adsorbing power: "A" Grade (natural clay) and "AA Grade" (extrusion improved clay).

Nonclay fractions are removed during processing so that the commercial products can contain up to 85 to 90% attapulgite.

THIELE KAOLIN CO., P.O. Box 1056, Sandersville, GA 31082

Kaoplate (Delaminated)

Chemical Type: Delaminated middle Georgia clay.

Chemical Analysis:

Ignition Loss	13.7–14.1%
Silica (SiO_2)	44.8–45.3%
Alumina (Al_2O_3)	37.5–39.7%
Ferric Oxide (Fe_2O_3)	0.3%
Titanium Dioxide (TiO_2)	1.0%
Manganese (Mn)	Nil
Copper (Cu)	Nil

Physical Properties:

Specific Gravity	2.63
Screen Residue Retained on 325 Mesh	0.001–0.02%
G.E. Brightness	88–90%
Oil Absorption (ASTM D281-31)	39
Refractive Index	1.56
% Solids Content	67–68
Bulking Value	0.0456
pH (20% Solids on Dry Clay and as Shipped Basis Slurry)	6.5–7.5

Key Properties: Recommended for aqueous applications only.

Thiele T-77

Chemical Type: East Georgia hybrid kaolin clay with fine particle size and high surface area.

Chemical Analysis:

Ignition Loss	13.4–14.1%
Silica (SiO_2)	43.5–44.5%
Alumina (Al_2O_3)	38.0–40.0%
Ferric Oxide (Fe_2O_3)	0.9–1.2%
Titanium Dioxide (TiO_2)	1.4–3.0%
Manganese (Mn)	Nil
Copper (Cu)	Nil
Free Moisture (Max)	1.0%

Physical Properties:

Specific Gravity	2.60
Screen Residue Retained on 325 Mesh	0.20%
Particle Size—Less than 2 Microns	90%
Surface Area (m^2/gram) B.E.T.	20–26
G.E. Brightness	79.5–82.5
pH (At 20% Solids on Dry Clay and as Shipped Basis Slurry):	
Dry	3.5–5.0
Slurry	6.5–7.5
Oil Absorption (ASTM D281-31)	36–41

Thiele B-80

Chemical Type: East Georgia hybrid kaolin clay with fine particle size and high surface area.

Chemical Analysis:

Ignition Loss	13.4–14.1%
Silica (SiO_2)	43.5–44.5%
Alumina (Al_2O_3)	38.0–40.0%
Ferric Oxide (Fe_2O_3)	0.9–1.2%
Titanium Dioxide (TiO_2)	1.4–3.0%
Manganese (Mn)	Nil
Copper (Cu)	Nil

Physical Properties:

Specific Gravity	2.60
Moisture (Max)	1.0%
Screen Residue Retained on 325 Mesh	0.03%
Particle Size—Less than 2 Microns	90%
G.E. Brightness	80–82%
pH	3.8–5.0
Oil Absorption (ASTM D281-31)	40

Key Properties (Both Grades): The adhesives industry has traditionally used water washed clays because of their extremely low levels of sand, quartz or other abrasive particles larger than 325 mesh. As a result of some new methods in air refining, Thiele can offer cost clays with 325 mesh screen residues approaching the best water washed clays. This new technology offers a potential savings of 70 to 100% available by switching from water washed clay to B-80.

There are other advantages in using B-80. The exclusive air refining process completely eliminates all clay agglomerates. This means B-80 is easier to disperse, especially in water systems. Also, since the clay has never been slurried in water, it is completely free of soluble salts which can (in some adhesives) result in lower water resistance. B-80 has an average particle size of 0.5 micron with 90% of its particles under 2 microns. This means that it will exhibit the same performance in thickening and strike-through as the finest grades of small particle size water washed clays.

For those applications which do not demand the extra low 325 mesh screen residues or the easier dispersion characteristics of B-80, T-77 is available. T-77 is also air refined but with higher 325 mesh screen residues. All of the other properties are quite similar to B-80.

R.T. VANDERBILT CO., INC., 30 Winfield St., Norwalk, CT 06855

Peerless China Clays

Chemical Type: "China Clay" produced by dry-processing the mineral kaolinite as found in the deposits of South Carolina. Available in many grades, according to dry brightness and fineness.

Product Name	Dry Brightness	Plus 200 Mesh (% Max)	Specific Gravity	pH (10% Solids Slurry)
No. 1	75	0.40	2.6	4.6
No. 1B	73	0.40	2.6	4.6
No. 2	70	0.40	2.6	4.6
No. 3	67	0.40	2.6	4.6
No. 4	74	3.0	2.6	4.6
No. 15	75	0.35	2.6	4.6

Note: Chemically treated to facilitate ease of dispersion in aqueous media.

No. 15B	73	0.35	2.6	4.6

Note: Chemically treated to facilitate ease of dispersion in aqueous media.

Chemical Analysis:

	Percent by Weight
Aluminum Oxide (Al_2O_3)	39.0
Silicon Dioxide (SiO_2)	43.8
Iron Oxide (Fe_2O_3)	1.6
Titanium Dioxide (TiO_2)	1.5
Ignition Loss	13.8
Manganese Oxide (MnO)	None
Magnesium Oxide (MgO)	Trace
Calcium Oxide (CaO)	Trace
Potassium Oxide (K_2O)	Trace
Sodium Oxide (Na_2O)	Trace

Physical Properties:

Brightness (G.E.)	80
Screen Residue (325 Mesh)	0.1% max
Specific Resistance	>60,000 ohms
Average Particle Size (By Sedimentation)	0.3 microns
Distribution (Percent by Weight):	
Smaller than 0.5 Micron	72%
0.5 to 2.0 Microns	22%
Larger than 2.0 Microns	6%
Moisture	1%
Oil Absorption (ASTM Rubout)	42

Key Properties: Series of clays for use as extender pigments in the coatings field. They have found application in traffic paints, barn paints, floor coverings, metal primers and in related materials, such as crack fillers and caulking compounds. Its color particle size, liquid demand and high specific resistance suggest its use in many types of coatings such as metal primers, emulsion paints, alkyd flats and traffic paints. It is particularly applicable in high PVC paints where dry opacity is desirable.

Section VII
Corn Starch and Starch Derivatives

CLINTON CORN PROCESSING CO., Clinton, IA 52732

Acid Modified Corn Starches

Chemical Type (All Grades): Clinton acid modified corn starches are prepared by treating an aqueous suspension of unmodified starch with a small quantity of strong acid such as hydrochloric acid. The degree of depolymerization which occurs depends on the reaction time. A mild alkali is introduced to neutralize the acid and stop hydrolysis when the proper fluidity has been reached. The starch is then washed to remove the salts of neutralization. The pH is adjusted to the desired range and the starch is filtered and dried.

Key Properties (All Grades): Acid modified starches are manufactured in various fluidities, up to and including 90. The relative fluidity is indicated by its product number. Clinton 290-B Industrial Modified Starch possesses a higher fluidity than Clinton 215-B Industrial Modified Starch. The hot pastes of acid modified starches are less viscous than those of unmodified starch. Their pastes are slightly clearer at elevated temperatures. It is desirable for most applications to hold the paste at high temperatures to prevent excessive set back. The temperature can be lowered if a thick opaque gel is desired. The gels of acid modified starches are cloudy and "short" or cohesive as compared to the stringier gels of oxidized starches, and their dry starch paste films are tough, horny and opaque.

These starches are extensively employed by the paper and textile industries for size applications, particularly at high temperatures. The starch solids concentrations possible with the higher acid modified starches make them applicable for use as adhesives in the paper industry. Also, the gel forming characteristics of acid modified starches are advantageous in many food and general adhesive applications.

Low Acid Modified Industrial Starches

215-B   220-B

Typical Analysis (as is or fresh basis):

Form	Pearl	Protein	0.30%
Moisture	12%	Ash	0.25%
pH	6.0	Chlorides	0.20%
Acidity	0.06%	Water Solubles	0.40%

Key Properties: These starches, having received the least acid modification of the series, are much like unmodified starch. On a comparative basis their hot pastes are progressively lower in viscosity than unmodified starch paste and their gels have less set back. In commercial applications where slightly higher percentages of starch solids are necessary, one of these starches is the logical choice to replace pearl corn starch.

Intermediate Acid Modified Industrial Starches

230-B   235-B   240-B   250-B   255-B
260-B   265-B   267-B   270-B

Typical Analysis (as is or fresh basis):

Form	Pearl	Protein	0.30%
Moisture	12%	Ash	0.28%
pH	6.0	Chlorides	0.20%
Acidity	0.06%	Water Solubles	0.40%

Key Properties: As their numbers indicate, these starches possess increasingly higher fluidity or can be cooked up to progressively lower viscosities. The starches in this series, considered in the intermediate modification range, exhibit the typical characteristics of all acid modified starches. They possess lower viscosities than regular corn starch, are more flowable and have less body and set back. The choice of which of these starches to use in a specific application depends upon the viscosity and the starch solids desired.

High Acid Modified Industrial Starches

277-B   285-B   290-B

Typical Analysis (as is or fresh basis):

Form	Pearl	Protein	0.30%
Moisture	12%	Ash	0.45%
pH	6.0	Chlorides	0.38%
Acidity	0.06%	Water Solubles	2.75%

Key Properties: These starches receive the highest acid modification. Clinton 290-B is the most highly modified starch of this series—its hot paste is the thinnest and it has the least body of any acid modified starch. It has the lowest set back upon cooling. The high acid modified starches are applicable in operations where high solids are required.

Clineo Modified Industrial Starches

700-D through **716-D**

Typical Analysis (as is or fresh basis):

Form	Ground Pearl	Color	White
Moisture	12%	Water Solubles	2.0%
pH	6.5		

Chemical Type: Hydroxyethyl starches, prepared by reacting ethylene oxide with corn starch to form derivatives which have the same physical appearance as the parent native corn starch.

Key Properties: However, when cooked, Clineo starches have decidedly different characteristics that are especially suitable for certain operations in the paper and textile industries.

Clinton Corn Dextrins

Chemical Type (All Grades): Dextrins are made by the modification of corn starch. This modification may be the result of the amount of catalyst which is added, or of the heat applied, or it may quite possibly be effected by both factors. [This "dextrinization" chemically reduces the larger (starch) molecules into smaller (dextrin) components.] There clearly exists an almost infinite number of dextrin products, as each variation in catalyst quantity and/or heat produces a different dextrin. By varying the time each product is roasted, an even greater range of dextrins is unveiled.

Key Properties (All Grades): Several general statements may be made concerning dextrins as a class.

(1) Dextrins are classified in terms of color, relative solubility, relative viscosity of their aqueous solutions, parent starch source and method of manufacture.

(2) Dextrins are more soluble than is the parent starch, in either hot or cold water. In addition, dextrins are less viscous and are better suited to adhesive applications than the parent starch.

(3) Dextrinization increases fluidity, while decreasing viscosity. This results in increasing solubility while the tendency to gel decreases. This allows the use of higher solids levels, improving the utility of the starch for use in adhesive functions.

(4) As a rule, "tack" and adhesion increase as the total solids of a dextrin solution is raised. Solutions of higher solids concentration tend to remain on the paper surface to which they have been applied, without wetting the paper to an undesirable degree or being absorbed by it.

These generalizations clearly show that dextrins possess certain inherent characteristics. These characteristics may be modified with a variety of additives in many ways to produce many different final adhesive characteristics.

Clinton White Dextrins

Typical Analysis (as is or fresh basis):

Form	Powder	pH	3.0
Color	White to	Ash	0.2%
	Off-White	Water Solubility	5.0–90.0%*

*Each dextrin has an individual solubility, accurate within 10%.

Chemical Type: White dextrins (called the "600" series) are, as an overall class, the least modified of the three dextrin types. Clinton can create any property desired in a corn dextrin, through custom roasting and blending.

Key Properties: The wide variety of properties available in these dextrins makes them adaptable to a wide range of applications. Some "600" dextrins are designed to be used at high viscosity and low solids, while others are ideally suited to the high solids, low viscosity functions. These dextrins are easily dispersed in water and form a relatively "short" paste. Machineability is excellent, making this group ideal for intricate high-speed operations.

The white dextrins offer a complete solubility range, from the short, thick paste of 5% soluble 635 (used primarily for coating applications and bag-bottom adhesives) to the 90% solubility of 652, the hard-sticking dextrin for difficult casesealing, tube-winding and carton-sealing applications.

Clinton Canary Dextrins

Typical Analysis (as is or fresh basis):

Form	Powder	pH	3.0
Color	Light Yellow	Ash	0.2%
	to Dark Buff	Water Solubility	85.0–95.0%*

*The majority of canary dextrins have water solubility of 90.0% and above.

Chemical Type: Canary dextrins (called the "700" series) are highly modified in order to attain high solubility and low viscosity. Clinton can create any property desired in a corn dextrin, through custom roasting and blending.

Key Properties: The complete modification and low viscosity at high solids make these dextrins ideal for rewettable applications such as envelope, flat gumming and gummed tape adhesives. The reduced setback properties of these dextrins contribute to high viscosity stability over long periods. Machineability is excellent and lengthy storage will not alter the prepared adhesive.

This series particularly lends itself to applications which require rapid drying and uniform films. Special processing insures quick tack. The canary dextrins, as a class, show good compatibility with chemicals added for improved adhesive performance. They also exhibit the highest adhesive qualities and the greatest versatility of the several dextrin categories.

Clinton British Gums

Typical Analysis (as is or fresh basis):

Form	Powder	Color	Light Brown to Brown
pH	5.5	Ash	0.2%
Water Solubility	42.0–95.0%*		

*Each dextrin has individual solubility accurate within 10%.

Chemical Type: British gum dextrins (called the "800" series) require the least catalyst and the longest roasting period of any dextrin class. Clinton can create any property desired, through custom roasting and blending.

Key Properties: The British gums offer an almost unlimited versatility of application because of a complete range of water solubilities and viscosities. Their high degree of dextrinization results in a smooth, free-flowing paste which exhibits remarkable viscosity stability over extended periods.

General uses to which British gums are ideally suited include: carriers which require high viscosity and low solids; extenders for more costly adhesives; rewettable adhesive applications (envelope seals, stamps, etc.); general adhesive uses.

Clinton Unmodified Corn Starches

Product Name	Moisture Content (%)	pH	Ash Content, max (%)	Protein Content (%)	Water-Solubles, max (%)
104-B	11.0	5.3	0.1	0.3	0.1

Chemical Type: Essentially refined native pearl white corn starch, basic unmodified, thick boiling.

Key Properties: Receives extensive refining to assure maximum purity.

105-B	12.0	5.5	0.1	0.3	—

Chemical Type: Regular pearl common corn starch, an unmodified (thick boiling) starch with a high rate of setback (viscosity increase of a starch paste upon standing).

Key Properties: Maximum quality and purity. Widely used where high viscosity, additional body, and/or nonflowing properties are desired. Available in four forms of moisture content: 12%, 10% and redried at 5% and 7½% at higher cost.

105-A	12.0	5.5	0.1	0.3	—

Chemical Type: An unmodified (thick boiling) starch, typically high in viscosity, with good body at both high and low solids.

Key Properties: Highly refined product. The basic powdered starch manufactured by the supplier. Set-back (viscosity increase upon aging of the starch paste) is characteristically high. Widely used in food applications. Available in four forms of moisture content: 12%, 10% and redried at 5% and 7½%.

105-A "M"	10.0	5.5	0.1	0.3	—

Chemical Type: Highly refined unmodified starch with a high, uniform gel strength.

Key Properties: Shows maximum uniformity of both body and set-back, in addition to the properties common to unmodified starches. Also, ideal for dry mixing applications.

106-B	11.5	6.0	0.1	0.3	—

Chemical Type: Fine white pearl unmodified corn starch.

Key Properties: Consistent product displaying predictable characteristics, typical of unmodified (thick boiling) starches. Has high viscosity, with good body at both high and low solids. Exceptionally well suited for use in chemical conversion processes. Its converted starch pastes have strong adhesive properties and can be utilized to great advantage in the surface sizing of paper.

Clincor 121-B	12.0	6.0	0.1	0.3	—

Chemical Type: Highly refined unmodified pearl starch, ultimately pure.

Key Properties: Shows the general properties of unmodified products. A unique manufacturing process develops an outstandingly uniform, stable viscosity which is highly resistant to breakdown, which makes it an ideal adhesive for high-speed corrugating operations. It is a highly versatile product which may be used in both the primary (carrier) and the secondary (raw) portions of adhesive formulations. This versatility promotes increased economy, as initial cost, inventory requirements and storage area are appreciably reduced. The high versatility makes operation with automated systems for the continuous preparation of adhesive pastes vastly simplified. Unsurpassed in quality, whether regular or specially treated components are being used, in terms of pin adhesion of the finished board and running efficiency.

Product Name	Moisture Content (%)	pH	Ash Content, max (%)	Protein Content (%)	Water- Solubles, max (%)
Clinvert 406-B	12	6.0	—	—	—

Chemical Type: Enzyme converting pearl white starch containing an effective buffer system for optimum pH control and efficient utilization of the enzyme during the conversion process.

Key Properties: Designed to be equally suitable for surface sizing or coating adhesive applications. Converts easily and is adaptable to a number of enzyme conversion time-temperature cycles. May be converted at concentrations up to 20% for surface sizing applications or up to 35% for coating color adhesives. The enzyme converted starch pastes can be readily diluted to any concentration desired at the size press. These diluted pastes have the viscosity characteristics required for surface sizing at many different application concentrations. The finished pastes are translucent in color.

When 406-B is enzyme converted for coating color adhesive operations it produces the thin viscosity normally required. It converts to the same viscosity each time, fulfilling another important property for coating adhesive applications. The converted pastes do not have an excessive reaction to the plasticizers normally used in coating formulations, so good control of the final coating color viscosity is maintained. The converted pastes from both low and high solids enzyme conversion cycles have strong adhesive properties. Therefore, the physical strength values are maintained as desired by mills making many different grades of surface-sized or coated paper.

Clinton 415-B	11.5	6.0	—	—	—

Chemical Type: Chemically treated unmodified fine white pearl corn starch. Contains an effective buffer system for optimum pH control and efficient utilization of the enzyme during the conversion process.

Key Properties: Designed specifically for enzyme conversion in the presence of pigment. The desired viscosity level is produced with minimum enzyme requirement for paper coating applications. Displays excellent uniformity and converts to the same viscosity range each time.

Clinvert 419-B	12	6.5	—	—	—

Chemical Type: Enzyme converting fine white pearl corn starch, produced by a special process.

Key Properties: Designed for use in size press or tub size applications. Displays such desirable characteristics as ease of conversion and good viscosity stability. Easily converted using normal conversion cycles at starch solids concentrations up to 25%. Extremely white and clean and designed to minimize accumulation of residue in the size tub and machine tanks. Displays little tendency to foam and has excellent stability under normal manufacturing and storage conditions. Wax pick requirements may be easily met or exceeded when it is used in the surface sizing of paper.

Clinvert 409-B	12	6.5	—	—	—

Chemical Type: Enzyme converting starch specially processed.

Key Properties: Increased ease of conversion and superior viscosity stability, advantageous for use in size press or tub size applications. Converts readily at starch solids concentrations up to 25% using normal time-temperature cycles. Wax pick requirements may be easily met or exceeded when used in the size press. Extremely white and clean. Designed to minimize the accumulation of residue in the size tub and demonstrates superior stability under normal manufacturing and storage conditions. Will not foam, even under agitation, and exhibits no adverse effects upon the retention of titanium dioxide.

Oxidized Corn Starches

Chemical Type (All Grades): Clinton manufactures oxidized starches by treating aqueous starch suspensions with a slightly alkaline solution of sodium hypochlorite. The oxidation is effected by reduction of the positive valent chlorine with the subsequent release of oxygen. The primary alcohol attached to the fifth carbon is converted into a carboxyl group along with the formation of water. Oxidized starches, in actual commercial practice, are finished at an alkaline pH so that the carboxyl group exists as a sodium salt. The oxygen not consumed in the formation of carboxyl groups probably assists in the depolymerization of the starch chains through oxidative alkaline degradation. This results in lower paste viscosities.

Key Properties (All Grades): A starch series of progressively higher fluidities may be manufactured since reaction time, temperature and pH can be varied. The greater the oxidation of a starch, the higher is its fluidity.

Oxidized starch retains the granule structure of unmodified starch, shows typical polarization crosses, is insoluble in cold water, and shows the typical starch iodine reaction. However, there is little similarity in other properties. Oxidized starches are much whiter, due principally to the bleaching action of the hypochlorite. They have a shorter cooking time, higher fluidity and are less opaque than their parent starches.

The films of oxidized starches are clear, tough and horny. Their gels show considerably less set back and are clear, more fluid and stringier than those of other starches.

Clinton oxidized starches are manufactured over a wide viscosity range, making possible the use of a high percentage of starch solids for improved adhesive strength. These starches are, by their properties, particularly adapted for use in the paper and textile industries where low viscosities and free-flowing characteristics at high solids are desired for surface and warp sizing.

The higher the product number of a Clinco starch, the greater is its degree of oxidation and the lower its viscosity. The most highly oxidized Clinco starches are referred to as "gums," a term which probably originated because their cooked pastes somewhat resembled those found with vegetable gums.

Higher Oxidized Modified Industrial Starches

Clinco 360-D
Clinco 370-D

Typical Analysis (as is or fresh basis):

Form	Ground Pearl (coarse powder)	Protein	Trace
		Ash	0.8%
Moisture	12%	Chlorides	0.3%
pH	7.5	Water Solubles	2.5%

Key Properties: Slightly more oxidized than 60 Gum, with a corresponding increase in the typical oxidized starch characteristics. Their clarity and paste viscosities are such that both are called gums. They are used in applications that require a higher concentration of starch than can be obtained with starches oxidized to a lesser degree.

Intermediate Oxidized Modified Industrial Starches

Clinco 345-D
Clinco 350-D
Clinco 351-D
Clinco 355-D
Clinco 357-D

Typical Analysis (as is or fresh basis):

Form	Ground Pearl (coarse powder)	Protein	Trace
		Ash	0.6%
Moisture	12%	Chlorides	0.2%
pH	7.5	Water Solubles	1.2%

Key Properties: These starches may be considered as intermediate between the lower oxidized starches and the gum starches. These oxidized starches may be used where clearer, more fluid pastes are needed or where higher starch solids are required. Clinco 355-D is the preferred starch when more surface holdup on the paper is required than can be obtained with the gum starches.

Least Oxidized Modified Industrial Starches

Clinco 315-D
Clinco 330-D

Typical Analysis (as is or fresh basis):

Form	Ground Pearl	Protein	Trace
	(coarse powder)	Ash	0.6%
Moisture	12%	Chlorides	0.5%
pH	7.5	Water Solubles	0.9%

Key Properties: These starches are the logical ones to use where an industrial application demands a starch paste that is less viscous than unmodified starch paste, and/or one that is clearer and shows less tendency to gel than an acid modified starch in a comparable fluidity range.

CORN PRODUCTS, CPC International, Inc., International Plaza, Englewood Cliffs, NJ 07632

Cerelose Dextrose 2001

Chemical Type: General purpose monohydrate dextrose.

Chemical Analysis and Physical Properties:

Form	White crystals	Screen test (Tyler mesh)	
Moisture	8.5%	% through 14	99.0% min.
Dextrose equivalent	99.5 min.	Color	White

Key Properties: Recommended as an adhesive base for the following:
Envelope adhesives (back gum): Provides flow control and lay flat.
Library paste: Increases open time.
Laminating adhesives: Provides flow control and lay flat.

FDA Status: GRAS (Substances that are generally recognized as safe.)

Corn Products Starch 3005

Chemical Type: General purpose, thick-boiling corn starch for industrial applications.

Chemical Analysis and Physical Properties:

Form	Powdered
Moisture	12.0%
pH	5.3
Wet grit @ 25°C	1.0% max.

Key Properties: Recommended as an adhesive base for the following:
Bag adhesives: Gives high viscosity and "buttery" paste consistency.
Library paste: Used in conjunction with low soluble white dextrin to give high viscosity and smooth paste.

FDA Status: Industrial starch-modified.

Corn Products Starch 3372

Chemical Type: Thick-boiling, pH-buffered corn starch specifically designed for enzyme conversion.

Chemical Analysis and Physical Properties:

Form	Powdered
Moisture	12.0%
pH	6.9
Wet grit @ 25°C	1.0% max.

Key Properties: Recommended as a base for laminating adhesives. An enzyme converting starch. When enzyme converted 3392 gives an internally plasticized adhesive which has good lay flat.

FDA Status: Industrial starch-modified.

Globe Corn Syrup 1132

Chemical Type: Regular conversion, normal viscosity syrup.

Chemical Analysis and Physical Properties:

Baumé	43.0	Carbohydrate composition	
Dry substance	80.3%	(average %–solids basis):	
Dextrose equivalent (DE)	43.0	Dextrose	19.0
Color	Water white to	Maltose	14.0
	very light straw	Maltotriose	12.0
pH	4.8	Higher saccharides	55.0
Starch	Negative	Weight per gallon @ 80°F	11.85 lb
Sulfite (SO$_2$)	<40.0 ppm	Brookfield viscosity	
Ash (sulfated)	0.3%	@ 80°F	56,000 cp

Key Properties: Recommended as a base for laminating adhesives. Increases open time and improves adhesive stability and machinability.

FDA Status: GRAS (Substances that are generally recognized as safe.)

PENICK & FORD, LTD., 920 First St., SW, Cedar Rapids, IA 52406

Gums

Astro Gum 3010 and 3020

Chemical Type: New patented starch derivatives.

Key Properties: Designed to be fully compatible in all proportions with highly hydrolyzed polyvinyl alcohol. Both have found application in special proprietary sizing and coating applications to gain properties generally associated with synthetic polymers used either alone, or in combination with starch. These qualities include improved grease, oil, or solvent holdout, higher binder strength with improved rheological properties, plus economies realized by combining the expensive PVA with the less expensive starch products. High volumes of the product are being used in textile sizings with PVA.

Astro Gum 3010 and 3020 may be used either as clear sizes or pigmented size systems on the size press or coater. In the latter case, the superior binding qualities make possible special functional coatings of low adhesive level, with improved flow characteristics, lower costs wherein surface strength may be held to acceptable levels while opacity and brightness may be improved. The combination of adhesives should be of particular value in overcoming deficiencies in grades made with a high content of reclaimed fibers.

The Astro Gums are completely stable when stored under normal conditions and may be cooked or handled in a similar manner as other modified starch products.

Essex Gums 1300, 1330, 1360 and 1390

Chemical Type: Starch derivatives produced as a result of chemical substitution into the starch molecule to produce a number of new and unique properties.

Key Properties: A general summary of the properties follows:

Minimized gelling and retrogradation tendency	Clear pastes and films
Improved water-holding ability	High gloss films
Lessened migration from applied films	High reactivity with resins
Good film-forming ability	Good chemical stability
Low gelatinization temperature; rapid cook-out	Wide range of viscosities available

The following applications can use Essex Gums:

Pigmented coatings	Remoistening adhesive for tape, labels
High gloss ink printing	and stamps
Grease, oil, wax and lacquer holdout	Total or partial replacement of hydro-
Tacky borated adhesives	philic film formers such as alginates,
Layflat or nonwrinkling adhesives	mannogalactans and CMC

Penford Gums 200 Series and 300 Series

Grades: The Penford Gums have been grouped in the following series:

	200 Series	300 Series
High viscosity	200	300
	220	330
	230	
	240	
Medium viscosity	250	360
	260	
	270	
Low viscosity	280	380
	290	

In addition to the above, the Penford Gum family includes the following products developed especially for specific applications:

Pen-Cote	Low viscosity, for high solids paper coating colors.
Pen-Cor	High viscosity, for corrugating adhesives.
Douglas Pen Sprae 3800	For spray applications.

Chemical Type: Hydroxy ethyl ether derivatives of corn starch. The hydroxy ethyl groups are attached directly to the starch molecule by strong ether linkages. The process for the manufacture of the Penford Gums is covered by U.S. Patents 2,516,632; 2,516,633; and 2,516,634.

Key Properties: The new chemical compound formed, while having the physical form and appearance of the original raw corn starch, has markedly changed properties which adapt it to a myriad of uses. The presence of these hydroxy ethyl ether groups on the starch molecule is very effective in reducing the associative forces between molecules, giving the following improved qualities:

 Lowered gelatinization temperature
 Improved viscosity stability during cooking
 Reduced gelling tendency, improved viscosity stability on
  cooling or aging
 Decreased retrogradation tendency
 Increased film solubility

These groups, coupled with resistance to reassociation, will provide:

 Improved water-holding ability
 More cohesive or glutinous paste characteristics
 Improved film-forming characteristics
 Clearer, more flexible films

Additional properties of the Penford Gums that are attributable to the special characteristics already discussed are:

 Increased reactivity to borax and resins
 Good compatibility
 Water hardness and pH not critical
 Good color stability
 Depolymerization by enzymes
 Viscosity reduction by high shear cooking

Starches

Douglas Pearl Starch

Chemical Type: Basic unmodified corn starch as it comes from the corn kernel. All other corn starch products are modified from it.

Key Properties: Often used at the wet end of the paper machine as an addition to improve internal bond.

Douglas E and EC Pearl Starches

Chemical Type: Unmodified starches designed for enzyme conversion.

Key Properties: Recommended for use in size press, calender stacks, coatings.

Douglas Corrugating Starch

Chemical Type: Basic pearl corn starch.

Key Properties: When used in the proper formulations performs as an adhesive for bonding liner-board to corrugating medium. A general formulation calls for the use of starch at 20% solids and also includes caustic and borax.

Crown Potato Starch

Chemical Type: Unmodified potato starch.

Key Properties: Suitable for use at the wet end of a paper machine and for enzyme converting operations. Recommended for use in furnish and wet end spray addition, calender stacks, coatings.

Douglas 3006 and CS-3018 Starches

Chemical Type: Specific starches for use with the P&F High Shear Cooker.

Key Properties: Recommended for size press use.

Crown Thin Boiling Starches
(X10 through XR,
Douglas Multiwall No. 3)

Chemical Type: Acid modified starches which range in viscosity from the very thick at low solids to the relatively thin in 10 to 12% solids range. Made by treating a suspension of pearl starch with acid. This results in a cleavage of some of the linkages within the starch molecule, producing a starch with shorter than normal chain lengths.

Key Properties: The result of this treatment is lower viscosity and the more severe the treatment the more fluid the product. Thin boiling starches are suitable for adhesives for the manufacture of bag pastes. They are also suitable for use at both the size press and calender stacks for improving surface sizing and internal bond on grades of paper such as kraft linerboard, bag stock and fine papers.

Douglas Clearsol Gums

Chemical Type: Frequently referred to as chlorinated starches. The chemicals used in their preparation, however, are oxidizing agents and the resulting products are actually oxidized starches. The process not only shortens the chain length, but also affects the molecule chemically.

Key Properties: These gums have the three following improvements:

> Better water-holding properties
> Better film-forming characteristics
> Reduced gelling tendencies

Also, higher solids with lower viscosities are possible than with the acid modified starches. One disadvantage of the oxidized corn starches is that in a highly pigmented sheet they act somewhat as a dispersant and so have a detrimental effect upon pigment retention.

Recommended for size press, calender stacks, coatings.

Penford Gum 200 and 300 Series
Pen-Cote
Pen Sprae 3800

Chemical Type (of the above): Hydroxyethyl ether derivatives of corn starch and are produced in a wide viscosity range. There are two series (200 and 300), with the 300 series having the higher degree of ethylation.

Key Properties (of the above): An ethylated starch has superior film-forming and water-holding properties. They do not affect pigment retention and their retrogradation tendencies are greatly reduced. For these reasons, they are excellent starches for use at the size press, calender stacks and in coatings where optimum results are desired. Since their inception, the Penford Gums have proven time after time their capability of highest possible performance. Recommended for size press, calender stacks, coatings.

Essex Gum 1300 Series

Chemical Type: Ethylated potato starches available over the same viscosity range as the Penford Gums.

Key Properties: The applications are much the same and do have some advantages over the Penford Gums. When cooked they produce a clear white sol and have proven exceptionally good for high gloss ink hold-out. They are more of a specialty starch where that little bit of extra test is required for surface sizing. Recommended for size press.

Astro Gum 21
Astro Gum 3010
Astro Gum 3020

Chemical Type: Astro Gum 21 is an anionic corn starch. Astro Gum 3010 and 3020 are midrange and low viscosity (respectively) anionic corn starch products.

Key Properties: Astro Gum 21 is designed for use as a wet end additive. It is retained by electrolytic attraction and the strength developed is not only a function of the physical strength of the starch, but also by its beneficial effect upon formation and higher degree of retention of both the starch and fines. It is recommended for use on all grades of paper and for furnish and wet end spray addition.

Astro Gum 3010 and 3020 are designed specifically for compatibility with polyvinly alcohol in formulations for size press, calender and coating purposes.

Astro X-100
Astro X-100B
Apollo 15
Apollo 20

Chemical Type: Cationic potato starches.

Key Properties: Have been found superior in all respects as wet end additives. They have been used in virtually all types of paper production and are suitable for developing mullen and increasing production speeds in kraft liner-board production, increasing pigment retention and wax pick in fine papers and improving sizing. To effect these improvements, only one-third as much of the Astro X-100 is required, as opposed to conventional additives.

Pen-Cor

Chemical Type: Ethylated starch derivative.

Key Properties: Designed to improve corrugating operations. Has superior water-holding and viscosity-stability properties as compared to pearl starch or other types of carrier starches. It is more efficient because of these properties and, consequently, a more economic operation.

Douglas Waterproof Adhesive No. 7

Chemical Type: Starch product containing the necessary resin to produce a waterproof adhesive for corrugated board.

Key Properties: The product is used in preparation of the carrier portion. By proper formulation, the adhesive can be adjusted for board for high humidity conditions or to pass government specifications. It can be used with pearl, corn, milo, tapioca or wheat starch.

Section VIII

Defoamers and Antifoams

CHEMICAL COMPONENTS, INC., 20 Deforest Ave., P.O. Box 291, East Hanover, NJ 07936

Product Name	Brookfield Viscosity (cp)	Specific Gravity
Hexafoam 100	800	1.10

Chemical Type: Nonsilicone organic defoamer, in straw yellow liquid form.

Key Properties: General purpose defoamer for use in systems where elimination of air in formulas, compounds, slurries and other mixes containing trapped air is desired. Recommended for use in paints, inks, paper coatings and castables.

COLLOIDS, INC., 394-8 Frelinghuysen Ave., Newark, NJ 07114

Product Name	Active Content (%)	Brookfield Viscosity (cp)	Pour Point (°C)	Flash Point (°C)	Weight per Gallon (lb)
Colloid 643	100	1,200	-17.8	182	7.0

Chemical Type: Light tan, opaque proprietary liquid defoamer.

Key Properties: Especially formulated to give outstanding foam control when used in the manufacture of paints and coatings. High efficiency for maximum economy. Quick bubble-break, resulting in reduction of foam build-up.

Meets the following food additives requirements of Subpart F:
Section 176.210—Defoaming agents used in the manufacture of paper and paperboard.
Section 176.200—Defoaming agents used in coatings.
Section 176.180—Components of paper and paperboard in contact with dry foods.

Toxicity: This product complies with the Toxic Substances Control Act PL94-469.

Product Name	Active Content (%)	Brookfield Viscosity (cp)	pH (5% Solids Slurry)	Flash Point (PMCC) (°C)	Weight per Gallon (lb)
Colloid 675	100	175	6.2	171	7.51

Chemical Type: Opaque, low viscosity proprietary liquid defoamer.

Key Properties: Especially designed to give outstanding foam control in various adhesive systems and latex paints. It has outstanding longevity when used in emulsion applications, such as water-based paints, paper coatings and adhesives. Dispersible in water and forms stable dilute dispersions. Effective foam knockdown and remains effective for longer periods than most defoamers.

Meets the following food additives requirements of Title 21 CFR Chapter I-Subchapter B:

Section 176.210—Defoaming agents used in the manufacture of paper and paperboard.

Section 175.105—Adhesives.

Section 176.170—Components of paper and paperboard in contact with aqueous and fatty foods.

Toxicity: This product complies with the Toxic Substances Control Act PL94-469.

Product Name	Active Content (%)	Brookfield Viscosity (cp)	pH (5% Solids Slurry)	Pour Point (°C)	Flash Point (PMCC) (°C)	Weight per Gallon (lb)
Colloid 681F	100	400	5.5	-17	179	7.34

Chemical Type: Off-white, opaque proprietary liquid defoamer.

Key Properties: Highly efficient liquid antifoam useful in a wide variety of industrial applications, such as:
Latex-based paints.
Stripping of resin emulsions.
Degassing of PVC suspension resins and emulsions.
General latex compounding.
Effluent from potato washing operations.

Its features are the following:
A homogeneous liquid with good shelf stability.
Highly efficient for maximum economy.
A low viscosity liquid, permitting application with spray nozzles where desired.
Effective in acid or alkaline media.
Highly persistent for long lasting protection.

Meets the requirements of CFR21, Subchapter B, Subparts:
175.105—Adhesives.
175.300—Resinous and polymeric coatings.
176.170—Components of paper and paperboard in contact with aqueous and fatty foods.
176.180—Components of paper and paperboard in contact with dry foods.
176.200—Defoaming agents used in coatings.
176.210—Defoaming agents used in the manufacture of paper and paperboard.

Colloid 681F is listed in:
Kosher Products Directory, by authority of U.O.J.C.A.
Meat and Poultry Inspection Program, U.S.D.A., Beltsville, Maryland.

Toxicity: This product complies with the Toxic Substances Control Act PL 94-469.

Colloid 691	100	500	7.8	-12	160	7.08

Chemical Type: Opaque, off-white proprietary liquid defoamer.

Key Properties: Has been especially designed to give outstanding foam control in paper coating colors. It has been found effective in a broad range of applications, including flexographic inks, latex paints, floor cleaning compounds and agriculture sprays.

Its features are:
Stable, uniform product.
Highly efficient.
Exceptionally long-lasting and remains effective for longer periods than most defoamers.

Meets the food additives requirements of Subpart F:
176.210—Defoaming agents used in the manufacture of paper and paperboard.
176.200—Defoaming agents used in coatings.

Toxicity: This product complies with the Toxic Substances Control Act PL 94-469.

EMKAY CHEMICAL CO., 319-325 Second St., Elizabeth, NJ 07206

Product Name	Active Content (%)	Viscosity (Gardner-Holdt)	Specific Gravity @ 77°F	Weight per Gallon (lb)
Emka Defoam BC	100	A-2	0.812	6.8

Chemical Type: Light amber proprietary composition versatile emulsion defoamer with bland odor.

Key Properties: Works equally well in all latex systems of the paint and adhesive trades to combat foam during manufacture. It also gives excellent can storage in emulsion paints to hold down foam during application, giving the best paint surface possible.

Its advantages are: effective, economical, silicone-free, mild odor, infinite stability, and versatile.

HENKEL CORP., 425 Broad Hollow Road, Melville, Long Island, NY 11746

Product Name	Active Content (%)	Flash Point (°F)	Specific Gravity @ 68°F
Defoamer C	100	329	—

Chemical Type: Mixture of special hydrophilic partial esters and aliphatic hydrocarbons (silicone-free), a cloudy, light yellow to light brown oil.

Key Properties: Highly effective defoamer for plastic dispersions and emulsion paints, particularly for fine particle dispersions. Easily dispersed in the emulsion and there is no danger of creaming, even at low viscosities. Especially recommended as a defoamer and as a foam inhibitor of PVC-acrylic, PVP, water-soluble alkyds and ethylene vinyl acetate. It is not suitable for styrene-butadiene emulsions.

Defoamer F	—	333	0.895

Chemical Type: Mixture of special fatty acid esters and aliphatic hydrocarbons, which are silicone-free.

Key Properties: Antifoaming agent for emulsion paints and other plastic dispersions. Can be used as a defoamer or foam preventive. Particularly suitable for use in plastic dispersions or emulsion paints based on polyvinyl acetate, polyacrylate and their copolymers. Not recommended for use in styrene-butadiene dispersions.

Defoamer G	—	333	0.925

Chemical Type: Mixture of special fatty acid esters, aliphatic hydrocarbons and a trace of silicone oil.

Key Properties: Highly active defoaming and foam preventive agent which can be used in practically all aqueous systems. Primarily intended for use in gloss and semigloss latex paint systems.

INTERSTAB CHEMICALS, INC., 500 Jersey Ave., P.O. Box 638, New Brunswick, NJ 08903

Product Name	Color (Gardner) (max)	Specific Gravity (75°F)	Viscosity (Gardner-Holdt) @ 77°F (max)	Weight per Gallon (lb)	Form
Anti-Foam Agent	White	0.811	F	6.76	Cream

Chemical Type: Proprietary 100% specially processed, silicone-free composition.

Key Properties: Designed to reduce or eliminate foam during preparation and application of water-based systems. A water-dispersible liquid, it can be added at any point during processing where foaming is most critical and troublesome. Will eliminate foam after it has formed. Specifically designed to be used in many types of water-based formulations, including oil, alkyd, acrylics, PVA, styrene-butadiene, vinyl acrylic, emulsion coatings, adhesives and rubber latex.

Product Name	Color (Gardner) (max)	Specific Gravity (75°F)	Viscosity (Gardner-Holdt) @ 77°F (max)	Weight per Gallon (lb)	Form
Anti-Foam 2079-87	Opaque White	0.875	T	7.29	Liquid

Chemical Type: Proprietary 100% active liquid, silicone free compound.

Key Properties: Designed to control or eliminate foaming during the preparation and application of water-based formulations. Effective in a wide variety of emulsion vehicles. Does not produce any adverse effects in a paint film. Retains excellent activity during long-term storage of the finished product.

Hevi-Duty Anti-Foam Agent	2	0.825	D	6.87	Liquid

Chemical Type: Proprietary 100% specially processed, silicone-free composition.

Key Properties: Exhibits high initial antifoam properties in the manufacture of emulsion formulations, regardless of composition. Retains excellent activity during long-term shelf storage of the finished product. Will not separate during storage. Exerts no adverse effect on paint film. Does not produce "fisheyes" when used in proper proportions. Specifically designed to eliminate or control foaming in many types of water-based formulations, including oil, alkyd, acrylics, PVA, styrene-butadiene, vinyl acrylic, emulsion paints, adhesives and rubber latex.

ISOCHEM, Cook St., Lincoln, RI 02865

Product Name	Active Content (%)	Solids Content (%)	pH	Specific Gravity	Weight per Gallon (lb)
IsoNofoam	100	100	7.0	0.90	7.1

Chemical Type: Proprietary composition nonionic chemically inert defoaming agent and air-release material (silicone-free).

Key Properties: High efficiency. Satisfactory for aqueous or nonaqueous foam control. Low surface tension to eliminate entrapped air. Completely stable and disperses in systems. No chemical reaction with foaming bodies. Excellent can stability on the shelf. No color interference. No fisheyes or craters. Low cost and small percentages required. Effective in acid and alkaline solutions. Water-free. Resistant to product separation and easy to handle in low viscosity fluids. Effective for removing foam from the following: adhesives and glues, paints, asphalts, waxes and polishes, textiles and paper processes, rubber latices, detergents and soaps, additive lube oils, latex compounding, insecticides, fermentation and filtration processes, monomer stripping, leather processing and glycol formulations.

MOONEY CHEMICALS, INC., 2301 Scranton Road, Cleveland, OH 44113

Defoamer 50

Physical Properties:

Chemical Activity	100%	Appearance	Slightly hazy liquid
Weight per Gallon @ 25°C	7.00 lb	Solubility	Dispersible in aqueous systems

Key Properties: Recommended for latex emulsion paints and particularly effective with vinyl acetate and acrylic latices. Defoamer 50 is equally suited for controlling foam during processing and application. Coatings containing Defoamer 50 can be stored for long periods of time without loss of antifoaming properties.

Defoamer 40

Physical Properties:

Chemical Activity	100%	Appearance	Hazy white liquid
Weight per Gallon @ 25°C	7.10 lb	Solubility	Dispersible in aqueous systems

Key Properties: Recommended for flat and semigloss latex paints, using acrylic, vinyl acrylic, polyvinyl acetate or styrene-butadiene emulsions. Coatings formulated with Defoamer 40 can be stored for long periods without loss of antifoaming properties. Defoamer 40 is considered by users to be the best defoamer available anywhere.

Defoamer 30

Physical Properties:

Chemical Activity	100%	Appearance	Slightly hazy liquid
Weight per Gallon @ 25°C	7.27 lb	Solubility	Dispersible in aqueous systems

Key Properties: Defoaming and antifoam agent designed for those emulsions where polyvinyl alcohol is used as the protective colloid. Particularly effective with polyvinyl acetate polymers and ethylene-vinyl acetate polymer emulsions. Defoamer 30 is suitable for latex adhesives as well as latex paints.

Defoamer 20

Physical Properties:

Chemical Activity	100%	Appearance	Slightly hazy liquid
Weight per Gallon @ 25°C	7.26 lb	Solubility	Dispersible in aqueous systems

Key Properties: Recommended for gloss and semigloss latex paint formulations in addition to being an effective defoamer and antifoam for many flat latex paints. Defoamer 20 is particularly effective with vinyl acrylics, vinyl acetates, 100% acrylics and styrene-butadiene emulsions. Coatings formulated with Defoamer 20 can be stored for long periods without loss of antifoaming properties.

Defoamer 10

Physical Properties:

Chemical Activity	100%	Appearance	Slightly hazy liquid
Weight per Gallon @ 25°C	7.33 lb	Solubility	Dispersible in aqueous systems

Key Properties: Recommended for polyvinyl acetates and acrylic emulsions. Defoamer 10 is equally suited for controlling foam during processing and application. Coatings formulated can be stored for long periods without loss of antifoaming properties.

Chemical Type (of all): Complete line of proprietary composition defoamers.

Key Properties (of all): Each of the defoamers has been formulated to meet specific and exacting criteria, with the following characteristics common to all of the line:

They are fluid and disperse easily. (Note—defoamers normally are added by volume.)
They do not separate upon aging.
They remain active over long periods of time.
They are used at low concentrations.
They do not emulsify with excessive agitation.
They do not interfere with surface tension properties or contribute to color problems.
They are silica and silicone free.

POLYMER RESEARCH CORP. OF AMERICA, 2186 Mill Ave., Brooklyn, NY 11234

Product Name	Active Content (%)	Specific Gravity (25°C)	Weight per Gallon (lb)
Silicone Defoamer #5037	15	0.980	8.2

Chemical Type: Highly efficient silicone defoamer.

Key Properties: Recommended for all types of aqueous foaming conditions. Can be used as received, but is usually diluted to the required working concentration for ease of addition to the foaming system. Can be readily dispersed in water to concentrations as low as 1% silicone and then added into the system as required. Excellent stability.

TRI-STAR CHEMICAL CO., P.O. Box 38627, Dallas, TX 75238

Tri-Star Antifoam 27

Chemical Type: Synergistic blend of organic chemicals in translucent to clear liquid form, with mild odor.

Physical Properties:

Activity	100%	Flash Point	240°F min
Specific Gravity @ 77°F	0.84	pH (1% dispersion)	4.8
Weight per Gallon	7.00 lb	Water Content	Nil

Key Properties: Very fast acting and is outstanding for rapid knockout of existing foam and the prevention of foam formation in systems containing detergents, adhesives, latex emulsions, industrial processes, paints, glues, paper coating formulations, water disposal systems, sewage plants, dye baths, textile printing, paper manufacturing, oil well drilling muds, refineries and petrochemical plants. Should be considered in all industrial processes where foam is a problem.

Regulatory: Complies with the following:

FDA 121.1519
FDA 121.2520

Section IX

Dispersing and Emulsifying Agents

ALLIED COLLOIDS, INC., 161 Dwight Place, Fairfield, NJ 07006

Dispex Dispersing Agents

Product Name	Active Content (%)	Viscosity Brookfield (cp)	pH	Specific Gravity
N40	40	20 @ 25%	7.25	1.30

Chemical Type: Sodium salt of a polymeric carboxylic acid.

Key Properties: Pale yellow liquid with slight odor. Standard product of the line.

A40	40	4.50 @ 15%	8.00	1.16

Chemical Type: Ammonium salt of a polymeric carboxylic acid.

Key Properties: Pale yellow liquid with ammoniacal odor. Available for applications where the presence of sodium ions is undesirable.

G40	40	600	7.50	1.20

Chemical Type: Sodium salt of a polymeric carboxylic acid.

Key Properties: Water white liquid with slight odor. Unique patented dispersing agent designed specifically for use in high sheen or glossy emulsion paints. Its use can considerably enhance the gloss level of a particular formulation.

Key Properties (All of the Above): The benefits of Dispex when used in latex compounds and adhesives can be summarized, as follows:

> High efficiency and low cost.
> Excellent viscosity stability.
> Water resistance unaffected as small quantities are used.
> High solids content can be achieved by rapid drying.
> Latex stabilization towards filler loading.
> Temperature and pH stability.

Strict molecular weight control during manufacture results in a consistent product of maximum dispersing efficiency. All grades are mobile liquids and, as such, are easy to handle and can be measured by volume. They are all miscible with water in all proportions. They are unaffected in efficiency by temperature fluctuations up to 100°C and are effective over a pH range of 5.5 to 14. They are nonfoaming in character.

CHEMICAL COMPONENTS, INC., 20 Deforest Ave., East Hanover, NJ 07936

Product Name	Viscosity Brookfield (cp)	Percent Solids Content	pH	Specific Gravity	Molecular Weight
Hexasperse 100	300	40	7.0	1.28	750

Chemical Type: Sodium neutralized polymer pigment dispersing agent, in water-clear liquid form.

Key Properties: Represents a technological break-through in the pigment dispersant field. Dramatically reduces the viscosity of compounds where high solids contents are desired. It imparts low viscosity stability to these compounds over a significant period of time. Recommended for paper coatings and other related applications. Excellent stability. Meets the following F.D.A. Regulations: 173.310, 176.170 and 176.180. The uses under these regulations are as boiler additive and in paper and paperboard in contact with aqueous and fatty and dry food respectively.

INTERSTAB CHEMICALS, INC., 500 Jersey Ave., New Brunswick, NJ 08903

Product Name	Color (Gardner) (max)	Specific Gravity (75°F)	Viscosity (Gardner-Holdt) @77°F (max)	Weight per Gallon (lb)	% Solids Content	Form
Intersperse	12	0.943	T	7.86	71.5	Liquid

Chemical Type: Proprietary composition highly active multifunctional dispersing agent.

Key Properties: Designed for solvent-based paints and high pigment concentrates and dispersions. Compatible with most oleoresinous varnishes, alkyds, oils, epoxy esters, chlorinated rubbers, nitrocellulose and other solvent-based coating vehicles. Some of the areas where Intersperse demonstrates distinct advantages are:

Minimizes color drifting because it allows complete color development in the dispersion operation.

Intersperse reduces solvent loss and production costs because of reduced heat build-up during high speed dispersion.

At equal cost, Intersperse yields any specified fineness of grind in less time than soya lecithin or other commercial grinding aids. This reduction in dispersion time means more production and lower overall cost.

On an equal cost basis in ball mill grinds, it consistently develops lower viscosity than soya lecithin. This increased fluidity permits fast, more complete draining at lower production cost.

It enables the formulator to incorporate the highest pigment concentration in the grind or base portion of any solvent based system.

It produces the maximum fineness of grind in any specified time at the lowest possible cost.

In most cases, Intersperse does such an effective job of wetting and dispersing the pigment that it eliminates the need for a suspending agent or other wetting agents.

Intersperse does not produce a yellow tint which requires compensating toners that reduce brightness.

SYNTHRON, INC., 44 East Ave., Pawtucket, RI 02860

Acrylon Dispersing Agents

Product Name	% Solids Content	Specific Gravity	Brookfield Viscosity (cp)	pH	Surface Tension (1% Solution @20°C) (dynes/cm)
Acrylon A06	25	1.15	60	8.50	70
Acrylon A10	40	1.28	1000	9.25	70
Acrylon A002	28	1.17	175	7.50	70

Chemical Type (All of the Above): Sodium salts of a low molecular weight carboxylated organic polymer, in hazy, straw-colored liquid form.

Key Properties (All of the Above): Particularly effective dispersing agents for mineral fillers and pigments in aqueous systems. They find widespread use in dispersing pigments and fillers in the paint and paper industries. Most small solid particles have a natural and marked tendency to agglomerate in water suspensions. This phenomenon results from attractive forces between the particles. As a result, high viscosities and/or settling occur rapidly. The Acrylons are absorbed on particles and cause repulsive forces preventing agglomeration. The use of Acrylons allow greater production rates because of their:

Low viscosity.

Properties not being affected by temperature changes.

Suitability for dispersing fillers in high concentration with a minimum of dispersing agent.

Properties not being affected by changing pH over a wide range, generally from 6 to 14. Their use is effective when used at rates as low as 0.2 to 0.5 on the weight of dry filler.

UNIROYAL CHEMICAL, Division of Uniroyal, Inc., Naugatuck, CT 06770

Polywet Anionic Oligomeric Dispersants

Product Name	Viscosity (Brookfield) (cp)	% Solids Content	pH	Specific Gravity	Molecular Weight
ND-1	65	25	7	1.13	1500

Chemical Type: Sodium salt of polyfunctional oligomer in water. Clear, colorless to slight yellow liquid.

ND-2	30	25	7	1.16	1000

Chemical Type: Sodium salt of polyfunctional oligomer in water. Clear, colorless to slight yellow liquid.

Key Properties (ND-1 and ND-2): Recommended for dispersions for paint, paper coating, latex compounding and dispersants for water treatment.

Effective as dispersants for most pigments, extenders and fillers, including rutile titanium dioxide, zinc oxide, kaolinite, bentonite, satin white, precipitated calcium carbonate, ground limestone, talc and iron oxides. Also, useful for zinc dust and manganese dioxide. They have the following characteristics:

(Polywet ND-1): Recommended for dispersion containing reactive pigments and biocides, like zinc oxide and phenyl mercury salts. Provides excellent stability at elevated temperatures and affords dispersions with exceptional fluidity.

Useful to impart thixotropy or false body to dispersions to inhibit settling without addition thickeners.

(Polywet ND-2): Recommended as an auxiliary dispersant with polyphosphates to impart stability to slurries at elevated temperatures or over prolonged storage. Useful in all types of latex paints, including gloss latex enamels. Recommended for use in shipments of pigment slurries. Complies with FDA Regulations governing coatings, adhesives, paper and paperboard and boiler water.

Toxicity (ND-1 and ND-2): Both are classified as not a "toxic substance" by ingestion as defined in Section 1500.3 of the Federal Hazardous Substance Act. Their acute oral LD_{50} (rats) is over 10 g/kg.

R.T. VANDERBILT CO., INC., 30 Winfield St., Norwalk, CT 06855

Darvan No. 1

Chemical Type: Sodium salts of polymerized alkyl naphthalene sulfonic acids, anionic. Buff granules.

Physical Properties:

Active Dispersing Agent	87% min
Inert Ingredients: Soluble	8.5% max
Inert Ingredients: Insoluble	0.05% max
pH (1% Solution)	8.0–10.5
Color (Gardner)(1% Solution)	4 max
Moisture	4.5% max
Apparent Density (Pounds per Cubic Foot)	35–37

Key Properties:

Pigment dispersing agent for aqueous systems.

While not a wetting agent, has a specific wetting action toward certain materials such as carbon black.

Stable in the presence of mild acids and alkalies.

Suggested amount to use is 1 to 4% based on pigment weight.

Readily soluble in warm water—a 20% solution is a convenient form for handling in the preparation of liquid compositions.

Darvan No. 7

Chemical Type: A polyelectrolyte dispersing agent in aqueous solution. Clear to slightly opalescent colorless liquid.

Physical Properties:

% Solids Content	25±1%
pH	9.5–10.5
Specific Gravity @ 25°C	1.16
Brookfield Viscosity @ 25°C	75 cp max

Key Properties:

Minimum-foaming pigment dispersing agent for aqueous systems.

Stable in the presence of mild acids and mild alkalies.

Very little tendency to foam; a 1% solution has about the same surface tension as water.

Suggested amount to use is 1 to 4% based on pigment weight.

Freezes at temperatures below –5°C. However, freezing does not affect the dispersing characteristics.

Darvan No. 31

Chemical Type: Sodium salt of a carboxylated polyelectrolyte. Yellow liquid.

Physical Properties:

% Solids Content	25% min
Specific Gravity @ 25°C	1.11
Brookfield Viscosity @ 25°C	75 cp max
pH	9–11

Key Properties:

Organic and inorganic pigment dispersing agent for aqueous system. Yellow liquid.

Stable in the presence of mild acids and mild alkalies.

Soluble in water systems.

Suggested amount to use is 1 to 4% on pigment weight.

Very effective in gloss as well as flat latex coatings.

Section X
Epoxy Resin Diluents

AZS CHEMICAL CO., Division of AZS Corp., 762 Marietta Blvd., N.W., Atlanta, GA 30318

Azepoxy Butyl Glycidyl Ethers

Product Name	Brookfield Viscosity (cp) (max)	Color (Gardner) (max)	Specific Gravity	Flash Point (Closed Cup) (°F)	Weight per Epoxide
B	4	2	0.91	>128	150

Chemical Type: Technical grade of butyl glycidyl ether, used as a monofunctional diluent to reduce the viscosity of some epoxy resin systems.

Key Properties: Will react with the resin/hardener system to become an integral part of the cured system. Can be cured with any of normal room temperature curing systems using amines, polyamides or Lewis acids. Will give maximum viscosity reduction with minimum loss of properties. Will permit increased filler loading.

C	25	4	1.08	250	193

Chemical Type: Technical grade of cresyl glycidyl ether, used as a monofunctional diluent to reduce the viscosity of some epoxy resin systems.

Key Properties: Exhibits low volatility and permits increased filler loadings. Resin systems using Azepoxy C exhibit low volatility, permit increased filler loading and show improved chemical resistance over the normal aliphatic diluents. Recommended for industrial flooring, laminating casting and tooling applications.

7	10	1	0.90	290	225

Chemical Type: Aliphatic glycidyl ether where the alkyl group is primarily C_8–C_{10} chain length.

Key Properties: Exhibits a low volatility and a very low order of toxicity. Recommended for efficient viscosity reduction of epoxy resins in casting, electrical, laminating, tooling and flooring applications where good impregnation or maximum filler loading is desired.

8	10	1	0.89	310	285

Chemical Type: Aliphatic glycidyl ether reactive diluent where the alkyl group is primarily C_{12}–C_{14} chain length.

Key Properties: Efficient, nontoxic, has low volatility and a good record of product safety which makes it of first consideration. Convenient from safety point of view as it is not considered to be a human skin sensitizer. Has low irritation potential for eye and skin effects and is practically nontoxic on ingestion.

17	5	3	0.92	200	225

Chemical Type: Technical grade of 2-ethyl hexyl glycidyl ether.

Key Properties: Classified as nontoxic and nonirritating as defined by the U.S. Government. Low volatile diluent for efficient viscosity reduction of epoxy resins. Recommended for casting, tooling, laminating and specialty coatings where low viscosity and high loading is desired. Blends of Azepoxy 17 and a standard liquid epoxy resin can be cured with amines, polyamides, anhydrides or other types of curing agents. The low volatility makes elevated temperatures cure cycles feasible.

Product Name	Brookfield Viscosity (cp) (max)	Color (Gardner) (max)	Specific Gravity	Flash Point (Closed Cup) (°F)	Weight per Epoxide
DA	15	2	0.89	310	285

Chemical Type: Inexpensive aliphatic glycidyl ether reactive diluent, where the alkyd group is primarily C_{10}–C_{14} chain length.

Key Properties: Efficiently nontoxic and has low volatility. It is typically used as a viscosity control to dilute liquid epoxy resins to the 700 cp range. It is not considered to be a human skin sensitizer and has low irritation potential for eye and skin effects.

Toxicity (all types): All of the above may be skin irritants and have other slight toxic effects. Consult the manufacturer for specific information.

PROCTER AND GAMBLE DISTRIBUTING CO., P.O. Box 599, Cincinnati, OH 45201

Epoxide No. 8

Chemical Type: Aliphatic glycidyl ether containing primarily n-dodecyl (C_{12}) and n-tetradecyl (C_{14}) alkyl groups, an epoxy resin diluent, in clear mobile liquid form.

Chemical Analysis:

Weight per Epoxide (WPE)	286	Specific Gravity @ 25°C	0.89
Oxirane Oxygen	5.6%	Surface Tension	33 dynes/cm
Hydrolyzable Chloride	0.1%	Viscosity, Brookfield @ 77°F	10 cp
Moisture	0.1%	Flash Point	310°F
Color (APHA)	25	Melting Point	35°F

Key Properties: When Epoxide No. 8 is used at the 15–18 phr level, resin viscosity is reduced to the preferred 500–700 cp range with minimal effects on cured resin properties. Less volatile than most commonly used diluents. A less volatile diluent reduces atmospheric contamination, making the workplace more pleasant.

R.T. VANDERBILT CO., INC., 30 Winfield St., Norwalk, CT 06855

Vanoxy Epoxy Diluents

Product Name	Brookfield Viscosity (cp)	Color (APHA)	Epoxide Equiv.	Flash Point (TOC) (°F)
RD-707	10	15	229	245

Chemical Type: Aliphatic glycidyl ether reactive diluent for epoxy resins.

Key Properties: Very low order of toxicity and low volatility. Efficient viscosity reducer.

RD-708	10	15	286	310

Chemical Type: Aliphatic glycidyl ether reactive diluent for epoxy resins.

Key Properties: Very low order of toxicity and low volatility. Efficient viscosity reducer.

Section XI
Fillers and Extender Pigments

COMMERCIAL MINERALS CO., 6899 Smith Ave., P.O. Box 363, Newark, CA 94560

Chemical Type: Finely ground crystalline calcium magnesium carbonate. Pigment grade white extender produced from natural dolomite (dry ground and air classified).

	. . . . Specification	
Product	*Mesh No.*	*Percent Pass*
D220	200	80
D202	200	98
D306 (D195)	325	94
D3002 (D-RM)	325	99.8

Chemical Analysis:

Calcium Carbonate ($CaCO_3$)	56.3 wt %	Aluminum Oxide and	
Magnesium Carbonate ($MgCO_3$)	42.8 wt %	Iron Oxide (R_2O_3)	0.2 wt %
Silica (SiO_2)	0.6 wt %	Specific Gravity	2.86

COMPOSITION MATERIALS CO., INC., 26 Sixth St., Stamford, CT 06905

Product Name	pH @ 25°C	Specific Gravity	Density (lb/ft³)	Charring Temperature (°F)	Moisture (%)	
Walnut Shell #108	4.5	1.33	40	470	5	

Mesh Analysis:

	Percent by Weight
Retained on 12 Mesh	0
Passing 12—Retained on 20 Mesh	0
Passing 20—Retained on 30 Mesh	0-2
Passing 30—Retained on 40 Mesh	14-22
Passing 40—Retained on 60 Mesh	25-35
Passing 60—Retained on 100 Mesh	35-47
Passing 100 Mesh	8-10

Product Name	pH @ 25°C	Specific Gravity	Density (lb/ft³)	Charring Temperature (°F)	Moisture (%)	
Walnut Shell Flour (Extra Fine)	4.5	1.33	39	470	4	60.0

Other Physical Constants:

Degree of Brightness (Measured Photometrically)	35-48%
Bulking Factor	0.0910
Specific Heat	0.43
Flash Point (Closed Cup)	380°F
Hardness (Rockwell)	91
Hardness (Brinell)	190

Particle Size Distribution (by U.S. Standard Screen Size):

	Percent by Weight	
	(max)	*(min)*
On 100 Mesh	0.0	0.0
Through 100 and on 150 Mesh	0.5	0.0
Through 150 and on 200 Mesh	1.5	0.0
Through 200 and on 250 Mesh	3.0	1.0
Through 250 and on 325 Mesh	5.0	3.0
Through 325 Mesh	96.0	90.0

Chemical Analysis (in percent by weight):

	#108	Extra Fine
Nitrogen	0.10	0.10
Furfural	9.00	10.00
Cellulose	60.00	60.00
Lignin	23.50	24.00
Cutin	5.00	5.00
Sugar	0.30	0.30
Ash	1.00	0.50
Oil	3.20	—
Methoxyl	—	6.50
Chlorine	—	0.10

Product Name	Acetone Soluble (% max)	Ash Content (%)	Oil Absorption (cc/10 g)	Ether Extract (% max)	Moisture (%)	Density (lb/ft³)
Woodflour #H-100	8.0	0.3	15	5.0	60	15.0

Chemical Type: Fibrous light brown kiln dry maple.

Typical Screen Analysis:

	Percent by Weight
On 40 Mesh	0.0
On 60 Mesh	0.0
On 80 Mesh	Trace
On 100 Mesh	10.0
On 150 Mesh	24.5
On 200 Mesh	24.5
Through 200 Mesh	41.0

GEORGIA MARBLE CO., 2575 Cumberland Parkway, N.W., Atlanta, GA 30339

Finely Ground Extender Pigments and Fillers

Product Name	Particle Size (μ)	Dry Brightness (Hunter Reflectometer) (% min)	Plus 325 Mesh (% max)	Hegman Fineness of Grind	Oil Absorption (lb/100 lb)
Gamaco	1.7	96	0.005	6	15

Key Properties: Recommended for use in paints and enamels, paper, plastics, rubber and wherever a fine extender pigment or filler is required.

| Calwhite | 2.25 | 96 | 0.008 | 5 | 13 |

Key Properties: An economical, high brightness general purpose extender pigment for use in paints, putties, rubber products, textile coatings, etc.

| Wingdale White | 2.5 | 94 | 0.200 | 3 | 12 |

Key Properties: Specifically recommended for use in all types of rubber products (except extruded), outside house paints, polyvinyl chloride products, putties, mastics and glazing compounds.

Product Name	Particle Size (μ)	Dry Brightness (Hunter Reflectometer) (% min)	Plus 325 Mesh (% max)	Hegman Fineness of Grind	Oil Absorption (lb/100 lb)
Gamaco II	1.7	95	0.005	—	15

Key Properties: Recommended for liquid polyester systems, BMC–SMC and similar application using polyester resin component.

Calwhite II	2.05	95	0.008	—	13

Key Properties: An economical filler for liquid polyester systems, BMC–SMC and similar application using polyester resin.

Chemical Type: All of these natural calcium carbonate pigments and fillers are produced from pure crystalline marble and manufactured under rigidly controlled conditions of wet or dry grinding. Classification is accomplished by centrifugal or air separation methods. These products differ in their particle size, as shown above.

Key Properties: An important advantage possessed by each of these products is the inherent economy, offering the highest quality calcium carbonate pigments and fillers at the lowest possible cost.

Medium and Coarse Ground Extender Pigments and Fillers

Product Name	Passing 325 Mesh (%)	Passing 200 Mesh (%)	Dry Brightness (Hunter Reflectometer) (% min)	Oil Absorption (lb/100 lb)
#10 White	99.5 min	—	92	8

Key Properties: Recommended for paints, rubber products, plastics, putties, etc., where cost is a factor and a minimum of +325 is permitted. Also, popular for latex rug backing compounds.

#9 White	93	—	92	*

Key Properties: Recommended for joint cements, rug backing compounds, etc., where even higher +325 is not detrimental.

#9NCS	93	—	**	*

Key Properties: Same uses as #9 White, but where color is not important.

80-325	80	—	90	*

Key Properties: For putties, mastics, adhesives, porcelain-type products, block sealers, etc., where still coarser materials are acceptable.

#8 White	75	—	90	*

Key Properties: A still coarser pigment for the same uses as 80-325.

Rock Dust	—	83	—	—

Key Properties: Recommended for coal mine dusting.

Industrial Filler	—	80	—	—

Key Properties: Recommended for synthetic marble, rice polishing.

RO-40	—	73	—	—

Key Properties: Recommended for asphalt filler, putty, ceramic material, foamed compounds.
*Too coarse to determine end point.
**No color specification.

Chemical Type: Pure white crystalline marble is the source of all these medium ground, classified products.

Key Properties: These products have found ready acceptance in the textile finishing field, heavily loaded rubber compounding, putty and mastic production, low sheen paint manufacturing, extension and strengthening of all types of adhesives and numerous other fields.

Low oil absorption, low binder requirement and relative high brightness typify these products that offer an excellent source of economical pigments and fillers.

Each of these products covers a broad range of particle sizes, including those retained on sieves and those which are subsieve.

GREFCO, INC., Minerals Division, 3450 Wilshire Blvd., Los Angeles, CA 90010

Dicalite Perlite Bulk-Aid Fillers

Product Name	Oil Absorption (lb/100 lb)	Retained on 325 Mesh (% max)	Surface Area (m²/g)	Loose Weight (lb/ft³)	Tamped Weight (lb/ft³)	Hegman Fineness of Grind
2	135	1.0	2.5	5.5	18.0	3
30	135	2.0	2.5	5.0	18.0	2
3	185	2.0 on 150 Mesh	1.9	5.0	16.0	—

Average Particle Size Distribution (Sedimentation Method):

	2	30	3
	. . . . (Percent by Weight).		
>40 μ	—	Trace	0.5
20-40 μ	1.0	1.0	3.5
10-20 μ	2.0	6.0	17.0
6-10 μ	11.0	13.0	29.0
3-6 μ	51.0	58.0	42.0
<3 μ	35.0	22.0	8.0

Characteristics of All Grades:

Color	White	Bulking Value (wt/gal)	19.4
Moisture	0.5% max	Porosity	88.0
Ignition Loss (Dry Basis)	1.5% (due to combined water)	pH (Range)	7.2-7.8
Specific Gravity	2.34	Brightness (G.E.)	83

Key Properties of All Grades:

2—Recommended for use in masonry paints, mastics and adhesives.

30—Recommended for use in paper, paints and match heads.

3—Recommended for use in paper, texture paints and insecticide carriers.

Chemical Type of All Grades: Produced from a siliceous material of volcanic origin known as perlite, which is a dense, glassy rock. It "pops" like popcorn when crushed and heated under proper conditions, expanding to 20 or more times its original volume. The exploded particles are milled and classified by an exclusive process. This special production method gives these materials a precise particle size distribution for maximum bulking characteristics. It is an amorphous mineral consisting of a fused sodium potassium aluminum silicate. The approximate analysis is 70% silica, 18% alumina, 7% potassium oxide and 4% sodium oxide. It is very light in weight and free from grit or organic matter.

Key Properties of All Grades: Four important characteristics make them very useful as fillers: (1) Fine particle size; (2) High brightness; (3) Very light weight; and (4) Mild abrasiveness.

MARTIN MARIETTA CHEMICALS, Executive Plaza II, Hunt Valley, MD 21030

MagChem 40

Chemical Type: High purity technical grade magnesium oxide processed from magnesium-rich brine.

Chemical Analysis:

	Percent by Weight
Magnesium Oxide (MgO), % Ignited Basis	98.00
Calcium Oxide (CaO)	0.80
Silica (SiO_2)	0.30
Ferric Oxide (Fe_2O_3)	0.20
Total R_2O_3	0.35
Chloride (Cl)	0.40
Sulfate (SO_3)	0.10
Loss on Ignition (LOI)	2.50

Physical Properties:

Bulk Density (lb/ft³) (kg/m³)	18 (288)
Mean Particle Size	5 μ
Surface Area per Gram (m²)	45
Iodine Number (meq/gram)	40
Activity Index (Seconds)	10
Screen Size, Wet (% Passing Minus 325 Mesh)	99

Key Properties: This fine white powder has a very high reactivity and a low bulk density. Because of its high surface area and fine particle size, MagChem 40 is well suited for many rubber formulations, particularly neoprene. It finds wide use as a filler, anticaking agent and pigment extender. It is extremely efficient in chemical processes where ease of conversion is a factor. It is also used in the production of oil additives and the desilication of water.

Toxicity: Not particularly hazardous. Suitable precautions in handling should include the practice of reasonable caution and personal cleanliness.

R.T. VANDERBILT CO., INC., 30 Winfield St., Norwalk, CT 06855

Vansil W Wollastonite

Chemical Type: A white, nonmetallic, natural mineral filler identified chemically as calcium metasilicate ($CaSiO_3$). Molecular weight of 116. White crystals with vitreous or pearly luster.

Chemical Analysis:

	Percent by Weight
Calcium Oxide (CaO)	46.10
Silicon Dioxide (SiO_2)	50.20
Aluminum Oxide (Al_2O_3)	0.21
Magnesium Oxide (MgO)	1.86
Iron Oxide (Fe_2O_3)	0.15
Sodium Oxide (Na_2O)	0.22
Ignition Loss	0.94
Total	99.68

Physical Properties:

Shape	Acicular
Specific Gravity	2.9
Hardness (Mohs Scale)	4.5
Coefficient of Expansion (mm/mm/°C)	6.5×10^{-6}
Transition Point to Pseudo Wollastonite	1200°C (2190°F)
Melting Point	1540°C (2800°F)

Key Properties: Useful as a functional filler in adhesives, abrasives, bonded abrasives, ceramics, elastomers, insulating materials, paint, plastics, resins, sealants and wallboards.

Section XII
Fire and Flame Retardants

DOVER CHEMICAL CORP., Davis at West Fifteenth St., Dover, OH 44622

Product Name	Chlorine Content (% by wt)	Softening Point (Ring and Ball)	Gardner Color	Weight per Gallon (lb)	Free Chlorine
Chlorez 700	70	103	<1	13.5	nil

Chemical Type: White powdered chlorinated paraffin resin.

Key Properties: Tasteless, odorless, essentially nontoxic resin, especially soluble in aromatic and chlorinated solvents. Compatible with most commonly used resins, rubbers, plasticizers, waxes and drying oils. Because of its very high active halogen content and low cost, it is widely used as a flame retardant additive in many applications.

Paroil 170HV	70	—	2	12.7	nil

Chemical Type: Liquid grade of chlorinated paraffin.

Key Properties: Very high viscosity, high chlorine content. Very high degree of tack.

Rez-O-Sperse 3	45	—	—	—	—

Chemical Type: Water emulsion of Paroil 170HV, containing 45+% chlorine.

Key Properties: Maximum flame retardant efficiency. It finds application in both cationic and anionic emulsions systems. In addition to their flame retardant contribution, they improve adhesion, impart chemical and water resistance and allow the user to formulate aqueous systems rather than solvent systems. Recommended for plasticizing and tackifying.

GAF CORP., Chemical Division, 140 West 51 St., New York, NY 10020

Flame Retardant Additive

Chemical Type: Dibromobutenediol in powder form.

Key Properties: Reactive flame retardant additive for polyurethane foams, films and elastomers and for polyester and epoxy systems. Effective in flexible foams and thermoplastics; maintains essentially linear structure of resin. Compatible with synergistic additives.

GREAT LAKES CHEMICAL CORP., Highway 52 NW, West Lafayette, IN 47906

Product Name	Freezing Point	Boiling Point
	................(°C)................	
BA-59 and BA-59P	180	316

Chemical Type: Tetrabromobisphenol A (TBBPA) [4,4'-isopropylidenebis(2,6-dibromophenol)] ($C_{15}H_{12}Br_4O_2$). Formula weight of 543.88. Bromine content of 58.8%. White, crystalline or powdered solid.

Key Properties: BA-59 (BA-59P) is used as a fire retardant in epoxy, polycarbonate and unsaturated polyester resin systems because of its structural compatibility, high bromine content and thermal stability. It is also used as an additive flame retardant in styrenic polymers.

Toxicity: BA-59 (BA-59P) has a very low order of toxicity by ingestion, inhalation or skin absorption.

Product Name	Melting Point (°C)	Specific Gravity	Volatiles (% max.)	Iron Content (ppm max.)
DE-79	105	2.66	0.25	10

Chemical Type: Octabromodiphenyl oxide ($C_{12}H_2Br_8O$). Formula weight of 801.4. Bromine content of 79.8%. Off-white powder.

Key Properties: Excellent fire retardant additive which can be used in many thermoplastics. The overall fire retardant performance can be improved when combined with antimony oxide. Can also be used for flame retarding urethanes, engineering plastics and fiber applications.

Toxicity: Toxicity information which is available suggests that DE-79 has a very low order of acute toxicity, whether by ingestion, inhalation, or skin absorption and a low order of chronic toxicity.

Product Name	Active Content (%)	Melting Point (°C)
DE-83R	97	308

Chemical Type: Decabromodiphenyl oxide ($C_{12}Br_{10}O$). Formula weight of 959.3. Bromine content of 83.3% theoretical. White powder. No detectable inorganic bromine.

Key Properties: Excellent fire retardant additive which can be used in many thermoplastics. The overall fire retardant performance can be improved when combined with antimony oxide. Can also be used for flame retarding, urethanes, engineering plastics and fiber applications.

Toxicity: Toxicity information which is available suggests that DE-83R has a very low order of acute toxicity, whether by ingestion, inhalation, or skin absorption and a low order of chronic toxicity.

HERCULES, INC., 910 Market St., Wilmington, DE 19899

Chlorafin Chlorinated Paraffins

Chemical Type: The Chlorafins are chlorinated paraffins, made by the reaction of chlorine and paraffin wax, under carefully controlled conditions. Effective stabilizers are also added to assure retention of good stability.

Chemical Analysis and Physical Properties:

	Chlorafin 40	Chlorafin 50
Chlorine content	40.0–43.5%	48.0–52.0%
Color (Gardner Scale)	6 max.	6 max.
Color (visual)	light amber	slightly light amber
Viscosity @ 25°C	26–39 poises	430–620 poises
Specific gravity @ 25°/25°C	1.16	1.26
Flash point	none	none
Fire point	none	none

Note: Special grades of any desired chlorine content can be made.

Key Properties: Volatility — The evaporation rate of Chlorafin 40 is comparable to that of tricresyl phosphate.

Chemical resistance — Chlorafin is quite resistant to the action of strong chemicals such as acids, alkalies and oxidizing agents.

Miscibility — Chlorafin is miscible with all classes of commonly used solvents with the exception of the lower aliphatic alcohols and water.

Compatibility — Chlorafin displays, generally, good plasticizing properties when it is used as the only plasticizer.

In general, the properties of Chlorafin 50, such as compatibility and solubility, are similar to those of Chlorafin 40 and its uses are also similar. Chlorafin 50 is useful where a product of higher chlorine content is desirable or where a viscous material is useful. The higher chlorine content of Chlorafin 50 makes it more effective than Chlorafin 40 in flame resistant compositions.

The Chlorafins should be considered for many applications because of their non-flammability, low vapor pressure and favorable plasticizing properties. The following are specific uses:

Flameproof and weatherproof textiles	Chemical resistant coatings and
Vinyl plastics	compounds
Vinyl coatings	Water repellent wood impregnants
Furniture finishes	Miscellaneous uses

HUMPHREY CHEMICAL CORP., P.O. Box 2, Edgewood Arsenal, MD 21010

Zinc Borates

Product Name	Zinc as Zinc Oxide	Boron as Boric Oxide	Ignition Loss
	........................(%)........................		
ZB-112	45	35	20
ZB-112 Special (Off-white color)	45	35	20
ZB-112R	47	34	19
ZB-237	33	41	26
ZB-325	52	29	19

Chemical Type: Zinc borates in white, nonhygroscopic, powdered form. They range from ZB-237 which is a low specific gravity, low zinc and high boron content to ZB-325, a higher specific gravity, lower boron and higher zinc content material.

Key Properties: Recommended for use as fire retardants, after-glow depressants and biostats. All grades have a lower specific gravity and offer a lower weight per unit of volume than antimony trioxide. All are easily dispersed in fire retardant systems. Zinc borate is being used successfully as an ingredient in fire retardant formulations containing halogens. In such formulations it is used either by itself or in conjunction with antimony trioxide. This combination reduces the total formulation cost and has, in most cases, increased fire retardance and decreased after-glow. The biocidal property increases resistance to mildew and fungi.

Performance at combustion temperatures — Promotes degradation and decomposition of halogens to form HCl. Loses water of hydration as steam (helps snuff out flames). Undergoes acid hydrolysis to form zinc chloride and free boric acid—both assist in fire retardation. Fuses and isolates substrate from atmospheric oxygen.

MAYCO OIL AND CHEMICAL CO., INC., Beaver and Canal Streets, Bristol, PA 19007

Mayco Chlorinated EP Additives

	DC-40	DC-50	DC-60	DC-70
Chlorine content, % by wt	43 min	50	60	70
Free chlorine	nil	nil	nil	nil
Viscosity @ 210°F, SSU	170	375	70	550
Viscosity @ 210°F, cs	—	80.4	13	118

(continued)

	DC-40	DC-50	DC-60	DC-70
Viscosity @ 100°F, SSU	–	25,000	2,800	–
Viscosity @ 100°F, cs	–	–	600	–
Gardner Color	3	3	3	4
Molecular weight	506	–	–	–
Specific gravity @ 60°/60°F	1.16	1.26	1.35	1.55
Weight per gallon @ 60°F, lb	9.75	10.5	11.3	12.9
Pour point (°F)	40	59	80	104
Flash point (°F)	none	none	none	none

Chemical Type: Chlorinated paraffin EP additives.

Key Properties: Designed to be used in a wide variety of industrial and metal working oil formulations. Higher chlorine levels.

NYACOL, INC., Megunco Road, Ashland, MA 01721

Nyacol Colloidal Antimony Oxide Dispersions

Product Name	Antimony Oxide (%)	Particle Size (μ)	pH (25°C)	Weight per Gallon (lb)	Stabilizer (% by wt)	Viscosity (cp)
A1510LP	10	15	2.5	9.17	none	2.2
A1510	10	15	5.0	9.17	none	2.2
A1530*	30	15	5.0	11.45	0.8	5.0
AI550**	50	15	5.0	15.1	2.8	20.0

*Contains 0.8% triethanolamine which allows higher product concentration.
**Contains 2.8% triethanolamine which allows higher product concentration.

Chemical Type (of all of the above): Anionic with negatively charged colloidal particles.

AP50   50

Chemical Type: Organic dispersible paste.

A1582   82

Chemical Type: Powder which is water redispersible to pH of 7.5 to 8.0.

A1588LP   88

Chemical Type: Water-insensitive powder.

Chemical Type (of all of the above): Extremely fine particle powders or dispersion of antimony pentoxide.

Key Properties: Less antimony oxide is required to achieve a given level of flame retardancy as compared to normal pigment antimony oxide. This affects the benefits of cost savings and availability.
Nonpigmenting and nondelustering effects are eliminated.
Nonsettling characteristics.
Complete impregnation of textiles, paper or open-celled foams.
Good adhesion to fibers.
Compatible with many latices.
Ease of compounding.

The spray dried versions of colloidal antimony oxide are round, free-flowing particles which are readily compounded into plastics and rubbers. They are easily mixed into plastisols and epoxy compounds without clumping or lumpiness.

VELSICOL CHEMICAL CORP., 341 East Ohio St., Chicago, IL 60611

Firemaster CA

Chemical Type: Chlorendric anhydride (technical grade), in white crystalline solid form.

Chemical Analysis and Physical Properties:

Assay, %	95.0
Chorendric acid, max %	4.0
Maleic anhydride, max %	1.0
Free volatiles, max %	1.0
Hexachlorocyclopentadiene, max %	0.14
Color (APHA), max	300
Molecular weight	371
Chlorine, %	54.5
Melting range, °C	235–250
Iron, ppm	5
Total volatile content, %	4

Key Properties: Solid, reactive flame retardant with an unusually stable chlorinated bicyclic structure. Being a difunctional acid anhydride it has many useful applications in the synthesis of flame retardant polymer systems, such as alkyds, urethanes, unsaturated polyester resins and epoxy resins. Besides being fire resistant, polyesters containing Firemaster CA often exhibit a significant degree of UV stability and corrosion resistance.

Toxicology: The data available indicate that Firemaster CA has a low order of toxicity. Refer to supplier's Material Safety Data Sheet for specific information.

Following are the results of acute toxicity tests conducted in general compliance with the Federal Hazardous Substances Act:

Test	Animal	Results
Acute Oral Toxicity	rat	LD_{50} = 1,138 mg/kg
Acute Dermal Toxicity	rabbits	LD_{50} = >2,000 mg/kg
Acute Inhalation Toxicity	rats (4 hr)	LC_{50} = >65 mg/ℓ

Eye irritation — Extremely irritating and corrosive to the eyes.
Skin irritation — Not a primary skin irritant, nor is it corrosive to the skin.

Section XIII

Latices

DOW CHEMICAL U.S.A., Midland, MI 48640

Dow Carboxylated Vinylidene Chloride-Butadiene Latices

Product Name	TSC (%)	Viscosity (cp)	pH	Weight per Gallon (lb)	Particle Size (μ)	Specific Gravity
XD-8609.01	48	85	6.5	9.2	0.15	1.13

Chemical Type: Soft, rubbery latex with approximately equal parts of butadiene and vinylidene chloride copolymerized in the molecule.

XD-8260.04	50	84	8.0	9.2	0.19	1.13

Chemical Type: Harder, more "boardy" latex having a significantly greater amount of vinylidene chloride than butadiene in the molecule. For that reason, it is used in formulating or compounding where ignition deterrence is required.

Key Properties: Both products have practical commercial value to two broad classes of customers:

(1) Companies who wet laminate one substrate to another.

(2) Formulators who compound "adhesives" used to wet laminate or bond various materials either to themselves or to other materials.

Both latexes offer the adhesives industry important property advantages in oil and solvent resistance, water resistance, mechanical stability and chemical stability.

Also, they can be formulated to have very low ignition potential due to the halogen introduced onto the molecule and to the effects of other additives. "Carboxylation" provides excellent stability with low surfactants, resulting in improved water resistance and increased formulating latitude. Carboxylation also means that film properties are largely controlled during production through control of the degree of cross-linking.

Dow Styrene-Butadiene Latices

Product Name	TSC (%)	pH	Weight per Gallon (lb)	Particle Size (μ)	Surface Tension (dynes/cm)	Tg (°C)
202	48	10.5	8.45	0.23	40	+6

Chemical Type: No carboxylation. Contains no antioxidant.

221	43	8.0	8.36	0.14	55	−30

Chemical Type: Carboxylated. Contains a nonstaining antioxidant.

238	50	8.0	8.52	0.13	50	+10

Chemical Type: Carboxylated. Contains no antioxidant.

Product Name	TSC (%)	pH	Weight per Gallon (lb)	Particle Size (μ)	Surface Tension (dynes/cm)	Tg (°C)
277	46	4.3	8.63	0.10	45	+20

Chemical Type: Highly carboxylated. Contains no antioxidant.

283	45	8.0	8.40	0.13	50	–27

Chemical Type: Carboxylated. Contains a nonstaining antioxidant.

285	53	8.0	8.53	0.20	48	+10

Chemical Type: Carboxylated. Contains no antioxidant.

Chemical Type (All of the Above): A new series of carboxylated styrene-butadiene latexes designed specifically for use in the adhesives industry.

Key Properties: Range from very soft and sticky types to very firm in this order, of increasing stiffness: 283-221-202-238-285-277. They all have excellent adhesion to most substrates. Excel when the price-volume relationship is considered. Recommended for a range of commercial bonding operations, including: paper to paper, foil to paper, aluminum to reinforced plastic, vinyl to glass, as adhesives for many food packaging structures, as binder and adhesive in the mastic and many other uses. These carboxylated styrene-butadiene latexes offer an exceptional range of performance attributes. Physical and film properties vary with choice of latex and are further influenced and controlled by choice of curing schedule and formulating additives. The carboxylation means they have excellent stability with low surfactants, resulting in improved water resistance and increased formulating latitude. Carboxylation also means that film properties can be enhanced at application stage by further curing through the available carboxyl group. Suggested curatives are heavy metal salts or melamine-formaldehyde resins, both of which readily react with carboxyl groups.

Product Name	TSC (%)	Viscosity (cp)	pH	Weight per Gallon (lb)	Particle Size (μ)	Chlorine in Polymer (%)
XD-30208.01	48	50	8.0	9.2	0.14	36

Chemical Type: Vinylidene chloride-butadiene latex.

Key Properties: A soft, tacky latex which is finding increasing use in ignition deterrent laminating applications. It is an excellent replacement for polychloroprene-type latexes, offering significant cost savings to the formulator and fabricator. This latex should be considered for these applications. This latex offers ignition resistance at an economical price. Excellent for bonding foil to fiberglass and foil/paper composites to fiberglass. The fiberglass can be of the rigid board type or of the more flexible wrap or home insulation construction. The latex imparts ignition deterrence to all of these structures.

ESSENTIAL CHEMICALS CORP., Merton, WI 53056

Product Name	Chemical Type/Key Properties
Eccotex 280	Latex copolymer raw material for compounding salt resistant pliable finishes—medium hard (38% active).
Eccotex 282	Acrylic latex copolymer raw material for compounding water emulsion floor sealers and undercoaters, semi-permanent (45% active).
Eccotex 287	Latex copolymer raw material for compounding semi-buffable floor finishes (43% active).
Eccotex 289	Latex copolymer raw material for compounding hard high gloss floor finishes (40% active).
Eccotex 2924	Wax copolymer raw material to formulate polishing, buffing and wire drawing compounds (24% active).

Product Name	Chemical Type/Key Properties

Ecco-Lok 40   Latex copolymer raw material (44% active).

Note: The supplier states that the above materials are basically used in the maintenance chemical field. They are unable to advise specifically which ones are used in the paints and adhesive fields.

FIRESTONE SYNTHETIC RUBBER & LATEX CO., 1200 Firestone Parkway, Akron, OH 44301

Hartex 103 Low-Ammonia Natural Latex

Chemical Type: High quality centrifuged latex. The preservation system consists of a low ammonia concentration in combination with 0.02% sodium dimethyldithiocarbamate and 0.02% zinc oxide. (Preservation concentrations are based upon wet weight of latex.)

Chemical Analysis and Physical Properties:

% Solids Content	61.5% min
Dry Rubber Content	60.0% min
% Solids Content minus Dry Rubber Content	1.75% max
Total Alkalinity (% NH_3 on Wet Weight)	0.21±0.02
KOH Number	0.55±0.05
Mechanical Stability	1400±300 seconds
Volatile Fatty Acid	0.05% max
pH	9.8
Sludge Content (% on Wet Weight)	0.03

Key Properties: Designed to offer important advantages over Firestone Hartex 104 and Hartex 102 natural latex in a wide range of applications. Processed in accordance with the very best commercial practices and is handled in large bulkings through a single factory to insure maximum uniformity.

Recommended for use in water-based pressure sensitive adhesives.

GAF CORP., 140 West 51 St., New York, NY 10020

GAF Latex Polymers
(Unmodified and Carboxylated Styrene-Butadiene Latexes)

Product Name	TSC (%)	Viscosity (cp)	pH	Weight per Gallon (lb)	Film Hand
1300 Series	49	3000	8.5	8.4	Soft
1400 Series	49	3000	8.5	8.4	Medium
1500 Series	49	3000	8.5	8.4	Stiff
2000 Series	48	3000	9.0	8.2	Medium

Key Properties (All of the Above): Recommended for tufted carpet laminating, backsizing and foam precoats. Very fast drying time and high tensile strength with low thickener demand. Good viscosity stability and filler wet out.

3000 Series	50	1400	8.5	8.34	Very Soft

Key Properties: Recommended for upholstery coatings and fabric sizings. These polymers have soft resilient hand, good tensile strength and water resistance.

4000 Series	51	200	6.4	8.4	Medium

Key Properties: Recommended for pigmented paper and paperboard coatings. Excellent binder strength, low odor and good rheology. Good compatibility with natural and synthetic binders. FDA approved.

Product Name	TSC (%)	Viscosity (cp)	pH	Weight per Gallon (lb)	Film Hand
5000 Series	49	200	3.5	8.47	Medium-Stiff

Key Properties: Polymers developed to serve as asbestos felt binders.

| 6000 Series | 49 | 84 | 8.5 | 8.4 | Soft |

Chemical Type: Carboxylated styrene-butadiene copolymers and styrene-butadiene-acrylonitrile terpolymers.

Key Properties: Designed as binders for nonwoven fabrics and saturated paper products. Products in series range from hydrophilic to hydrophobic and adhere to a wide variety of substrates.

| 7000 Series | 48 | 200 | 9.3 | 8.47 | Medium to Soft |

Key Properties: Binders for fibrous padding applications.

| 8000 Series | 38.5 | 50 | 9.3 | 8.37 | Very Soft |

Chemical Type: Reactive copolymers and terpolymers.

Key Properties: Designed for use as binders in pigment printing applications.

| 9000 Series | 49 | 100 | 8.5 | 8.4 | – |

Key Properties: Recommended for use as reinforcing agent in latex formulations. Increases tensile strength, hand and compression sets in latex foams and backing. Used as plastic pigments in paper to improve brightness, gloss, wet rub, and K&N receptivity.

W.R. GRACE & CO., 55 Hayden Ave., Lexington, MA 02173

Daran Polyvinylidene Chloride Coatings

Product Name	TSC (%)	Viscosity (max cp)	pH	Weight per Gallon (lb)	Particle Size (μ)
220	61	75	3.5	11.1	0.12

Key Properties: For all-purpose use. Coatings exhibit high barrier protection against water, moisture, grease, oils and gases. Can be applied as supplied, using conventional multi-purpose coating equipment. Heat sealability is 285°-350°F. (These same properties apply to all the Daran product line.)

| 229 | 61 | 50 | 2.5 | 11.1 | 0.17 |

Key Properties: High adhesion low heat seal. Designed to provide both low heat seal temperatures and optimum adhesion to paper and board. Offers high barriers as well as non-blocking and excellent slip characteristics. Coatings produced are odor free. Heat sealability range is 250°-350°F.

| 820 | 50 | 30 | 2.0 | 10.45 | 0.14 |

Key Properties: Primerless. Combines the barrier advantages of other Daran coatings with primerless adhesion to many plastic films. It also offers low foaming, improved whiteness and gloss, excellent flow and leveling. Should be tested on specific substrates. Heat sealability range is as low as 215°F.

| X-805 | 50 | 30 | 2.0 | 10.35 | 0.13 |

Key Properties: Water based barrier PVDC adhesive and primer providing very high initial and aged bond strengths to a variety of films. An activation temperature as low as 180°F and excellent machinability produce superior film laminations.

Physical Properties (All Darans have the following properties):

Color   Cream white

Freezing Point	36°F
Oxygen Barrier	0.5
Grease Barrier	Excellent
Freeze-Thaw Stability	Excellent
Mechanical and Storage Stability	Excellent
Film Appearance	Clear and glossy

Daratak Adhesive Emulsions

Product Name	TSC (%)	Viscosity (cp)	pH	Weight per Gallon of Latex (lb)	Weight per Gallon of Latex Solids (lb)	Particle Size (μ)
52L	55	450	6.3	9.1	9.93	1.5

Chemical Type: Low molecular weight, low viscosity PVAc emulsion.

Key Properties: Fast breaking. Useful in adhesives, textiles, surface coatings and heat sealing applications.

| 56L | 55 | 450 | 6.3 | 9.1 | 9.93 | 1.0 |

Chemical Type: Very high molecular weight, low viscosity PVAc emulsion.

Key Properties: Fast drying, excellent adhesion. Finds wide application in adhesives, primer-sealers, textile finishes and paper saturation.

| 61L | 55 | 3000 | 5.0 | 9.1 | 9.93 | 1.0 |

Chemical Type: High viscosity, borax tolerant PVAc homopolymer.

Key Properties: Films exhibit excellent water resistance. Used in heat sealing, paper and foil adhesives and in fire retardant mastics.

| 62L | 55 | 2500 | 6.0 | 9.1 | 9.93 | 1.0 |

Chemical Type: High molecular weight, high viscosity PVAc homopolymer.

Key Properties: Characterized by excellent adhesion, fast setting and good penetration into coated surfaces.

| 65L | 55 | 1700 | 5.0 | 9.1 | 9.93 | 1.5 |

Chemical Type: Medium molecular weight and viscosity PVAc homopolymer.

Key Properties: Strong wet tack. Compatible with fully hydrolyzed polyvinyl alcohol. Ideal for paper packaging adhesives.

| 71L | 55 | 1850 | 5.5 | 9.1 | 9.93 | 1.5 |

Chemical Type: Medium molecular weight PVAc homopolymer emulsions.

Key Properties: Imparts fast grab to adhesive formulations. Tolerant of large amounts of solvents and plasticizers. Used for packaging adhesives.

| 74L | 55 | 3000 | 5.0 | 9.0 | 9.35 | 1.0 |

Chemical Type: Acrylic terpolymer emulsion.

Key Properties: Dries to a permanently tacky film. Designed for adhering to a wide range of non-porous substrates.

| 78L | 55 | 4000 | 5.0 | 9.1 | 9.93 | 1.0 |

Chemical Type: Alkali soluble high reactive carboxylated PVAc emulsion.

Key Properties: Excellent adhesion to aluminum foil and many coated surfaces. May be cross-linked with formaldehyde resins to provide glossy coatings.

| 79L | 56 | 1150 | 4.5 | 9.0 | 9.9 | 1.0 |

Chemical Type: A unique emulsion polymer.

Key Properties: Possesses excellent compatibility with various dextrins, increasing speed and reducing stringing in remoistenable glues for envelopes and labels.

| 84L | 49 | 2000 | 5.0 | 8.6 | 8.8 | 1.0 |

Chemical Type: High molecular weight pressure sensitive laminating adhesive base, supplied as an aqueous emulsion.

Key Properties: This polymer dries to form an aggressively tacky film with good resistance to cold flow.

Product Name	TSC (%)	Viscosity (cp)	pH	Weight per Gallon of Latex (lb)	Weight per Gallon of Latex Solids (lb)	Particle Size (μ)
17-200	55	1750	5.3	9.1	9.93	1.5

Chemical Type: Medium molecular weight, low-medium viscosity, general purpose PVAc homopolymer emulsion.

Key Properties: Noted for its superior compounding and machining stability.

| 17-230 | 55 | 2600 | 5.3 | 9.1 | 9.93 | 1.5 |

Chemical Type: Medium-high viscosity version of 17-200.

Key Properties: Noted for its superior compounding and machining stability.

| 17-300 | 55 | 3500 | 5.3 | 9.1 | 9.93 | 1.5 |

Chemical Type: High viscosity version of 17-200.

Key Properties: Noted for its superior compounding and machining stability.

| SP-1065 | 56 | 2250 | 5.0 | 9.3 | 9.93 | 1.0 |

Chemical Type: Externally plasticized PVAc emulsion with exceptional compounding stability.

Key Properties: For highly pigmented mastics, adhesives, coatings, drywall cement and patching compounds.

| SP-1500 | 52 | 2000 | 6.8 | 8.5 | 8.7 | 0.5 |

Chemical Type: A unique emulsion.

Key Properties: Meets critical requirements for many film and foil adhesive applications.

| A | 55 | 1200 | 4.5 | 9.0 | 9.76 | 1.5 |

Chemical Type: A copolymer emulsion similar to Daratak B.

Key Properties: Combines excellent compounding properties with extremely good machining. Greater hardness and less flexibility than Daratak B. Of particular value in bonding flexible, nonporous substrates.

| B | 55 | 1200 | 4.5 | 9.0 | 9.57 | 1.5 |

Chemical Type: A copolymer emulsion.

Key Properties: Combines excellent compounding properties with extremely good machining. Of particular value in bonding flexible, nonporous substrates.

All of the above emulsions have the following properties:

Color	White
Odor	Slight, characteristic
Mechanical Stability	20+ min

Product Name	Emulsion. Borax Stability	Clarity	Gloss	Water Resistance	Tg (°C)
52L	No	Clear	High	Good	+31
56L	No	Slight Haze	Low	Good	+37
61L	Yes	Clear	Medium	Excellent	+40
62L	No	Slight Haze	Medium	Good	+35
65L	No	Slight Haze	Medium	Good	+31
71L	No	Slight Haze	Medium	Good	+35
74L	Yes	Clear	High	Excellent	−18
78L	Yes	Clear	Medium	Good	+40
79L	Yes	Slight Haze	High	Poor	+23

Product Name	*Emulsion* Borax Stability	Clarity	Gloss	*Film Properties* Water Resistance	Tg (°C)
84L	Yes	Clear	High	Excellent	−16
17-200	No	Slight Haze	Medium	Good	+31
17-230	No	Slight Haze	Medium	Good	+31
17-300	No	Slight Haze	Medium	Good	+31
SP-1065	Yes	Clear	High	Excellent	+25
SP-1500	Yes	Slight Haze	Medium	Excellent	−15
A	No	Clear	High	Excellent	+12
B	No	Clear	High	Excellent	+8

All of the above films have the following properties:

Color	Water White
Odor	None
Light Stability	Excellent (SP-1500-Good)
Aging Characteristics	Excellent (SP-1500-Good)

PVO INTERNATIONAL, INC., World Trade Center, San Francisco, CA 94111

Polidene Vinylidene Chloride Copolymer Emulsions

Product Name	Viscosity (Brookfield) (cp)	% Solids Content	pH	Specific Gravity 25°/25°C	Film Forming Temp. (°C min)
33-075	62.5	55	3.75	1.17	−

Key Properties: Principal binder in fire retardant coatings and moisture vapor barriers. Effective in intumescent coatings at low loading weight. Excellent durability. Recommended for industrial maintenance, textile backings, fire retardant adhesives and building products.

33-943	30	55	6.25	1.15	2

Key Properties: Binder developed for flexible, high-build, fire retardant coatings. Excellent moisture resistance and exterior durability. Recommended for industrial maintenance, textile backings, fire retardant adhesives and building products.

Texigel Acrylic Copolymer Emulsions and Solutions

13-011	4	41	3.0	1.07	2

Chemical Type: Carboxylated acrylic emulsion.

Key Properties: Binder in water-based flexographic inks. Imparts good rub resistance and transfer properties. Also used as a pigment dispersing medium in latex paints. Recommended for interior and exterior trade sales paints, product finishes, industrial maintenance and inks.

17-0088	15,000	40	8.8	1.06	−

Chemical Type: Acrylic solution.

Key Properties: Dispersing medium for gloss latex paints. Supplied in a water/isopropanol mixture. Recommended for interior and exterior trade sales paints, product finishes and industrial maintenance finishes.

23-500	500	40	11.0	1.30	−

Chemical Type: Acrylate salt solution.

Key Properties: Pigment dispersing agent for a wide range of dry pigments. Low foaming and efficient at low addition rates. Recommended for interior trade sales, exterior trade sales, product finishes, industrial maintenance, building products and paper coatings.

Product Name	Viscosity (Brookfield) (cp)	% Solids Content	pH	Specific Gravity (25°/25°C)	Film Forming Temp. (°C min)
23-555	2,000	30	8.75	1.05	—

Chemical Type: Acrylic solution.

Key Properties: Pigment dispersing medium for latex coatings. Designed to improve gloss, flow, wet-edge and dispersion stability. Cross links in baking finishes. Recommended for interior and exterior trade sales paints, product finishes, industrial maintenance and paper coatings.

Product Name	Viscosity (Brookfield) (cp)	% Solids Content	pH	Specific Gravity (25°/25°C)	Film Forming Temp. (°C min)
13-300	4	40	3.0	1.08	2

Chemical Type: Carboxylated acrylic emulsion.

Key Properties: Alkali soluble high viscosity thickening agent designed for most synthetic rubber latices and their compounds. Recommended for textile backing compounds.

13-302	4	35	3.0	1.08	2

Chemical Type: Carboxylated acrylic emulsion.

Key Properties: Alkali soluble high viscosity thickening agent designed for latex paints. Both 13-300 and 13-302 are useful in thickening acrylic emulsions for print-bonding adhesives in nonwoven fabrics. Recommended for interior and exterior trade sales paints, industrial maintenance, textile backings and building products.

13-935	4	25	2.5	1.06	—

Chemical Type: Carboxylated acrylic emulsion.

Key Properties: Alkali-soluble thickener giving high viscosity at high shear in clay coatings by trailing blade. Effective in latex paints. Recommended for interior and exterior trade sales paints, building products and paper coatings.

Texigel Polyacrylate Gels

23-005	60,000	15	8.5	1.06	—

Chemical Type: Sodium polyacrylate.

Key Properties: Thickening agent for natural and synthetic rubber latices, stabilizers, protective colloids and binders. Recommended for textile backings.

23-018	37,500	12.5	9.5	1.07	—

Chemical Type: Sodium polyacrylate.

Key Properties: Combines higher efficiency and ease of handling. Recommended for textile backings.

ROHM AND HAAS CO., Independence Mall West, Philadelphia, PA 19105

Rhoplex Crosslinkable* Acrylic Emulsions

Product Name	% Solids Content	Weight per Gallon (lb)	pH (As Packed)	Viscosity (Brookfield) (cp)	Ti** (°C)	Emulsifying System
Emulsion E-1612	45	8.8	9.7	100 max	+7	Anionic

Key Properties: Designed as a spray insulation adhesive/binder. Broad range of wet and dry specific adhesion.

Rhoplex B-15	46	8.8	6.4	50	-4	Nonionic

Key Properties: Forms soft, tacky films with excellent adhesion to metal, cellulosic and plastic substrates.

Product Name	% Solids Content	Weight per Gallon (lb)	pH (As Packed)	Viscosity (Brookfield) (cp)	Ti** (°C)	Emulsifying System
Rhoplex LC-40	55	8.9	4.5	70	-4	Anionic

Key Properties: Forms soft, flexible films with rapid water release. Has excellent adhesion to cellulosics, plastics, vinyl films, aluminum and other metal surfaces.

Rhoplex LC-67	63	8.7	5.0	250	-38	Anionic

Key Properties: Suited for formulating laminating and pressure sensitive adhesives. Displays very high shear resistance, good tack and peel, and resistance to plasticizer migration.

Rhoplex N-495	58	9.0	4.5	2500	-3	Anionic

Key Properties: High speed packaging and laminating emulsion. Has excellent adhesion, "wet-grab," fast water release and quick rate of set. Good adhesion to aluminum, vinyl films, wood, hardboard, paper and steel.

Rhoplex N-560	57	8.6	7.7	150	-32	Anionic

Key Properties: Recommended for laminating and permanent pressure sensitive adhesives, displays high peel, tack and excellent adhesion to vinyls, metals and polyolefins.

Rhoplex N-619	57	8.9	7.7	185	-30	Anionic

Key Properties: Similar to Rhoplex N-560, but offers shear resistance at a slightly higher price.

Rhoplex N-580	55	8.9	7.7	80	-40	Anionic

Key Properties: Suggested for removable pressure sensitive adhesives where high tack, high shear resistance and excellent adhesion are required.

Rhoplex N-1031	54	8.6	7.5	100	-32	Anionic

Key Properties: Offers very high shear resistance with good tack and peel strength. Retains adhesion and pressure sensitive properties over plasticized vinyl even after aging.

Note: All of the above meet F.D.A. Regulation 175.105.

   *Can be crosslinked by the addition of nitrogen resins or certain other crosslinking agents.

   **Ti: The temperature at which the Torsional Modulus of an air-dried film is 300 kg/cm^2.

Rhoplex Self-Crosslinking* Acrylic Emulsions

Rhoplex CA-12	53.5	9.0	5.9	350	-4	Anionic

Key Properties: Designed for use in formulating latex contact adhesives. Excellent adhesion to various substrates, including plastics and metals. Outstanding heat resistance.

Emulsion E-1126	60	8.8	4.0	250	-18	Nonionic

Key Properties: An acrylic polymer which, after drying, remains pressure sensitive up to 8 hours and achieves full cure at room temperature. Displays excellent adhesion and water resistance.

Rhoplex K-14	46	8.6	3.0	200	-47	Nonionic

Key Properties: Softest of the self-crosslinking Rhoplex emulsions. Forms very soft, tacky films. Suggested for fabric laminating applications where softness and low temperature flexibility are required. Excellent durability to washing. Fair durability to dry-cleaning.

Rhoplex K-3	46	8.6	3.0	50	-32	Nonionic

Key Properties: Forms soft, tacky films. Suggested for fabric laminating applications where a soft resilient adhesive binder is required. Excellent durability to washing. Fair durability to dry cleaning.

Rhoplex K-87	46	8.6	3.0	200	-18	Nonionic

Key Properties: Similar in properties to Rhoplex K-3, but is slightly harder and has a higher level of durability to dry cleaning. Excellent durability to washing.

Product Name	% Solids Content	Weight per Gallon (lb)	pH (As Packed)	Viscosity (Brookfield) (cp)	Ti** (°C)	Emulsifying System
Rhoplex HA-8	45.5	8.7	3.0	550	-14	Nonionic

Key Properties: Similar in properties to Rhoplex K-87, but is slightly harder. Excellent durability to washing. Good durability to dry cleaning.

Rhoplex HA-24	44.5	8.7	2.9	400	-7	Anionic

Key Properties: Similar in properties to Rhoplex HA-8, but is slightly harder and exhibits better running properties. Excellent durability to washing. Good durability to dry cleaning.

Rhoplex E-32	46	8.8	3.0	200	-2	Nonionic

Key Properties: Forms fairly firm, tack-free films. Suggested for fabric laminating and flocking applications. Excellent durability to dry cleaning and washing.

Rhoplex E-358	60	9.0	7.0	1500	0	Nonionic

Key Properties: A high solids emulsion suggested for formulating high solids adhesives for fabric laminating and flocking applications. Excellent durability to washing and dry cleaning.

Rhoplex E-269	46	8.8	3.2	600	+7	Anionic

Key Properties: An emulsion which self-thickens to a selected viscosity by the careful addition of ammonium hydroxide to the proper pH. Suggested for fabric laminating and flocking applications. Excellent durability to washing and dry cleaning.

Rhoplex RA-90	46	8.8	3.5	65	+12	Nonionic

Key Properties: Of particular interest in compounding laminating and flocking adhesives and pigment binder systems. Exhibits excellent solvent resistance, good flexibility and good pigment loading capability. Excellent durability to washing and dry cleaning.

Rhoplex HA-12	45	8.8	3.0	400	+17	Nonionic

Key Properties: Forms firm, tack-free films. Suggested for fabric laminating. Excellent durability to washing and good durability to dry cleaning.

Rhoplex HA-20	46	9.8	2.2	100	+29	Anionic

Key Properties: Forms hard, brittle, flame resistant films. Suggested for formulating flame resistant adhesives or bonding systems for nonwoven webs. Excellent durability to washing. Good durability to dry cleaning.

Rhoplex HA-16	45.5	8.8	3.0	450	+33	Nonionic

Key Properties: Hardest and most rigid of the Rhoplex self-reactive emulsions. Suggested for formulating adhesives for fabric laminating and flocking applications where a firm hand is desired. Excellent durability to washing. Good durability to dry cleaning.

Rhoplex TR-407	45.5	8.9	3.0	30	+30	Anionic

Key Properties: Designed as a fiberfill binder. Will not discolor in high temperature molding. Imparts excellent loft. Useful as binder for nonwovens for shirt interliners. Excellent adhesion to synthetic and natural fibers. Excellent durability to washing. Good durability to dry cleaning.

Rhoplex TR-520	50	8.9	3.0	60	-8	Anionic

Key Properties: Acrylic emulsion with excellent tolerance to large quantities of flame retardant salts. Excellent binder for disposable, flame retardant and nonwoven fabrics. Excellent water release. Excellent durability to washing. Good durability to dry cleaning.

*Crosslinks without the addition of an external crosslinking agent.

**Ti: The temperature at which the Torsional Modulus of an air-dried film is 300 kg/cm^2.

Regulatory: The following Rhoplex Emulsions meet F.D.A. Regulation 175.105: HA-8, HA-24, E-32, E-358, E-269, RA-90, HA-12, HA-20, HA-16, TR-407, TR-520.

Rhoplex Noncrosslinking* Acrylic Emulsions

Product Name	% Solids Content	Weight per Gallon (lb)	pH (As Packed)	Viscosity (Brookfield) (cp)	Ti** (°C)	Emulsifying System
Rhoplex AC-19	44.5	8.8	9.5	950	+16	Nonionic

Key Properties: The least expensive Rhoplex acrylic emulsion. It has slightly less wet adhesion than Rhoplex AC-22 to vinyl film, hardboard, paper and paperboard substrates.

Rhoplex AC-22	44.5	8.9	9.8	500	+16	Nonionic

Key Properties: A low cost Rhoplex emulsion. Suggested for formulating general purpose laminating adhesives. Has excellent adhesion to vinyl film, hardboard, paper and paperboard substrates.

Rhoplex AC-33	46.5	8.9	9.0	1500	+16	Nonionic

Key Properties: One of the lower cost Rhoplex emulsions. Suggested for general purpose laminating applications and has excellent adhesion to vinyl film, hardboard, paper and paperboard substrates. Meets F.D.A. Regulation 175.105.

Rhoplex AC-61	46.5	8.9	9.8	60	+16	Anionic

Key Properties: Similar to Rhoplex AC-234 in specific adhesion, but gives harder, more abrasion resistant, higher cohesive strength films.

Rhoplex AC-73	46.5	8.9	9.5	250	+32	Nonionic

Key Properties: Forms films at elevated temperatures which are hard and grease resistant. Suggested primarily as a modifier for the softer Rhoplex emulsions.

Rhoplex AC-234	46.5	8.9	9.5	1200	+7	Nonionic

Key Properties: Related to Rhoplex AC-33 in film properties, but has a higher level of specific adhesion to a variety of substrates.

Rhoplex AC-235	46.5	8.8	9.0	500	+12	Nonionic

Key Properties: Similar to Rhoplex AC-234 in adhesion and film properties.

Rhoplex B-60A	46.5	8.9	9.0	1500	+16	Nonionic

Key Properties: Similar to Rhoplex AC-33, but supplied at a lower viscosity. Meets F.D.A. Regulation 175.105.

Rhoplex B-85	38	8.8	9.8	10	+103	Anionic

Key Properties: The hardest acrylic emulsion polymer offered for adhesive use. Suggested primarily as a modifier for the softer Rhoplex emulsions.

Rhoplex LC-45	65	8.8	5.0	300	-7	Anionic

Key Properties: Suited for formulating water-based laminating adhesives. Outstanding adhesion to wood and plastic substrates, mastics, panel and flooring adhesives.

*Regardless of curing conditions crosslinking does not occur.

**Ti: The temperature at which the Torsional Modulus of an air-dried film is 300 kg/cm^2.

CHAS. S. TANNER CO., Box 1848, Greenville, SC 29602

Dur-O-Cryl 520

Chemical Type: Self-crosslinking high molecular weight acrylic polymer emulsion, obtained with a newly developed emulsion polymerization technique.

Physical Properties:

Appearance	Milky emulsion
% Solids Content	45.0–47.0
Ionic Nature	Essentially nonionic

Viscosity, Brookfield (#2/60 RPM)	200-700 cp
pH	4.2-5.0
Weight per Gallon	9.1 lb
Film	Very soft, clear
Tg	-38°C

Key Properties: As a result of the increased molecular weight, finishes, coatings and adhesives based on it exhibit increased durability to laundering and dry cleaning, reduced tackiness of films, increased elongation and toughness of films.

Recommended applications are:

Lamination and flocking adhesives.

Nonwoven binders.

Pigment binders.

Fabric finishing.

Dur-O-Cryl 620

Chemical Type: Self-crosslinking high molecular weight acrylic polymer emulsion, obtained with a newly developed emulsion polymerization technique.

Physical Properties:

Appearance	Milky emulsion
% Solids Content	45.0 - 46.0
Viscosity, Brookfield (#2/50 RPM)	100-300 cp
pH	5.7-6.3
Weight per Gallon @ 25°C	8.7 lb
Film	Soft, clear
Tg	-18°C

Key Properties: As a result of the increased molecular weight, finishes, coatings and adhesives based on it exhibit increased durability to laundering and dry cleaning, reduced tackiness of films, increased elongation and toughness of film.

Recommended applications for Dur-O-Cryl 620 are:

Lamination and flocking adhesives.

Backcoating of upholstery and drapes.

Nonwoven binder.

Pigment binder.

Fabric finishing.

Dur-O-Cryl 720

Chemical Type: Self-crosslinking proprietary emulsion.

Physical Properties:

Appearance	Milky emulsion
% Solids Content	45
Ionic Nature	Nonionic
Viscosity, Brookfield (#2/20 RPM)	250-500 cp
pH	4.5-5.5
Weight per Gallon	8.75 lb
Film	Soft, clear
Tg	+2°C

Key Properties: Finishes incorporating Dur-O-Cryl 720 are durable to launderings and dry cleanings when properly cured. This product does not require the addition of a nitrogenous or other external crosslinking agent, although, for maximum durability, the addition of an acid or latent acid catalyst is recommended to catalyze the crosslinking reaction.

Dur-O-Set E-200

Chemical Type: Vinyl acetate-ethylene emulsion polymer, with a protective colloid of partially acetylated polyvinyl alcohol.

Physical Properties:

% Solids Content	55% min
Viscosity, Brookfield (#3/20 RPM)	2100-2800 cp
pH	4.0-5.0
Residual Monomer	0.5% max
Particle Size	$0.2 - 1.0\,\mu$
Weight per Gallon @ 72°F	8.8 lb
Appearance	Milky white emulsion
Emulsification System (PVOH Protected)	Nonionic
Mechanical Stability	Excellent
Ethanol Tolerance	Good
Chlorinated Solvent Tolerance	Excellent
Aromatic Solvent Tolerance	Excellent
Thickening Characteristics	Excellent
Borax Stability	Unstable
Wet Tack	High

Typical Film Properties:

Adhesion to Glass	Good
Flexibility	Excellent
Clarity	Slightly hazy
Glass Transition Temperature	-2° to +2°C

ASTM D638-68:

Modulus 100%	300 approx.
Tensile Strength at Break	300 approx.
Elongation at Break	800% approx.

Key Properties: Particularly effective in bonding polyvinyl chloride films. Evaluations have been conducted that show bonding strength is excellent, even with aging. Laminations of cellulosic materials to 2 mil polyvinyl chloride film have been measured on an Instron Tester initially and after rapid aging. No reduction in peel strength was found after several days at 50°C.

Dur-O-Set E-200 aggressively bonds films of a wide variety of materials, including polyvinylidene chloride, cellulose acetate, polystyrene, PVDC-coated cellophane and acrylic film to porous substrates.

Effectively bonds such widely varied substrates as paper, cellulosics, nylon fabric, wood, hard board, urethane foam and specific types of coated paperboard. It works well as a base for metallic foil laminating adhesives.

Dur-O-Set E-200 meets the requirements of the following FDA Regulations:

175.105: Adhesives
175.300: Resinous and polymeric coatings.
176.170: Components of paper and paperboard in contact with aqueous and fatty foods.
177.1350: Vinyl acetate-ethylene copolymers.

Dur-O-Set H-100

Chemical Type: Polyvinyl acetate homopolymer emulsion.

Physical Properties:

% Solids Content	54.5-55.5%
pH	4.5-5.5
Viscosity, Brookfield (#2/20 RPM)	1000-2000 cp
Particle Size	$2-10\,\mu$
Free Monomer	<1%
Ionic Nature	Nonionic
Odor	Slight, characteristic
Mechanical Stability	Excellent

Film Properties: Odorless, tasteless, slightly cloudy, hard brittle film. Properties of the film can be modified to suit specific applications when necessary by proper compounding of the emulsion prior to use.

Key Properties: Dur-O-Set H-100 may be used in textile finishing as a binder, in compounding adhesives and in paint formulations. As a fabric finish, it imparts a firm crisp full hand. This finish is semipermanent through several launderings and will not retain chlorine from bleaches. The very low B.O.D. presents no waste disposal problems.

Dur-O-Set SB-321

Chemical Type: Polyvinyl acetate homopolymer emulsion, unmodified and unplasticized.

Physical Properties:

% Solids Content	55.0-57.0
pH	4.0-6.0
Viscosity, Brookfield	600-1200 cp
Free Monomer	1.0% max
Weight per Gallon	9.2 lb
Resin per Gallon	0.495
Ionic Nature	Nonionic
Appearance	Viscous white milky fluid
Odor	Slight, pleasant
Storage Stability	Excellent at 40°-120°F
Borax Tolerance	Excellent

Film Properties:

Film Hardness	Hard and brittle
Water Resistance	Fair
Appearance	Smooth, no graininess, slightly hazy.

Key Properties: Greater chemical stability is obtained, because of its borax stability. The larger particle size gives it added advantages for stiffening of textile fabrics and generally its nonsticking qualities seem to be better than Dur-O-Set H-100.

Possible applications are in paper coatings, adhesives and in paint. It has been found to be especially good in indoor paint formulations and is recommended for this use.

UNION CHEMICALS DIVISION, 1345 N. Meacham Road, Schaumburg, IL 60196

Amsco-Res Non-Carboxylated Latexes

Product Name	Viscosity (Brookfield) (cp)	% Solids Content	Residual Monomer (max %)	pH	Weight per Gallon (lb)	Tg (°C)
4000	50	41.5	0.07	10.7	8.3	-23

Chemical Type: Styrene/butadiene latex, freeze-thaw unstable and borax stable with anionic particle charge.

4040	150	50.0	0.07	6.8	8.3	-8

Chemical Type: Styrene/butadiene latex, freeze-thaw unstable and borax stable with nonionic particle charge.

Key Properties: Recommended for use in adhesives, caulks and mastics.

Amsco-Res Polymer Emulsions

Product Name	Viscosity (Brookfield) (cp)	% Solids Content	pH	Weight per Gallon @25°C (lb)	Tg (°C)	Residual Monomer (% by Weight)
1145	200	45.5	4.5	9.1	35	0.5

Chemical Type: Self cross-linking vinyl copolymer, freeze-thaw and borax stable with anionic particle charge.

Key Properties: Recommended for textile finishing and nonwoven binders.

1159	600	50.0	4.5	9.1	35	0.5

Chemical Type: Self cross-linking vinyl copolymer, freeze-thaw unstable, borax stable with anionic particle charge.

Key Properties: Recommended for textile finishing and nonwoven binders.

3001	2,000	58.0	4.5	9.2	26	0.5

Chemical Type: Vinyl acetate, freeze-thaw and borax stable with nonionic particle charge.

Key Properties: Recommended for adhesives, caulks, mastics, interior paints and textile finishing.

3001P	3,500	60.0	4.5	9.2	16	0.5

Chemical Type: Vinyl acetate, freeze-thaw unstable and borax stable with nonionic particle charge.

Key Properties: Recommended for adhesives, caulks and mastics.

3004	1,000	55.0	5.0	9.1	28	0.5

Chemical Type: Vinyl acetate, freeze-thaw stable and borax unstable with nonionic particle charge.

Key Properties: Recommended for adhesives, caulks, mastics and textile finishing.

3006	3,500	51.5	5.0	9.1	22	0.5

Chemical Type: Vinyl acetate, freeze-thaw stable and borax unstable with nonionic particle charge.

Key Properties: Recommended for adhesives.

3007	1,600	55.0	5.0	9.1	28	0.5

Chemical Type: Vinyl acetate, freeze-thaw stable and borax unstable with nonionic particle charge.

Key Properties: Recommended for adhesives, caulks, mastics and textile finishing.

3008	2,500	55.0	5.0	9.1	28	0.5

Chemical Type: Vinyl acetate, freeze-thaw stable and borax unstable with nonionic particle charge.

Key Properties: Recommended for adhesives, caulks and mastics.

3011	1,650	55.0	3.0	8.8	4	0.7

Chemical Type: Vinyl acrylic, freeze-thaw unstable and borax stable with nonionic particle charge.

Key Properties: Recommended for adhesives, caulks, mastics, exterior paints, interior paints, textile finishing and paper coating.

3016	1,650	55.0	3.0	8.8	14	0.7

Chemical Type: Vinyl acrylic, freeze-thaw unstable and borax stable with nonionic particle charge.

Key Properties: Recommended for interior paints, textile finishing, nonwoven binders and paper coating.

3034	2,500	57.0	5.0	9.1	18	0.5

Chemical Type: Vinyl acetate, freeze-thaw unstable and borax unstable with nonionic particle charge.

Key Properties: Recommended for adhesives.

3151	1,200	55.0	5.0	9.1	26	0.5

Chemical Type: Vinyl acetate, freeze-thaw unstable and borax unstable with nonionic particle charge.

Key Properties: Recommended for adhesives.

Amsco-Res Polymer Emulsions Carpet and Upholstery Latexes for Textile Backcoating

Product Name	Viscosity (Brookfield) (cp)	% Solids Content	Residual Monomer (max %)	pH	Weight per Gallon (lb)	Tg (°C)
4125	300	50.0	0.07	9.2	8.3	–32

Chemical Type: Styrene/butadiene (45/55) latex, freeze-thaw unstable and borax stable with anionic particle charge.

4150	300	50.0	0.07	8.7	8.3	–16

Chemical Type: Styrene/butadiene (55/45) latex, freeze-thaw unstable and borax stable with anionic particle charge.

4151	300	50.0	0.07	9.2	8.3	–30

Chemical Type: Styrene/butadiene (45/55) latex, freeze-thaw unstable and borax stable with anionic particle charge.

4170	300	50.0	0.07	9.2	8.3	–7

Chemical Type: Styrene/butadiene (65/35) latex, freeze-thaw stable and borax stable with anionic particle charge.

4171	300	50.0	0.07	9.2	8.3	–7

Chemical Type: Styrene/butadiene (65/35) latex, freeze-thaw stable and borax stable with anionic particle charge.

4176	300	50.0	0.07	8.7	8.3	–21

Chemical Type: Styrene/butadiene (50/50) latex, freeze-thaw unstable and borax stable with anionic particle charge.

Amsco-Res Polymer Emulsions (Specialty Latexes)

6190	300	50.0	0.07	9.2	8.3	–7

Chemical Type: Styrene/butadiene (65/35) latex, freeze-thaw stable and borax stable with anionic particle charge.

Key Properties: Recommended for use in adhesives and exterior paints.

4104	300	50.0	0.07	9.2	8.3	–7

Chemical Type: Styrene/butadiene (65/35) latex, freeze-thaw stable and borax stable with anionic particle charge.

Key Properties: Recommended for use in caulks and mastics.

6180	300	50.0	0.07	9.2	8.3	–6

Chemical Type: Styrene/butadiene (65/35) latex, freeze-thaw stable and borax stable with anionic particle charge.

Key Properties: Recommended for use in paper coatings.

Amsco-Res Polymer Emulsions (Specialty Latexes for Nonwovens)

5400	300	50.0	0.07	9.2	8.3	–35

Chemical Type: Styrene/butadiene (45/55) latex, freeze-thaw unstable and borax stable with anionic particle charge.

5500	300	50.0	0.07	9.2	8.3	–28

Chemical Type: Styrene/butadiene (45/55) latex, freeze-thaw unstable and borax stable with anionic particle charge.

5600	300	50.0	0.07	8.7	8.3	–24

Chemical Type: Styrene-butadiene (50/50) latex, freeze-thaw unstable and borax stable with anionic particle charge.

Product Name	Viscosity (Brookfield) (cp)	% Solids Content	Residual Monomer (max %)	pH	Weight per Gallon (lb)	Tg (°C)
5700	300	50.0	0.07	8.7	8.3	-15

Chemical Type: Styrene/butadiene (55/45) latex, freeze-thaw unstable and borax stable with anionic particle charge.

| 5900 | 300 | 50.0 | 0.07 | 9.2 | 8.3 | -5 |

Chemical Type: Styrene-butadiene (65/35) latex, freeze-thaw stable and borax stable with anionic particle charge.

WITCO CHEMICAL , 277 Park Ave., New York, NY 10017

Witcobond Urethane Latices

	W-160	W-170	W-180	W-210	W-290H
% Solids Content	35	35	35	35	35
Odor	. . . . Characteristic (Solvent Free)				
Particle Charge	Anionic	Anionic	Anionic	Cationic	Anionic
Viscosity, Brookfield @ 77°F (cp)	30	30	30	150	200
pH @ 77°F	7.5	7.5	7.5	7.6	7.5
Surface Tension (dynes/cm)	40	40	40	46	42
Weight per Gallon @ 77°F (lb)	8.5	8.5	8.5	8.6	8.91
Specific Gravity @ 77°F	1.03	1.03	1.03	1.03	1.07
Glass Transition Temperature (Tg)(°F)	-49	-53	-60	-45	–
Stability:					
Mechanical (High Shear)	. . . . Excellent				
Freeze-Thaw (-18°C)(Cycles)	5+	Fails 1	Fails 1	–	6+
Typical Film Properties:					
Tensile Strength (psi)	4300	1200	*	5000	4500
Ultimate Elongation (%)	725	1200	*	500	720
Modulus at 100% (psi)	350	200	*	900	260
Modulus at 300% (psi)	700	350	*	1900	540
Modulus at 500% (psi)	1250	500	*	–	1550

Hydrolytic Stability at 70°C (158°F)
(After 24 hours immersion) Strong with slight whitening**   Excellent

*Films are too soft to determine accurate figures.
**No whitening should occur with the addition of at least 3 parts Cymel 373.

Note 1: Witcobond 210—Good film properties are dependent on film fusion, which is achieved between 130° to 150°C (266° to 302°F). If substrates to be coated or surface treated are sensitive to temperatures in this range, it is recommended that about 5% of a water-miscible, coalescing solvent be added, based on latex weight.

Note 2: For optimum film properties, it is recommended that 6.5 parts Witcobond XW latex adjunct be added to 100 parts Witcobond W-290H urethane latex. Witcobond XW latex adjunct is a 50% solids water dispersion of epoxy. The pot life of this mixture typically is in excess of one week at room temperature. Drying can be achieved at 88° to 107°C (190° to 225°F).

Witcobond W-160, W-170, W-180

Chemical Type: Urethane latices which are low viscosity, stable, aqueous emulsions of fully reacted polyurethane. No solvent is present.

Key Properties: Suggested for use as high performance adhesives or coatings. Because of their composition, they may be crosslinked for added performance.

Witcobond W-210

Chemical Type: Low viscosity, light-stable, aqueous emulsion of fully reacted polyurethane. No solvent is present.

Key Properties: Strong, cohesive films can be formed simply by evaporation of water. This latex is suggested for use as a coating or surface treatment where the properties of a light-stable, water based urethane are suitable, primarily with textiles, nonwovens, paper, wood, urethane foam or other porous substrates. Witcobond urethane latex can also exhibit good adhesion to fiberglass and other glass materials.

Films typically exhibit excellent water and solvent resistance. Upon exposure such films characteristically exhibit good retention of their original film properties and do not blanche.

Meets FDA Regulation 21 CFR 175.105

Witcobond 290H

Chemical Type: Low-viscosity, high-solids, stable, aqueous system.

Key Properties: Offers light-stable films and may be considered for applications such as:

Coating or surface treatments for porous or nonporous substrates.

Adhesive for difficult substrate combinations.

Frothable interlayer or coating.

Nonwoven saturant and binder.

Toxicity: Independent laboratory tests indicate that these latices are not an eye or primary skin irritant.

Section XIV
Oils

ARIZONA CHEMICAL CO., Wayne, NJ 07470

Actinol Distilled Tall Oils

Product Name	Color (Gardner) (max)	Acid Value (min)	Saponi- fication Value	Specific Gravity @ 25°C	Viscosity	Flash Point (O.C.) (°F)
D30LR	7	183	190	0.94	90 cp*	400

Chemical Type: Distilled tall oil, consisting of 67.8% fatty acids, 30% rosin acids and 2.2% unsaponi-fiables.

D40LR	7	178	188	0.95	60 cp*	400

Chemical Type: Distilled tall oil, consisting of 61.5% fatty acids, 36% rosin acids and 2.5% unsaponi-fiables.

EPG	2	196	200	0.90	A**	400

Chemical Type: Tall oil fatty acid, consisting of 38% linoleic (nonconjugated) acid, 7% linoleic (conjugated) acid, 51% oleic acid, 2% stearic acid and 2% other fatty acids.

FA1	5	193	198	0.90	A**	400

Chemical Type: Special tall oil fatty acid, consisting of 35% linoleic (nonconjugated) acid, 9% linoleic (conjugated) acid, 47% oleic acid, 3% saturated acids and 6% other fatty acids.

FA2	4	195	199	0.90	A**	400

Chemical Type: Tall oil fatty acid, consisting of 37% linoleic (nonconjugated) acid, 7% linoleic (conjugated) acid, 50% oleic acid, 2% saturated acids and 4% other fatty acids.

FA3	3	196	200	0.90	A**	400

Chemical Type: Tall oil fatty acid, consisting of 38% linoleic (nonconjugated) acid, 7% linoleic (conjugated) acid, 50% oleic acid, 2% stearic acid and 3% other fatty acids.

*Brookfield. **Gardner-Holdt.

PVO INTERNATIONAL, INC., World Trade Center, San Francisco, CA 94111

Castor Oil

Product Name	Acid Value (max)	Viscosity (Gardner-Holdt)	Color (Gardner) (max)	Weight per Gallon (lb)	Iodine No.
#3	6	U	8	8.08	85

Key Properties: Lubricant and plasticizer. Passes JJJ-C-86, Grade 3.

Product Name	Acid Value (max)	Viscosity (Gardner-Holdt)	Color (Gardner) (max)	Weight per Gallon (lb)	Iodine No.
#1	2	U	3	8.08	—

Key Properties: Lubricant and plasticizer. Passes JJJ-C-86, Grade 1.

Product Name	Acid Value (max)	Viscosity (Gardner-Holdt)	Color (Gardner) (max)	Weight per Gallon (lb)	Iodine No.
Tung Oil	8	H-J	12	7.85	165

Key Properties: Meets Federal Spec. TT-T-775. Fast cook and fast dry.

Linseed Oil

Product Name	Acid Value (max)	Viscosity (Gardner-Holdt)	Color (Gardner)	Weight per Gallon (lb)	Iodine No.
Raw	1.5	A	11	7.75	177

Key Properties: Meets Fed. Spec. TT-L-215A.

Alkali Refined	0.2	A	5	7.75	177

Key Properties: Refined and bleached for kettle bodying. Meets MIL-L-15180A.

White Refined	5	A	5	7.75	177

Key Properties: Mixing and grinding oil with good wetting properties.

Nonbreak	0.5	A	11	7.75	177

Key Properties: For alkyd cooking. Heat bleach. 5-7 color.

Boiled	3.5	A	11	7.82	170

Key Properties: Driers added. Clear coat on wood. Passes TT-L-190A.

Blown Z-3	7	Z3-Z4	10	8.25	125

Key Properties: Good wetting and flow properties. Recommended for caulks and putties.

Pavolin KB Q	4	Q	4	7.80	140

Key Properties: Medium acid for paints and enamels.

Pavolin KB Z2	6	Z2	4	8.0	130

Key Properties: Same as Pavolin KBQ, except heavier.

Safflower Oils

Product Name	Acid Value (max)	Viscosity (Gardner-Holdt)	Color (Gardner)	Weight per Gallon (lb)	Iodine No.
Nonbreak	1	A	9	7.72	144

Key Properties: Nonyellowing. Recommended for house paints, varnishes, alkyds.

Blown Z4	7	Z3-Z4	7	8.10	100

Key Properties: Good wetting and good flow. Recommended for mastics and caulks.

Alkali Refined	0.3	A	4	7.72	144

Key Properties: Lighter color than Nonbreak.

KB Z-2	9	Z2	5	8.00	101

Key Properties: Good flexibility and color. Recommended for enamels and nonyellowing paints.

KB HA Z-7	22	Z7½	6	8.20	90

Key Properties: Cold liming and puffing vehicle.

Conj. 122-G	3	G	2	7.80	—

Key Properties: Conjugated to give fast dry. Exhibits good color retention and flexibility.

Product Name	Acid Value (max)	Viscosity (Gardner-Holdt)	Color (Gardner)	Weight per Gallon (lb)	Iodine No.
Conj. 122 Z-3	5	Z3	3	8.00	–

Key Properties: Same as Conj. 122-G plus high gloss, good wetting, fast dry.

Conj. 122 Z-8	6	Z8	4	8.2	–

Key Properties: Same as Conj. 122-G with higher viscosity.

Enamel Oil	4	Z2½	6 max	–	–

Key Properties: Painter type gloss interior and exterior.

Saff-White #1	3	M-O	6 max	7.80	120

Key Properties: Excellent brushing and leveling. Includes puffing agent. Recommended as exterior house paint vehicle.

Saff-White #3	3+	M-O	3	7.80	120

Key Properties: Same as Staff-White #1 with no puffing agent.

Methyl Linoleate	0.5	6 cts.	9	7.42	143

Key Properties: Equivalent use to fatty acids in alkyd manufacturing. Fast drying. Excellent penetrant for rust surfaces.

Limed 1011	0.8	Gel	6	7.10	–

Key Properties: Low cost. 40% nonvolatile. Recommended for interior flat wall paints.

Walnut Oil

Product Name	Acid Value (max)	Viscosity (Gardner-Holdt)	Color (Gardner) (max)	Weight per Gallon (lb)	Iodine No.
Enamel Oil	4	Z2½	6	–	–

Key Properties: Painter type gloss for interior and exterior.

1011-W	1.5 max	Z6½	8	–	–

Key Properties: 40% nonvolatile. Low cost. Recommended for interior flat wall paints.

Walwhite #1	3	M-O	7	7.80	120

Key Properties: Excellent brushing and leveling. Recommended as exterior house paint vehicle.

Walwhite #3	3	M-O	7	7.80	120

Key Properties: Excellent brushing and leveling. Recommended as exterior house paint vehicle.

Blown Z-4	7	Z3-Z4	7	8.10	100

Key Properties: Recommended as plasticizer for caulks and putties.

Alkali Refined	0.2	A	5+	7.67	152

Key Properties: Light color for alkyd cooking.

REICHHOLD CHEMICALS, INC., RCI Bldg., White Plains, NY 10603

Gloss Oils

Product Name	Color (Gardner) (max)	Acid No. (Solution)	Viscosity (Gardner-Holdt) @ 25°C	Percent Solids Content (max)	Lime Content (% on Solids)
N-GLO-5Y (35-702)	10	48	Z	65	5

Chemical Type: Gloss oil in mineral spirits with Pensky Martens Closed Cup Flash Point of 109°F. Meets Rule 442.

Key Properties: A rosin gloss oil having pale color, a high nonvolatile content and a high viscosity. It is a general purpose rosin gloss oil.

Product Name	Color (Gardner) (max)	Acid No. (Solution)	Viscosity (Gardner-Holdt) @ 25°C	Percent Solids Content (max)	Lime Content (% on Solids)
T-GLO-8Y-210 (35-705)	10	35	Z	66	9

Chemical Type: Tall Oil gloss oil in mineral spirits with Pensky Martens Closed Cup Flash Point of 114°F. Meets Rule 442.

Key Properties: A pale tall oil gloss which contributes outstanding sag resistance to a paint. Made from a special Newport pale tall oil which has an inherent 6.7 gallon oil length. This feature offers a real savings in the amount of other drying oils used in low cost paint and varnish formulations.

Key Properties (both of the above): These gloss oils are noted for the following:

Excellent Gloss—Show more brilliantly glossy films than similar products of equivalent lime content made in the conventional varnish kettles.

Pale Color—Both are clarified and are pale colored materials. Tints and light shades become more lively when these gloss oils are used.

Low Acid Value—Both gloss oils have low acid values with the result that the tendency of gloss oil to "liver" (react slowly with the basic pigments, is reduced to a minimum).

Lack of Water Sensitivity—It has become fairly common practice to produce thixotropic paints for a quite wide variety of uses. These paints contain a certain quantity of water or alkaline solution. Newport gloss oils are particularly recommended for this work, since their great uniformity plus their method of manufacture prevent uncontrollable thickening and make it possible to obtain a uniform consistency.

Drying Rates—The drying rates of gloss oils are excellent and will be found to be as fast or faster than gloss oils of similar composition which have been made in the ordinary open varnish kettle.

Suggested Uses: These gloss oils, in conjunction with drying oils and/or alkyds, are the basis of low cost paints and enamels.

T-GLO-8Y-210, in addition to its applications in paints and enamels, is used in low cost caulking compounds and wax-based concrete curing compounds.

Section XV

Plasticizers

ARGUS CHEMICAL, 633 Court St., Brooklyn, NY 11231

Drapex Epoxy Plasticizers

Physical Properties:

	3.2	4.4	6.8	10.4
Color	1	1	<1	>1
Specific Gravity @ 77°F	0.899	0.922	0.992	1.0385
Oxirane Oxygen (%)	3.8	5.1	7.25	9.6
Iodine Value (Hanus)	1.8	2.2	1.3	2.0
Refractive Index @ 77°F	1.4540	1.4580	1.4720	1.4788
Acid Value	0.7	0.5	0.5	0.5
Moisture (%)	1.05	0.05	0.08	0.08
Pour Point (°F)	56	–4	32	23
Odor	Faintly Fatty	Faintly Fatty	Faintly Fatty	Faintly Fatty
Viscosity, Brookfield @ 77°F (cp)	19.9	20	320	1,000
Flash Point (C.O.C.) (°F)	—	428	554	554
Water Solubility @ 68°F (% by wt)	—	<0.0	<0.01	<0.01
Molecular Weight (Approximate)	—	420	1,000	1,000

Drapex 3.2

Chemical Type: Low-temperature monoester epoxy plasticizer.

Key Properties: Imparts heat and light stability, outstanding low-temperature flexibility along with low volatility.

When used in vinyl dispersions, Drapex 3.2 imparts low initial viscosity and because of its low solvating action, excellent viscosity stability for extended periods of time.

Synergistic stabilizing action is provided when this octyl epoxy stearate is used with the Mark barium/cadmium stabilizers.

Drapex 4.4

Chemical Type: Octyl tallate.

Key Properties: Imparts low temperature flexibility and excellent heat and light stability to vinyl compounds.

In combination with certain vinyl stabilizers, such as Mark barium/cadmium types, the use of Drapex 4.4 results in a synergistic improvement of this stability at low cost.

Drapex 4.4 also provides vinyl plastisol and organosol formulations with low initial viscosity and extended viscosity stability by virtue of its low solvating action.

Drapex 6.8

Chemical Type: Liquid epoxidized soybean oil.

Key Properties: Specifically designed to provide optimum heat and light stabilizing performance together with maximum compatibility in vinyl compounds.

Drapex 10.4

Chemical Type: Epoxidized linseed oil.

Key Properties: Plasticizer having the highest oxirane oxygen available anywhere, using the latest peracetic acid oxidation technology. Designed specifically to provide nontoxic vinyl compounds with optimum heat stabilizing action when used with primary metallic stabilizers such as Argus' line of calcium-zinc additives.

BASF CORP., 491 Columbia Ave., Holland, MI 49423

Palamoll

Product Name	Viscosity (mPa-s)	Specific Gravity (20°C)	Refractive Index	Saponification Value	Boiling Point (°C)	Color (Hazen)
632	2,750	1.14	1.463	620	Decomposes	200
644	7,000	1.13	1.469	563	–	300
646	11,500	1.13	1.469	563	–	300

Key Properties: Suitable for coatings that have to withstand oil fractions.

Physical Constants (all of the above):

Acid Value	2.5 max
Light Transmittance (Iodine Scale)	2-5
Flash Point (Pensky-Martens)	200°C
Rate of Saponification	~50
Volatiles (2 hours @ 150°C)	0.2% max

Chemical Type: Polyadipate polymer plasticizers.

Key Properties (of all grades): Clear, yellowish liquids at room temperature. They display good stability to light and resistance to oil fractions. Their volatility is very low and they are compatible with many coatings raw materials. The main difference between them is their viscosity, a property that largely governs their application.

The Palamolls are soluble in conventional paint solvents, with some exceptions. The main application for the Palamolls is as nonsolvent plasticizers for cellulose nitrate. They yield coatings with particularly high resistance to gasoline. These coatings are also weather-resistant and have good flexibility retention. These plasticizers are also incorporated in paints based on nonhydrolyzable binders, where their low volatility and resistance to migration offer particular advantages. They are hydrolyzable and excessive proportions will reduce the resistance to alkalies of the coatings.

Palatinol

Product Name	Viscosity (mPa-s)	Specific Gravity (20°C)	Refractive Index	Saponification Value	Boiling Point (°C)	Color (Hazen)
A	13	1.121	1.5023	505	177	7.5

Key Properties: Used solely as auxiliary plasticizer, due to its high volatility.

AH	82.5	0.083	1.4865	288	232	25.0

Key Properties: Cellulose nitrate lacquers and binders derived from chlorinated binders containing this yield flexible coatings that are resistant to low temperature and high-temperature aging.

C	20.5	1.048	1.4930	404	183	25.0

Key Properties: Cellulose nitrate systems containing this yield extremely flexible yet fairly hard coatings with good resistance to outdoor exposure.

Product Name	Viscosity (mPa-s)	Specific Gravity (20°C)	Refractive Index	Saponification Value	Boiling Point (°C)	Color (Hazen)
DN	165	0.978	1.5015	270	248	25.0
IC	42.5	1.041	1.4900	404	174	25.0

Key Properties: Cellulose nitrate systems containing this yield extremely flexible yet fairly hard coatings with good resistance to outdoor exposure.

M	17.5	1.192	1.5155	575	136	7.5

Key Properties: Used solely as auxiliary plasticizer, due to its high volatility.

O	57.5	1.173	1.5040	400	200	25.0

Key Properties: Very suitable for cellulose nitrate lacquers that have to withstand oil fractions.

Z	122.5	0.966	1.4855	251	259	30.0

Key Properties: Suitable secondary plasticizer for binders that dry by physical means.

911	75	0.966	1.4850	250	272	60.0

Key Properties: Suitable for use at low temperatures, because of flexibility.

Chemical Type: Light-stable phthalate plasticizers with good solvent properties.

Key Properties: Compatible with many binders that dry by physical means and are used in a wide variety of applications in the coatings sector and for polyvinyl chloride. They are low viscosity, clear, colorless, practically odorless, anhydrous liquids that are readily soluble in conventional organic solvents, but insoluble in water. The main application is for cellulose nitrate lacquers. They can also be used in paints derived from nonhydrolyzable binders, chlorinated rubber and cellulose ethers, if the demands on the resistance to alkalies and acids are not very high.

Palatinol AH, C, DN, IC and O can be ground with pigments and are suitable for the preparation of color pastes, etc.

Plasticizer 9	330	1.165	1.465	0.2	Decomposes	−37

Chemical Type: A glycerol ether alcohol that is less hygroscopic than glycerol.

Key Properties: Clear, odorless and oily liquid that decomposes on distillation. Soluble in water and most paint solvents with the exception of aliphatic, aromatic and terpene hydrocarbons. Compatible with gelatin and ethanol-soluble resins, such as Suprapal AP, Phthalopal resins, shellac, etc. Suitable for gelatin film, aqueous solutions of resins and cellulose nitrate coatings that are resistant to aliphatic and aromatic hydrocarbons, but not resistant to water.

Plastomoll

Product Name	Viscosity (mPa-s)	Specific Gravity (20°C)	Refractive Index	Saponification Value	Boiling Point (°C)	Color (Hazen)
DOA	14	0.924	1.447	304	215	25.0
DIDA	29	0.918	1.453	263	241	25.0
NA	28.5	0.928	1.453	284	231	25.0
632	2,750	1.14	1.463	620	Decomposes	200.0
644	7,000	1.13	1.469	563	Decomposes	300.0
646	11,500	1.13	1.469	563	Decomposes	300.0

Chemical Type: Low viscosity adipate-type plasticizers with good stability to light and good solvent properties for cellulose nitrate.

Key Properties: Suitable for coatings that have to withstand low temperatures and are compatible with many binders that dry by physical means. They are also suitable for plasticized PVC with a low brittle temperature. Clear, colorless, anhydrous, practically odorless, neutral liquids. Very efficient plasticizers for many polymers. They are very popular in coating systems that have to withstand low temperatures, by virtue of their good compatibility with many coatings raw materials. They are soluble in conventional organic solvents, but insoluble in water. They are mainly used for cellulose nitrate lacquers, in the coatings sector, if the demands on the resistance to chemicals are not very severe. They are also used for paints derived from nonhydrolyzable binders and for grinding pigments. The films they yield are quite hard and flexible. They can also be used in road marking paints.

EAST COAST CHEMICALS CO., P.O. Box 160, Cedar Grove, NJ 07009

Product Name	Specific Gravity	Flash Point (°F) (min)	Color (ASTM)	ASTM Distillation (ASTM D86) (°F)
Escoflex Plasticizer 435	1.0	300	1.5	545–700

Chemical Type: Completely aromatic distillate with exceptional compatibility with nitrile rubber.

Key Properties: Can be used to soften a nitrile stock without sacrifice in oil resistance. Provides good processability and contributes to improved temperature performance. Primary plasticizer and can, therefore, be used as the sole plasticizer in nitrile rubber rather than the more expensive ester plasticizers.

Escoflex 416 Dibenzoate

Chemical Type: Proprietary composition plasticizer.

Key Properties: Comparable to DOP or TXIB in initial physical properties, except for low temperature flexibility.

Superior to TXIB and comparable to DOP in retention of properties or aging.

Superior with respect to weight loss and Shore Hardness, indicating the lower volatility of Escoflex 416 Dibenzoate.

Comparable to DOP and TXIB in retention of physical properties after oil immersion.

Offers the advantage of better performance in the nitrile rubber formulations tested.

VELSICOL CHEMICAL CORP., 341 East Ohio St., Chicago, IL 60611

Benzoflex 9-88 Plasticizer

Chemical Type: Dipropylene glycol dibenzoate, in clear oily liquid form, with mild ester odor.

Chemical Analysis and Physical Properties:

% Assay (as Benzoate Ester)	98.0%
Color (APHA)	100 max
Acidity (as Benzoic Acid)	0.1% max
Molecular Weight (Theoretical)	342.3
Boiling Point (5 mm Hg)	232°C
Freezing Point (Becomes a Glass)	−40°C
Pour Point	−19°C
Specific Gravity @ 25°/25°C	1.120
Weight per Gallon @ 25°C	9.346 pounds
Moisture	0.07%
Refractive Index @ 25°C	1.5282
Flash Point (COC)	414°F
Vapor Pressure @ 25°C (mm Hg)	2.29×10^{-7}
Dielectric Constant @ 60 Hz (1,000 Hz)	7.6
Volume Resistivity	1×10^{10}

Key Properties: High solvating monomeric plasticizer. It is an exceptionally good solvator for PVC, an energy-saving property which improves processability by lowering gel and fusion temperatures. This unusual affinity for PVC is also reflected in tests for permanence. Vinyls containing Benzoflex 9-88 show good resistance to solvent extraction and perform well in long term volatility tests.

Benzoflex 9-88 is also used in latex adhesives to give improved flexibility, tack and bond properties. In PVA homopolymer emulsion adhesives, it has been shown to enhance film formation and surface wetting. Where the dried solids of a PVA adhesive are later fused to form a bond, as in foil to foil laminations, Benzoflex 9-88 will reduce the temperature necessary for heat sealing. Other uses of Benzoflex 9-88 are as a plasticizer in elastomers, such as Buna-N and SBR, latex caulks and sealants based on PVA, EVA and acrylates; and in formulating color concentrates for PVC.

Hydrolytically stable under normal conditions.

Various volatility tests have shown Benzoflex 9-88 to be somewhat more volatile than dioctyl phthalate, but comparable to other high solvating plasticizers such as butyl benzyl phthalate.

Has a medium to low viscosity at normal handling temperatures.

Toxicity: The following are the results of acute toxicity tests conducted in general compliance with the methods outlined in the Federal Hazardous Substances Act and CFR Title 49:

Test	Animal	Results
Acute Oral Toxicity	Rat	LD_{50} = 4,673 mg/kg
Acute Dermal Toxicity	Rabbit	LD_{50} = >2,000 mg/kg
Acute Inhalation Toxicity	Rat	4 hr LC_{50} = >200 mg/l
Eye Irritation	Rabbit	Not an eye irritant
Skin Irritation (DOT Protocol)	Rabbit	Not a primary skin irritant

Regulatory: Benzoflex 9-88 is not considered a hazardous substance as defined by either the Federal Hazardous Substances Act or by the Department of Transportation under CFR Title 49.

Approved for use in Food-Packaging Adhesives subject to the limitations defined in 21 CFR 175.105. Refer to the manufacturer's Material Safety Data Sheet for specific handling instructions.

Benzoflex 50

Chemical Type: Mixed dipropylene glycol dibenzoate and diethylene glycol dibenzoate.

Chemical Analysis and Physical Properties:

% Assay (Ester Content)	98.0% min
Color (APHA)	100 max
Acidity (as Benzoic Acid)	0.1% max
Molecular Weight-Average	328.3
Boiling Point (5 mm Hg)	240°C
Freezing Point	0°C
Refractive Index @ 25°C	1.535
Weight per Gallon @ 77°F	9.6
Flash Point (Tag Open Cup)	414°F

Key Properties: Properties of this plasticizer with PVC include the following:

Very strong solvent action, permitting lower processing temperatures.
Superior silica gel migration resistance.
Higher tensile strength.
Higher hot strength in extruded film.
Improved lacquer marring resistance.
Good compatibility with aromatic, aliphatic and chlorinated hydrocarbons.
Increased clarity, brilliance and hardness.
Benzoflex 50 is of interest in calendering, extrusion, molding and plastisol-like processes, and adhesives, especially PVAc emulsion type.

In addition to polyvinyl chloride and polyvinyl acetate, Benzoflex 50 is compatible with ethyl cellulose, cellulose acetate butyrate and cellulose acetate, and is a plasticizer for Buna-N rubber and/or phenol formaldehyde resins.

Section XVI

Polybutenes

AMOCO CHEMICALS CORP., 200 East Randolph Drive., Chicago, IL 60601

Amsco Polybutenes

Product Name	Viscosity (ASTM D445) @ 100°F (cs)	Viscosity (ASTM D445) @ 210°F (cs)	Flash Point (C.O.C.) (°F min)	API Gravity (ASTM D287) @ 60°F	Color (APHA) (Haze Free) (max)	Color (APHA) (Haze) (max)
L-14	30	—	280	37.5	70	15
L-50	109	—	280	34.5	70	15
L-100	219	—	285	33.5	70	15
H-25	—	52	300	30.5	70	15
H-35	—	77	310	29.5	70	15
H-50	—	117	310	28.5	70	15
H-100	—	215	380	27.5	70	15
H-300	—	663	440	26.5	70	15
H-1500	—	3204	470	25.5	70	15
H-1900	—	4226	470	24.5	70	15

Note: All of the above have no visual appearance of foreign material and their odor passes testing.

Chemical Type: These butylene polymers are composed predominantly of high molecular weight mono-olefins with a minor isoparaffin content. Viscous, nondrying liquids.

Key Properties: Chemically stable, colorless and resistant to oxidation by light and heat. Volatilization or decomposition at sufficiently high temperatures leaves no residue. Another important characteristic is tackiness which increases with increased molecular weight. These unique properties have led to their use in formulations for a wide variety of applications. The interesting structure of Amoco polybutenes also offers opportunities to synthesize new chemical derivatives with special advantages. Some areas in which polybutenes are currently being used include: (1) Adhesives, as a tackifier, strengthener and extender. (2) Caulks, sealants and glazing compounds as the primary vehicle or a modifier. (3) SBR, butyl and natural rubbers as a plasticizer and low cost extender. (4) Ashless lubricants. (5) Surface treatments for moisture and vapor-proof papers and carbon paper. (6) Electrical oils and insulating materials. (7) Reactive intermediates for specialty chemicals, surfactants and various adducts.

Toxicity: Generally considered nontoxic and require no special handling when used as directed.

COSDEN OIL & CHEMICAL CO., 8350 North Central Expresssway, Dallas, TX 75221

Polyvis Polybutenes

Chemical Analysis (All Polyvis Polybutenes)

Neutralization Value (mg KOH/g)	0.01	Carbon Residue (%)	None
Total Chlorides (ppm)	25	Water Content (ppm)	15
Total Sulfur (%)	Nil		

Product Name	MW	Viscosity (SUS) @ 210°F	Viscosity Index	Weight per Gallon (lb)	Flash Point (C.O.C.) (°F)	Fire Point (C.O.C) (°F)	Pour Point (°F)
0SH	340	41	65	6.97	270	290	-60
06SH	450	63	68	7.09	310	325	-40
09SH	500	91	95	7.14	310	330	-30
015SH	570	158	85	7.15	320	340	-10
025SH	650	250	85	7.19	330	360	-10
5SH	800	554	104	7.35	375	410	+5
10SH	940	1090	110	7.41	420	480	+15
30SH	1350	3070	115	7.46	450	510	+40

Note: Gardner Color of all grades, -1°F

EXXON CHEMICALS, 3020 West Market St., Akron, OH 44313

Vistanex Polyisobutylenes

Product Name	Brookfield Viscosity @ 350°F (cp)	Flory Molecular Weight	Staudinger Molecular Weight	Needle Penetration @ 25°C	Nonvolatile Content @ 300°F	Specific Gravity @ 23°C
LM-MS	30,500	44,350	10,650	17.7	97.0% min	0.914
LM-MH	58,000	53,000	11,950	12.0	97.0% min	0.914

Chemical Type: Vistanex LM is a low molecular weight polyisobutylene, which is a highly paraffinic hydrocarbon polymer composed of a straight chain molecule having only terminal unsaturation. Water-white to pale yellow color with clear, free of suspended matter appearance. Soft, very viscous, permanently tacky, completely amorphous, nonpolar and somewhat elastomeric in color. Will soften gradually when heated, but does not exhibit a thermoplastic melting point. They contain no stabilizer.

Key Properties: The molecular structure makes Vistanex LM relatively inert and resistant to chemical and oxidative attack and soluble in hydrocarbon solvents. The molecular structure permits close, unstrained molecular packing and to this is attributed the unique low air, moisture and gas permeability of the polymer. Their inertness affords exceptional storage stability. Vistanex LM polyisobutylene is an important ingredient in a diversity of applications. End uses particularly capitalize on the unique and attractive properties of this material. This polymer provides permanent tack in pressure sensitive adhesives, hot melt adhesives, cements and sealants. Additionally, it contributes softness, flexibility (particularly at low temperatures) and low permeability characteristics to many compounds in which it is used as, for example, in sealants, caulks and wax and asphalt blends. Vistanex LM is used for adhesiveness and high-temperature stability in a variety of greases and as an oil additive to improve viscosity index.

Toxicity: The low order of toxicity and favorable FDA regulatory position of Vistanex LM has resulted in its use as a soft, stable ingredient in chewing gum base and as a tackifier in surgical adhesives.

Section XVII
Polyvinyl Acetates

AZS CHEMICAL CO., 762 Marietta Blvd., N.W., Atlanta, GA 30318

Seycorez Polyvinyl Acetates

Product Name	Solids Content (%)	Brookfield Viscosity (cp)	pH	Weight per Gallon (lb)	Particle Size (μ)	Residual Monomer (%)
C-79	55	1,000	.5.0	9.2	2.5	<0.5

Chemical Type: Homopolymer polyvinyl acetate dispersion stabilized with partially hydrolyzed polyvinyl alcohol.

Key Properties: General purpose base for wood adhesives and packaging adhesives. Fast setting speed and high wood compression strength. Can be used in bookbinding, case sealing, cellophane bonding, bag sealing and paper laminating. Excellent solvent tolerance and mechanical stability.

Product Name	Solids Content (%)	Brookfield Viscosity (cp)	pH	Weight per Gallon (lb)	Particle Size (μ)	Residual Monomer (%)
C-79-H	55	1,600	5.0	9.2	2.5	<0.5
C-80	55	1,800	5.0	9.2	2.5	<0.5
C-80-H	55	2,200	5.0	9.2	2.5	<0.5
C-91	55	5,000	5.0	9.2	2.5	<0.5

Chemical Type (of all of the above): Homopolymer polyvinyl acetate dispersion stabilized with partially hydrolyzed polyvinyl alcohol.

Key Properties (of all of the above): Excellent solvent tolerance and mechanical stability. General purpose base for wood adhesives and packaging adhesives. Fast setting speed and high wood compression strength. Also, can be used in bookbinding, case sealing, cellophane bonding, bag sealing and paper laminating.

Product Name	Solids Content (%)	Brookfield Viscosity (cp)	pH	Weight per Gallon (lb)	Particle Size (μ)	Residual Monomer (%)
G-16	46	—	5.0	9.0	—	—

Chemical Type: Alkali-swellable vinyl acetate copolymer.

Key Properties: Used in coating applications. Often used as a replacement for casein. Compatible with the usual paper coating binders and other additives.

Product Name	Solids Content (%)	Brookfield Viscosity (cp)	pH	Weight per Gallon (lb)	Particle Size (μ)	Residual Monomer (%)
H-52	46	450	6.5	8.8	—	—

Chemical Type: Homopolymer polyvinyl acetate dispersion stabilized with partially hydrolyzed polyvinyl alcohol.

Key Properties: Excellent pigment binding ability and adhesion to various surfaces when applied as a coating resin. The rate of crosslinking can be adjusted by the addition of only acid or latent-acid catalyst. pH can be adjusted by ammonium hydroxide or weak acids.

Section XVIII

Preservatives and Fungicides

COSAN CHEMICAL CORP., 400 Fourteenth St., Carlstadt, NJ 07072

Product Name	PMA Content (% min)	Mercury Metal Content (% min)	Particle Size (μ)	Weight per Gallon (lb)
Cosan JTA-20	20	12	2.1	2.8

Chemical Type: Highly functional, soluble, phenylmercury acetate (20% active) which has been deposited on an inert, synthetic, hydrous calcium silicate to produce a free-flowing, relatively nondusting powder which is simple to incorporate into dry formulations.

Key Properties: When a product contains water or is ultimately applied by dispersion in water, it is subject to attack by bacteria unless a functional preservative is used at the proper concentration. Joint cements, acoustical plasters, textures, adhesives and other powdered products are commonly applied out of water solutions or dispersions. One of the major problems in the use of these products is microbial spoilage. Cosan JTA-20 offers:

>Immediate preservation upon wetting of the formulation which is to be protected.

>Immediate solution in water since the active ingredient, phenylmercuric acetate, is present in its soluble state. This minimizes the opportunity for entrapment of undissolved particles which ultimately could result in pockmarks.

>Elimination of the danger of photographing of joints and nail heads, yellowing and browning, since it is a nonphenolic.

EPA Registration Number 8489-3.

Toxicity: All mercury compounds are toxic and irritant and may produce delayed chemical burns. Cosan JTA-20 is a mercury compound and should be handled with care. When handling the concentrate, rubber gloves, goggles, apron and other protective garments should be worn. In case of accidental contact, flush copiously with water and follow instructions on label.

Product Name	Active Content (%)	Melting Point (°C)	Boiling Point @ 760 mm (°C)	Boiling Point @ 10 mm (°C)	Specific Gravity
Cosan PCMC	100	64	230.5	111.0	1.215

Chemical Type: Parachlorometacresol (PCMC) is a high purity chemical.

Key Properties: Suggested as an antimicrobial agent, for industrial water-based and emulsion compositions, such as adhesives and coatings, to impart package stability.

Toxicity: This product is dangerous when heated to decomposition (235°C), liberates highly toxic fumes of phosgene. May be allergenic; avoid skin contact. May be fatal if swallowed or absorbed through skin. May produce severe burns. Do not get in eyes, on skin or on clothing. Avoid breathing spray mist. Wash thoroughly after handling.

Product Name	Assay (%)	Melting Point (°C)	Flash Point Open Cup (°F)	Weight per Gallon (lb)
Cosan S	95	99.5	280	11.6

Chemical Type: Nonmetallic organic chemical (3,5-dimethyltetrahydro-1,3,5,2H-thiodiazine-2-thi-one), a highly effective antimicrobial preservative. In white crystalline powder form with light, distinctive odor.

Key Properties: Recommended for use in water systems, including adhesives, dispersed colors, protein colloids and resin emulsions. Chemically cleared for broad use as an anti-microbial, in "Cumulative Pocket Supplement—Code of Federal Regulations, Title 21–Food and Drug," as of January 1, 1962. The exact reference is as follows, in these paragraphs:

121.2505C	Simicides
121.2520C5	Adhesives
121.2529:	Paper, Paperboard, Foodboard, Coatings

EPA Registration No. 8489-4.

Product Name	Assay (%)	Gardner Color (max)	Weight per Gallon (lb)
Cosan 158	15	6	7.92

Chemical Type: Yellow liquid of low viscosity with an active content of tributyltin benzoate. Also contains 21.2% alkyl (derived from fatty acids of coconut oil) amine hydro-chlorides.

Key Properties: Has been designed to insure preservation against in-can bacterial attack and provid-ing the necessary antifungal ingredient for mildew resistance of the applied coating. Recommended for the following end uses: ready-mixed joint cements and spack-ling compounds, where the control of both bacteria and fungi is required.

EPA Registration No. 8489-21.

Toxicity: Danger! Keep out of reach of children. Corrosive. Causes eye damage and skin irritation. Do not get in eyes, on skin or on clothing. Wear goggles or face shield and rubber gloves when handling. Harmful or fatal if swallowed. Avoid contami-nation of feed, food or drinking water.

Product Name	Active Content (%)	Gardner Color (max)	pH	Weight per Gallon (lb)
Cosan 635-W	25	3	7.0	7.95

Chemical Type: Liquid complexed alkyl amine (5.0% caprylyl, 7.0% capryl, 56.0% lauryl, 18.0% myristyl, 7.0% palmityl, 5.0% stearyl, 2.0% linoleyl) hydrochlorides, in water solution.

Key Properties: Suggested for use as an antibacterial agent in water-containing compositions, such as adhesives, paints and their components, to impart package stability. Will provide stability to bacterial attack for susceptible water and emulsion compositions. These may include resin emulsions, such as polyvinyl acetate, acrylic, polyvinyl acetate-ethylene copolymer and vinyl acrylic types. Latex paints and adhesives prepared from the above resin emulsions will be similarly preserved.

EPA Registration No. 8489-13.

Toxicity: Danger! Concentrated solutions are corrosive and cause severe eye and skin dam-age. Do not get in eyes, on skin or on clothing. May be fatal or harmful if swal-lowed. Precautions should include the use of eye shields, rubber gloves and pro-tective clothing. In the event of skin contact, wash thoroughly with soap and wa-ter. If in contact with the eye, flush with copious amounts of water. In case of severe burns, consult a physician promptly.

DOW CHEMICAL U.S.A., Midland, MI 48640

Product Name	Active Content (%)	Inert Content (%)	Gardner Color (max)	Sieve Analysis (% through No. 5 U.S.S.)	Specific Gravity
Dowicil 75	67.5	32.5	4	100	1.54

Chemical Type: 1-(3-chloroallyl)-3,5,7-triaza-1-azoniaadamantane chloride (67.5%), as an active ingredient. Inert ingredients are 32.5%, including 25% sodium bicarbonate (as stabilizer) and 9.5% other raw materials.

Key Properties: The adhesives industry can use this product to protect its products against deterioration and spoilage during manufacture, storage and service. Dowicil 75 is used to preserve water-based latex paints in the can while having no effect on the viscosity, stability and other physical properties of the formulation or on the color fastness of the paint film. It can also be used as a preservative in many other products.

Regulations: EPA Registration No. 464-403.

Meets the requirements of Food Additive Regulation CFR121.2520 for use as a preservative in adhesives; CFR121.2522 for preservation of polyurethane resins in contact with dry bulk foods; and CFR121.2526 for use only as a preservative at a level of 0.3% in the latexes used as pigment binders in paper and paperboard intended for use in contact with dry and fatty foods.

ICI AMERICAS, INC., Concord Pike and New Murphy Road, Wilmington, DE 19897

Product Name	Active Content (%)	Brookfield Viscosity @ 25°C (cp)	pH	Specific Gravity @ 25°C	Retained on 300 Mesh British Standard Sieve (% max)
Proxel CRL	32.5	22	10.5	1.12	0.1

Chemical Type: A microbiostat preservative, with 1,2-benzothiazolin-3-one, as the active agent. Dark brown liquid, noncrystallizing above 14°F, with an amine-like odor.

Key Properties: Antimicrobial preservative for adhesives, binders, paper coatings and various aqueous compositions. Effective preservative for aqueous compositions such as solutions, emulsions and dispersions, providing protection during storage. Proxel CRL has found wide acceptance in adhesives, binders and paper coatings. It is rapidly being adopted as a replacement for phenolic preservatives because of its many significant performance advantages. The benefits of Proxel CRL are as follows:

Economy – Use levels are significantly lower than for many other products.
Liquid form – Can be pumped, metered and easily handled.
Stability – Dilute aqueous solutions are stable. Will not decompose or separate during storage. May be thawed after freezing.
Long life – Maintains effectiveness if exposed to high temperatures.
Solubility – Miscible in water in all proportions.
Effective preservative against a wide spectrum of bacteria and fungi.

EPA Registration Number 10182-3.

INTERNATIONAL DIOXCIDE, INC., 136 Central Ave., Clark, NJ 07066

Product Name	Specific Gravity (min)	Boiling Point (°F)	Freezing Point	pH
Anthium Dioxide	1.063	214	22	9.0

Chemical Type: Anthium Dioxide complex is a combination of oxygen and chlorine joined as ClO_2 in aqueous solution.

Chemical Analysis:

Sodium carbonates (Na_2CO_3 and $NaHCO_3$)	3.65 % by wt
Chlorine dioxide (ClO_2)	5.00 % by wt
	(50,000 ppm)
Water	91.35 % by wt

Key Properties: Chlorine dioxide has long been recognized as an effective antimicrobial and odor control agent. Now in convenient solution form with control of sustained release,

Anthium Dioxide provides a means of using ClO_2 in products and processes where previously available unstable solutions of ClO_2 gas were not practical. It is easy to use and safe to store.

Recommended for use in water-based adhesives and polyvinyl-acetate-based paints and adhesives as a shelf-life preservative.

OTTAWA CHEMICAL DIVISION, Ferro Corp., 700 North Wheeling St., Toledo, OH 43605

Ottasept Extra   (Purified for use in fine cosmetics and pharmaceuticals.)
Ottasept Technical   (For other applications.)

Chemical Type: Ottasept has several synonyms — PCMX®, p-chloro-m-xylenol; 3,5-dimethyl-4-chlorophenol, 2-chloro-m-5-xylenol, and 2,6-dimethyl-4-hydroxy-chlorobenzene.

Chemical Analysis and Physical Properties:

	Ottasept Extra	Ottasept Technical
Physical characteristics	. . . . crystalline solid	
Crystal size	. . . . . . 30 mesh, approximately	
Odor	faint	characteristic
Color	white	white to off-white
Molecular weight	156	156
Melting point	114°C min	112°C min
Bulk density	5 lb/gal	5 lb/gal
Water suspension	. . . . neutral to litmus	
Caustic solubility	10 g dissolves completely in 100 cc	
	. . . . .5% NaOH at 25°C	
Specific assay	98.5% min	97.0% min

Key Properties: Potent antimicrobial which is a safe and effective preservative covering a broad spectrum.

Toxicity: The safety of Ottasept has been proven by a wealth of available toxicity data. In particular, a large number of acute and chronic animal studies have shown the toxicity potential of Ottasept to be almost nonexistent, with one exception. When Ottasept was applied to the eyes of test animals, mild conjunctivitis developed.

THIOKOL/VENTRON DIVISION, 150 Andover St., Danvers, MA 01923

Cunilate 2174-NA
Cunilate 2174-NO

Chemical Type: Copper-8-quinolinolate (10%)—completely solubilized. 1.80% minimum copper on metal.

Basic carriers —

Cunilate 2174-NA	2-ethylhexoic acid and VM&P naphtha
Cunilate 2174-NO	2-ethylhexoic acid and odorless mineral spirits

Physical Properties:

Solids content, %	33.0–38.0
Specific gravity @ 77°F	0.883–0.888
Flash point, °F	
Cunilate 2174-NA	35
Cunilate 2174-NO	125
pH	5.8
Acid value	60
Appearance	greenish yellow liquid

Key Properties: Both are recommended for formulation into paints, protective coatings and adhesives to make products which are highly resistant to fungus growth. Allowed by the Food and Drug Administration in adhesives for food packaging.

EPA Registration Numbers:

Cunilate 2174-NA	2829-17
Cunilate 2174-NO	2829-44

Cunimene D-2629

Chemical Type: 40% dehydroabietylamine pentachlorophenate in 60% aromatic solvents.

Physical Properties:

Specific gravity	1.01–1.03
Flash point (C.O.C.), °F	125–130
Appearance	amber liquid

Key Properties: Fungistatic treatment for direct addition to rubber compounds, adhesives, caulking compounds, textiles, and other related materials where microbiological preservation is required.

Toxicity: Care should be taken in handling. Follow precautions described by manufacturer.

EPA Registration Number: 2829-58

Cunimene D-2747

Chemical Type: Unique formulation of dehydroabietylamine pentachlorophenate in a special formulation.

Physical Properties:

Specific gravity @ 60°/60°F	1.035
Solids content, %	55
Flash point, °F	117
pH	5.1

Key Properties: Special formulation which enhances fungicidal and bactericidal activity through synergistic action. Exceedingly effective as a preservative for water-based paints, glues and adhesives, and cellulosic materials.

Water-based paints – The use of 0.35% Cunimene D-2747, based on the weight of a finished water paint, will impart excellent preservative qualities throughout the shelf life of the paint.

Glues and adhesives – The preservation of many glues and adhesives can be readily accomplished by the use of 0.30 to 0.60% of Cunimene D-2747, based on the weight of the finished adhesive.

EPA Registration Number: 2829-59

Cuniphen 2778-I

Chemical Type: Special formulation of 2,2'-methylene-bis(4-chlorophenol), supplied as a liquid concentrate.

Physical Properties:

Solids content, %	70
Specific gravity	1.086
Flash point, °F	180
pH	5.7
Gardner color	11

Key Properties: Especially designed for use in the preservation of glues and adhesives. Recommended where an effective preservative is required and is easily dispersible in glues, adhesives and resin emulsions.

The active ingredient of Cuniphen 2778-I has been cleared for use for adhesives and for paper and paperboard used in contact with aqueous and fatty foods and dry foods.

Toxicity: Care should be taken in handling. Follow precautions described by manufacturer.

EPA Registration Number: 2829-36
Canadian Registration Number: 12315

Durotex Antimicrobials

Chemical Type: Active ingredient of 10,10'-oxybisphenoxarsine.

Physical Properties:

	7599	7603	7604
Active content, %	2	2	2
Specific gravity @ 25°C	1.00	1.05	0.99
Solids content, %	73	25.5	24.5
Color	. . . . .off-white		
Type of emulsifier system	nonionic	anionic	cationic
pH of aqueous solution	6.5	6.5	3.5

Key Properties: Fungistatic and bacteriostatic emulsions for protecting textile and adhesive products. Durotex antimicrobials are highly effective emulsion concentrates formulated for incorporation into textiles, nonfood-grade glues and adhesives, latexes and similar products. The series offers a unique range of emulsion systems for maximum compatibility with a wide range of textile treatment baths and adhesive formulas. Durotex treated products provide protection against deterioration, staining and unsightly changes in surface appearance caused by bacterial or fungal attack.

Many commercial water repellent formulations and binders such as polyvinyl acetate or acrylics are compatible with the Durotex antimicrobials and will enhance the durability of the system. Susceptible products successfully protected with Durotex antimicrobials include: adhesives and glues (nonfood grade); ink bases; latex emulsions; wallpaper constructions; and many types of fabrics.

Toxicity: Care should be taken in handling. Follow precautions described by manufacturer.

EPA Registration Numbers:

Durotex 7599 = 2829-89; Durotex 7603 = 2829-90; Durotex 7604 = 2829-91

R.T. VANDERBILT CO., INC., 30 Winfield St., Norwalk, CT 06855

Vancide Mold Inhibitors

Product Name	Specific Gravity
PA	1.40

Chemical Type: Trans-1,2-bis(n-propylsulfonyl)ethene.

Key Properties: A very effective mold inhibitor in all paints except those normally formulated with ammonia or other alkaline materials. Will not become discolored during exposure to sulfide atmospheres. Essentially insoluble in water, oils or solvents commonly used in paint products. Stable under common storage conditions.

PA Dispersion	1.01

Chemical Type: Dispersion of Vancide PA (50% active) in 37% air pollution exempt mineral spirits and 13% other diluents.

Key Properties: Easily incorporated into a paint formulation.

51Z

Chemical Type: Contains 87.0% zinc dimethyldithiocarbamate and 7.5% zinc 2-mercaptobenzothiazole as active ingredients.

Key Properties: Has been shown to inhibit mold growth in many commercial caulking compounds. Readily compatible with formulations of either butyl rubber or acrylic copolymers containing petroleum or alcohol solvents without affecting performance or color.

Vancide Preservatives

Product Name	Active Assay (%)	Specific Gravity	Melting Range (°C)	Molecular Weight
89	90	1.69	171	300.6

Chemical Type: N-trichloromethylthio-4-cyclohexene-1,2-dicarboximide industrial preservative.

Key Properties: Can be properly formulated to work as an efficient fungistat in paints, latex adhesives, polyethylene, lacquer, soap and wallpaper flow paste. May be stored indefinitely in closed containers.

51	30	1.15	—	—

Chemical Type: Contains 27.6% sodium dimethyldithiocarbamate and 2.4% sodium 2-mercaptobenzothiazole as active ingredients.

Key Properties: Can be properly formulated to work as an efficient fungistat in adhesives, paper mill production, cooling towers, paper and paperboard, cotton fabrics, wood veneer and cutting oils.

Product Name	Active Assay (%)	Specific Gravity	Boiling Range (°C)	Molecular Weight
TH	–	0.89	208	–

Chemical Type: Hexahydro-1,3-5-triethyl-s-triazine preservative for latex paints.

Key Properties: Prevents attack by microorganisms while the paint is in storage. This nonmetallic liquid, which is completely soluble in water, can be added at any point during paint preparation.

Section XIX
Resins–Acrylic

AZS CHEMICAL CO., 762 Marietta Blvd., NW, Atlanta, GA 30318

Seycorez Acrylics

Product Name	Viscosity (Brookfield) (cp)	Percent Solids Content	pH	Weight per Gallon (lb)	Specific Gravity
K-25	2,500	52.0	9.5	8.9	1.07

Chemical Type: Acrylic emulsion polymer.

Key Properties: Emulsion polymer with high tack properties which facilitates use as an adhesive for nonporous substrates. Compatible with other adhesive base polymers and common additives included in adhesive applications. Excellent adhesive for bonding polyethylene to kraft, used alone or in a compound. Excellent machinability. Valuable where plasticizer migration must be considered.

ROHM AND HAAS CO., Independence Mall West, Philadelphia, PA 19105

Acryloid Solid Grade Thermoplastic Acrylic Resins

Product Name	Weight per Gallon (lb)	Solubility Parameter	Tg (°C)	Tukon Hardness
A-11	9.8	9.4	100	18–19

Key Properties: Excellent adhesion, hardness exterior durability.

A-21	9.8	9.4	105	21–22

Key Properties: Hardest of the Acryloid thermoplastic resins, resistant to alcohol and water.

A-30	9.8	9.4	86	21–22

Key Properties: Abrasion resistance, toughness and flexibility. Higher molecular weight.

B-44	9.6	9.4	60	15–16

Key Properties: Hardness, flexibility and adhesion.

B-48N	9.2	9.3	50	11–12

Key Properties: Toughness, exterior durability, adhesion to bare metal.

B-50	9.2	9.3	50	11–12

Key Properties: Similar to Acryloid B-48N, but lower cost and adhesion to primed or treated metals.

Product Name	Weight per Gallon (lb)	Solubility Parameter	Tg (°C)	Tukon Hardness
B-66	9.1	9.0	50	12-13

Key Properties: Fast air-dry and excellent compatibility with other film formers.

B-67	8.8	8.6	50	11-12

Key Properties: Improve properties of medium and long-oil alkyd and oleoresinous resins.

B-72	9.6	9.3	40	10-11

Key Properties: Compatible with vinyl and silicone resins. Stable, durable, transparent.

B-82	9.7	9.4	35	10-11

Key Properties: Low cost with typical acrylic properties.

Section XX

Resins–Epoxy

AZS CHEMICAL CO., 762 Marietta Blvd., NW, Atlanta GA 30318

Azepoxy Epoxy Resins

Product Name	Viscosity (Brookfield) (cp)	Color (Gardner) (max)	Weight per Epoxide	Specific Gravity	Flash Point (Closed Cup) (°F)
113	600	4	195	1.15	>200

Chemical Type: Low viscosity cresyl glycidyl ether modified epoxy resin. The modifying diluent is mono-functional and is fully reactive with the hardener.

Key Properties: Recommended for applications requiring room temperature curing. Hardener suggestions are Azamide 900, Azamide 450 and Azamide 340. Can be cured with amines, polyamides and other room temperature hardeners.

115	600	3	186	1.15	178

Chemical Type: Low viscosity butyl glycidyl ether modified epoxy resin. The modifying diluent is fully reactive with the hardener to become chemically combined in the product.

Key Properties: Recommended for applications requiring room temperature cure. Has broad acceptance for electrical, laminating and casting applications. Hardener suggestions are Azamide 900, Azamide 450 and Azamide 360. Can be cured with amines, polyamides and other room temperature hardeners.

118	750	3	200	1.11	>200

Chemical Type: Low viscosity epoxy resin made by the modification with a nontoxic aliphatic monofunctional diluent. Properties of butyl glycidyl ether modified resin without its toxicity.

Key Properties: Recommended for flooring, laminating, casting and decorative applications. Hardener suggestions are Azamide 900, Azamide 450, Azamide 360 and Azamide 340. Can be cured with normal room temperature hardener systems, as well as with anhydrides and other hardeners requiring elevated temperatures.

126	8,250	2	183	1.16	>200

Chemical Type: Low viscosity, unmodified bisphenol A type epoxy resin.

Key Properties: Excellent physical and electrical properties. Exhibits excellent chemical resistance and maximum heat resistance. Readily mixes with modifiers, fillers and liquid hardeners at room temperature. Hardener suggestions are Azamide 900, Azamide 450, Azamide 360 and Azamide 325. Can be cured with amines, polyamides, anhydrides and other hardeners.

128	13,000	3	188	1.16	>200

Chemical Type: Medium viscosity, unmodified bisphenol A type epoxy resin.

Key Properties: Considered as the basic type of resin and is recommended for adhesives, castings, coatings, electrical, flooring, laminating and all other applications. Pourable at room temperature. Highly versatile. Easily mixed with modifiers, fillers and liquid hardeners at room temperature. Can be mixed with solid hardeners, if heated. Hardener suggestions are: Azamide 900, Azamide 450, Azamide 360 and Azamide 325. Can be cured with amines, polyamides, anhydrides, Lewis acids and many other types of hardeners.

Product Name	Viscosity (Gardner-Holdt)	Color (Gardner) (max)	Weight per Epoxide (100% solids basis)	Specific Gravity	Flash Point (Closed Cup) (°F)
601x75	Z_2-Z_6	5	500	1.08	80

Chemical Type: 75% solution of Azepoxy 601 in xylene.

Key Properties: Most frequently used solid epoxy resin solution for coating applications. Exhibits excellent physical properties and chemical resistance. Readily compounded. The most frequently used hardener is Azamide 215x70.

Toxicity (all types): May cause skin irritation following frequent or prolonged contact. In the event of skin contact, wash well with soap and water. In the event of eye contact, flush with water and consult a physician.

CIBA-GEIGY RESINS, Saw Mill River Road, Ardsley, NY 10502

Araldite Liquid Epoxy Resins

Product Name	Viscosity (Brookfield) (cp)	Color (Gardner) (max)	Weight per Epoxide	Weight per Gallon (lb)	Flash Point (Open Cup) (°F)
502	2,850	3	241	9.45	—

Chemical Type: Modified liquid epoxy resin.

Key Properties: Especially useful in formulating high-solids masonry coatings with excellent adhesion and water resistance. These coatings provide excellent protection for low-cost building blocks, such as concrete and cinder block and many other masonry surfaces.

506	600	3	180	9.45	195

Chemical Type: Liquid epoxy resin, modified with a reactive diluent.

Key Properties: Amine cured systems show excellent overall physical and electrical properties, good chemical resistance, moderate heat resistance and good adhesion. This resin is recommended for casting, electrical, laminating and tooling applications where very low viscosity is required for good impregnation or maximum filler loading. Cures can be effected at room or slightly elevated temperatures.

Product Name	Viscosity (Brookfield) (cp)	Color (Gardner) (max)	Weight per Epoxide	Weight per Gallon (lb)	Epoxy Value
507	600	7	—	9.50	0.53

Chemical Type: Modified liquid epoxy resin. The modification, accomplished during manufacture, results in a liquid (epichlorohydrin-bisphenol A) base epoxy resin in combination with a proprietary mono-epoxide reactive diluent.

Key Properties: The pot life, cure time and other cure properties are quite similar to unmodified resins. Has low vapor pressure. This resin exhibits these advantages over similar viscosity modified epoxy resins: superior chemical resistance, pleasant odor, less skin sensitizing, lower vapor pressure and no premium price. Recommended for use in adhesive, electrical, flooring, laminating and tooling applications.

508	3,500	5	428	9.42	0.24

Chemical Type: Flexible liquid epoxy resin which serves as a flexible modifier for conventional liquid epoxies.

Key Properties: Compatible with conventional epoxy resins in all proportions. Excellent working properties and does not affect the color of cured systems. Elongation in various ranges can be obtained. Enables obtaining the increased flexibility, improved impact resistance and high elongation required for many applications, such as adhesives, caulking and sealing, electrical potting, high solids coatings, impregnating, etc. Does not produce the undesirable effects on properties of other modifiers.

Product Name	Viscosity (Brookfield) (cp)	Color (Gardner) (max)	Weight per Epoxide	Weight per Gallon (lb)	Epoxy Value
509	600	1.5	195	9.26	0.52

Chemical Type: Low toxicity epoxy resin modified with Epoxide No. 7, a reactive modifier.

Key Properties: The outstanding characteristics of amine cured systems are excellent overall physical and electrical properties, practically nontoxic, good chemical resistance, moderate heat resistance, good adhesion, long pot life and low exotherm. Recommended for casting, electrical, laminating, tooling and flooring applications where very low viscosity is required for good impregnation or maximum filler loading.

6004	5,750	2	187	9.7	0.54

Chemical Type: Liquid epoxy resin based on bisphenol A, for general purpose use.

Key Properties: Can be cured with amine, anhydride, polyamide and other hardeners, depending on the application and desired properties. Easy handling and can be highly filler loaded. Good choice for clear coating applications. Excellent mechanical and electrical properties, good chemical resistance and outstanding versatility. Recommended for adhesives, castings, coatings, flooring, laminating, filament winding and tooling.

6005	8,500	2	189	9.7	0.35

Chemical Type: General purpose liquid epoxy resin based on bisphenol A.

Key Properties: Can be cured with amine, anhydride, polyamide and other hardeners, depending on the applications and desired properties. Easy to handle without reducing the curing properties. Light color makes it a good choice for higher build clear coatings. Excellent mechanical and electrical properties, good chemical resistance, highly versatile and can be highly filler loaded. Recommended for adhesives, castings, coatings, flooring, laminating, filament winding and tooling.

6010	14,000	2	187	9.7	0.54

Chemical Type: Unmodified liquid epoxy resin, based on bisphenol A and epichlorohydrin.

Key Properties: For general purpose use to be applied in cold or heat cured systems. Superior mechanical and electrical properties, excellent chemical and heat resistance, excellent adhesion, outstanding versatility. Easy to cure with a variety of different type hardeners. Compatible with many different fillers, diluents and accelerators. FDA regulated. Can be used in paints, coatings, electrical and construction applications.

6020	18,000	3	200	9.7	0.50

Chemical Type: Unmodified, liquid bisphenol A epoxy resin, which can be cured with amines, anhydrides, polyamides and other hardeners, depending on the application and desired properties.

Key Properties: Light color, excellent adhesion and mechanical properties, good chemical resistance, good electrical properties and FDA regulated. Can be used for adhesives, castings, coatings, laminating, filament winding and tooling.

GY9513	–	3	209	9.1	0.47

Chemical Type: Modified liquid epoxy resin based on bisphenol A.

Key Properties: Characterized by minimal handling hazards and very slight odor. When used with amine hardeners, compositions demonstrate excellent physical properties, adhesion, appearance and surface properties. Especially good for applications where low viscosity is necessary for good impregnation or maximum filler loading or where reduced odor and diminished toxicity is required. Applications are casting, electrical, laminating, tooling, flooring, coating and impregnation.

Product Name	Viscosity (Brookfield) (cp)	Color (Gardner) (max)	Weight per Epoxide	Weight per Gallon (lb)	Epoxy Value
GY9533	5,000	2	—	9.6	0.51

Chemical Type: Precatalyzed liquid epoxy resin for advancement to solid resin solutions and esters.

Key Properties: This specially catalyzed resin can be readily reacted with bisphenol A to yield solutions of conventional epoxy resins, epoxy esters and high molecular weight epoxy resins. This easily controlled "fast" reaction provides rapid plant processing, high performance characteristics and reduces raw material costs. Based on a permanently active, nonlabile catalyst and bisphenol A, assuring greater control over the reaction, ease of adjustment during manufacture and consistent reactivity in the finished products. Suitable for coating and laminating applications.

GY9573	8,500	300	183	9.7	0.55

Chemical Type: Unmodified general purpose epoxy resin. Can be cured with amine, anhydride, polyamide and other hardeners depending on the application and desired properties.

Key Properties: Can be easily handled and highly filler loaded. Lighter color makes it a good choice for clear coating formulations. Has good chemical resistance, outstanding versatility, is noncrystallizing and meets FDA regulations. Recommended for adhesives, castings, flooring, laminating, filament winding and tooling.

GY9574	5,000	75 max*	172 min	9.7	0.58
	*APHA				

Chemical Type: General purpose epoxy resin based on bisphenol A.

Key Properties: Can be cured with amine, anhydride, polyamide and other hardeners depending on the applications and desired properties. Easily handled and can be highly filler loaded. Lighter color makes it a good choice for clear coating formulations. Frequently crystallizes at room temperature. Crystallization may be induced by chilling, seeding by dust particles or incorporation of filler. Warming to 50°C restores resin to a liquid state. Has good chemical resistance, outstanding versatility and meets FDA regulations. Recommended for adhesives, castings, flooring, laminating, filament winding and tooling.

Toxicity (all grades): Some of the above are slightly toxic and are skin irritants. Please refer to manufacturer's Material Safety Data Sheets.

HENKEL CORP., 425 Broad Hollow Road, Melville, Long Island, NY 11746

Genepoxy Resins

Product Name	Epoxide Equiv. Weight	Color (Gardner) (max)	Viscosity (Brookfield) (cp) @ 25°C	Percent Resinous	Specific Gravity @ 25°C	Flash Point
M200	200	5	8,750	100	1.16	140
M205	222.5	4	1,250	100	1.12	158
370-H55	201*	Milky	4,250	55	1.09	100

*Based on percent resinous in total composition

REICHHOLD CHEMICALS, INC., RCI Bldg., White Plains, NY 10603

Epotuf Liquid Epoxy Resins

Product Name	Viscosity (Brookfield) (cp)	Epoxide Equivalent	Color (Gardner) (max)	Weight per Gallon (lb)
37-127	750	200	3	9.20
Chemical Type: Rigid.				
37-128	750	200	8	9.35
Chemical Type: Low viscosity.				
37-130	550	185	3	9.45
Chemical Type: 100% reactive (low viscosity).				
37-135	5,750	190	3	9.50
Chemical Type: 100% reactive (medium viscosity).				
37-137	600	193	3	9.45
Chemical Type: 100% reactive (low viscosity).				
37-139	8,500	185	3	9.65
Chemical Type: 100% reactive (medium viscosity).				
37-140	12,500	188	3	9.65
Chemical Type: Medium viscosity.				
37-141	19,000	203	3	9.65
Chemical Type: Undiluted high viscosity.				
37-151	50,000	500	—	8.95
Chemical Type: Permanently flexible.				
37-200	750,000	255	5	13.70
Chemical Type: Self-extinguishing.				

Epotuf Hard Epoxy Resins

Product Name	Viscosity (Gardner-Holdt)	Epoxide Equivalent	Color (Gardner) (max)	Weight per Gallon (lb)
37-001	D-G	488	2	10.0
37-002	G-K	650	2	10.0
37-004	Q-U	950	2	10.0
37-006	X-Z	1825	3	10.0
37-007	Z-Z_2	2250	5	10.0
37-009	Z_2-Z_5	3250	5	10.0

Epotuf Hard Epoxy Resin Solutions

Product Name	Viscosity (Gardner-Holdt)	Color (Gardner) (max)	Percent Solids Content	Weight per Gallon (lb)
38-501	Z_1-Z_4	3	75	9.00

Chemical Type: Base Resin 37-001 in methyl isobutyl ketone and xylol (2:1 ratio).

Product Name	Viscosity (Gardner-Holdt)	Color (Gardner) (max)	Percent Solids Content	Weight per Gallon (lb)
38-503	Z_1-Z_5	3	55	9.10

Chemical Type: Base Resin 37-007 in Cellosolve acetate.

38-504	Z-Z_3	3	55	8.85

Chemical Type: Base Resin 37-007 in toluol and Cellosolve acetate (1:1 ratio).

38-505	Z_3-Z_6	3	75	9.10

Chemical Type: Base Resin 37-001 in xylol.

38-507	Z_2-Z_5	3	75	9.10

Chemical Type: Base Resin 37-001 in toluol.

38-521	Z_4-Z_6	4	75	9.10

Chemical Type: Base Resin 37-001 in Cellosolve.

Note: Special cuts of Epotuf Hard Resin, in any solvent or combination of solvents, can be made based on an initial 20 drum order.

Chemical Types (all of the above): Epoxy resins are resins which contain two or more epoxide groups per molecule. The most common epoxy resins are produced by reacting bisphenol A with epichlorohydrin. Other starting materials can be used.

Key Properties (all of the above): Epoxy resins can be crosslinked to form hard films with good solvent resistance. Because there is relatively wide spacing between the crosslinking sites at the end of the molecule, a high degree of flexibility is realized in the cured film. It follows that as the molecular weight of an epoxy resin increases, the adhesion and flexibility achieved in a formulation will also increase. The structure of the epoxy resins explains their unusual chemical resistance, solvent resistance, adhesion, hardness and flexibility.

R.T. VANDERBILT CO., INC., 30 Winfield St., Norwalk, CT 06855

Vanoxy Liquid Epoxy Resins

Product Name	Viscosity (Brookfield) (cp)	Color (Gardner) (max)	Weight per Epoxide	Weight per Gallon (lb)	Flash Point (O.C.) (°F)
115	600	5	185	9.5	240

Key Properties: Yields products with high physical strength, excellent chemical resistance and good electrical properties.

120	7,000	5	188	9.7	175

Key Properties: Used for surface coatings, laminating, casting and potting of electrical equipment.

125	5,000	1	175	9.7	400

Key Properties: Tends to crystallize due to high purity, but can be reliquefied by heating. For laminating and filament winding.

126	8,000	2	182	9.7	350

Key Properties: For products requiring ultimate physical strength, good chemical resistance and electrical properties.

128	13,000	3	189	9.7	450

Key Properties: Laminating, casting and coatings.

129	5,000	3	198	9.6	450

Key Properties: Precatalyzed resin for manufacture of solid epoxy resins and epoxy resin esters.

Product Name	Viscosity (Brookfield) (cp)	Color (Gardner) (max)	Weight per Epoxide	Weight per Gallon (lb)	Flash Point (O.C.) (°F)
130	19,750	9	194	9.7	480

Key Properties: Laminating, casting, surface coatings, adhesives and potting of electrical equipment.

| 134 | 690 | 5 | 255 | 9.7 | 450 |

Key Properties: Laminating, casting, surface coatings, adhesives and potting of electrical equipment. Color and viscosity measured on 70% by weight solution in Butyl Dioxitol.

| 171 | 6,500 | 12 | 430 | 8.2 | 150 |

Key Properties: Improves impact strength, flexibility and toughness without reducing other properties below useful limits.

Vanoxy Semi-Solid Epoxy Resins

Product Name	Viscosity (Brookfield) (cp) or (Gardner- Holdt)	Color (Gardner) (max)	Weight per Epoxide	Weight per Gallon (lb)	Flash Point (O.C.) (°F)
136	A₁-B (G-H)	5	313	9.9	450

Key Properties: Laminating, casting, surface coatings, adhesives and potting of electrical equipment. Color and viscosity measured in a 40% by weight solution in Butyl Dioxitol.

| 172 | 2,650 | 10 | 700 | 9.0 | 200 |

Key Properties: For flexible surface coatings. The color and viscosity is measured on a 75% by weight solution in xylene.

Vanoxy Solid Epoxy Resins

Product	Viscosity (Brookfield)	Color (Gardner)	Weight per Epoxide	Weight per Gallon	Viscosity (Gardner- Holdt)
201	135	4	500	9.9	D-G

Key Properties: Recommended for use in amine-cured surface coating systems and dry lay-up laminations.

| 202 | 235 | 4 | 650 | 9.9 | G-K |

Key Properties: Recommended for use in amine-cured surface coating systems and dry lay-up laminations.

| 204 | 530 | 4 | 913 | 9.6 | P-U |

Key Properties: Recommended for use in amine-cured surface coating systems, dry lay-up laminations and in preparing ester-type surface coating vehicles for floor varnishes, metal finishes, appliance primers, etc.

| 207 | 2,300 | 5 | 2,250 | 9.6 | Y-Z |

Key Properties: Recommended for use in bake-type surface coating applications, appliance primers and tank, drum and can linings.

| 209 | 6,900 | 5 | 3,250 | 9.9 | Z₂-Z₅ |

Key Properties: Suitable for use in bake-type surface coating applications for tank, drum and can linings and wire coatings.

Note: Color and viscosity of all grades are measured on a 40% by weight solution in Butyl Dioxitol.

Section XXI
Resins–Ester Gums

CROSBY CHEMICALS, INC., 600 Whitney Bldg., New Orleans, LA 70130

Crosby Ester Gum Resins

Chemical Type: A full line of Ester Gum resins beginning with tall oil rosin, Resin 721, Polros A or Crosdim as the basic raw materials and modifying them generally with glycerine or pentaerythritol.

Physical Properties:

	Acid Value	Softening Point (Ring & Ball)(°C)	Viscosity* (Gardner)
Ester Gum	7	85	A
Ester Gum HV	7	93	D
Ester Gum HE	8	78	B
PE Ester Gum	14	102	G
Ester Gum CP	16	165	U
Ester Gum 721	8	73	A
PE Ester Gum 721	14	98	F

*60% in mineral spirits.

Key Properties: Ester Gums made with Resin 721, generally denoted by the number 721 in the name, have a somewhat lower melt and viscosity than those made with tall oil rosin.

Conversely, ester gums made with Polros A generally are higher in melt and viscosity than the standard ester gums. The ester gums made from Polros A are generally designated by an "H" in the name.

Ester Gum CP is the only ester gum made from Crosdim.

Ester Gum CP is known for its high melt and viscosity.

These resins have a wide application in the adhesive industry with some resins being used in the coating and ink fields.

FRP CO., P.O. Box 349, Baxley, GA 31513

Ester Gums

Product Name	Softening Point (Ring & Ball) (°C)	Acid Value (max)	Viscosity (Gardner- Holdt)*	Color (Gardner)* (max)	Weight per Gallon (lb)	Flash Point (O.C.) (°F)
HMP	95	8	D-G	10	9.05	405

Chemical Type: Glycerol ester of gum rosin.

133

Key Properties: Offers good viscosity and color retention. Improves brushability, gloss and resistance to sagging in quality interior coat formulations. Typical uses are in: interior coatings, varnishes, phenolic modifier, lacquers and hot melt adhesives.

Product Name	Softening Point (Ring & Ball) (°C)	Acid Value (max)	Viscosity (Gardner-Holdt)*	Color (Gardner)* (max)	Weight per Gallon (lb)	Flash Point (O.C.) (°F)
HMP 2010	92	8	E-H	10	9.05	490

Chemical Type: A premium lacquer grade ester gum.

Key Properties: Especially designed for good resistance to film hazing at high resin-nitrocellulose ratios. Exhibits high alcohol tolerance and lower acid value than standard grade ester gums. Typical uses are in: interior lacquers, exterior lacquers, metal coatings and polishing type lacquers.

Filtrez 13	101	12.5	E-H	10	9.1	505

Chemical Type: Pentaerythritol ester of rosin widely soluble in drying oils, especially raw, cooked and blown.

Key Properties: Compatible with hydrocarbon waxes, ester waxes and chlorinated rubber. Heat-stable, resistant to oxidation and exhibits a high refractive index, desirable for use in gloss finishes. Typical uses are in adhesives, antifouling paints, emulsion paints, oleoresinous varnishes and wax compositions.

Filtrez 15	110	12.5	H-K	10	9.1	440

Chemical Type: Pentaerythritol ester gum offering a high melting point, viscosity and low acid value.

Key Properties: It improves brushability, nonsagging and resistance to water spotting in interior enamel and gloss paint formulations. Typical uses are in: interior enamels, gloss paints, varnishes, phenolic modifier, hot melt adhesives and waxes.

*60% solution in mineral spirits.

REICHHOLD CHEMICALS, INC., RCI Bldg., White Plains, NY 10603

Synthe-Copal and Pentacite Ester Gums

Product Name	Softening Point (Ring & Ball) (°C)	Color (USDA) (max)	Acid Number (max)	Specific Gravity	Weight per Gallon (lb)
43-201	100	N	12	–	–
43-204	90	M	10	1.08	9.00
43-218	78	WW	10	–	–
43-221	103	N	15	–	–
43-245	89	N	12	1.08	9.00
43-406	105	M	17	1.07	8.90

Key Properties: Ester Gums are widely used in general purpose varnishes, nitrocellulose lacquer sealers, and as hardeners for alkyds when incorporated in solvent cuts. Ester gums are also essential in chewing gums and in some rubber adhesives. They contribute to emulsion stability and the tack properties of rubber emulsion adhesives, and are of value in solvent-type as well as vinyl-acetate emulsion adhesives.

Section XXII
Resins–Ethylene/Vinyl Acetate

ALLIED CHEMICAL, FIBERS AND PLASTICS CO., P.O. Box 2332R, Morristown, NJ 07960

A-C Vinyl Acetate-Ethylene Copolymers

Product Name	Softening Point (Ring & Ball) (°C)	Needle Penetration	Specific Gravity	Brookfield Viscosity @ 140°C (cp)	Vinyl Acetate (%)
400 & 400A	95	9.5	0.92	550	14
402 & 402A	102	7.0	0.91	275	2
403 & 403A	106	4.0	0.92	400	2
405	96	8.5	0.91	550	11
430	60	80	0.93	600	26
440	Viscous	Liquid	0.94	350	40

DU PONT CO., Wilmington, DE 19898

Elvax 40 and 150 Ethylene/Vinyl Acetate Copolymer Resins

Properties:

	Elvax 40	Elvax 150
Viscosity @ 30°C (0.25 g/100 ml toluene)	0.70	0.78
Melt Index (ASTM D1238)	57	43
Softening Point (Ring & Ball) (ASTM E28) (°F)	220	230
Tensile Strength (psi)	650	850
Elongation at Break (%)	1,450	1,050
Elastic (Tensile) Modulus (psi)	300	700
Hardness, Shore A-2 Durometer (ASTM D2240)	40	65
Vinyl Acetate Content (% by weight)	40	33
Ethylene Content (% by weight)	60	67
Residual Vinyl Acetate Monomer (% maximum)	0.3	0.3
Odor	Slight	Slight
Antioxidant (ppm BHT)	750	750
Specific Gravity	0.965	0.957
Bulk Density (lb/ft^3)	30	33
Refractive Index (n_D^{25})	1.476	1.482

Chemical Type: Copolymers of ethylene and vinyl acetate, containing more than 30% vinyl acetate.

Key Properties: Adhesives: Elvax 40 and 150 resins are suggested for use in—

Laminating adhesives for a variety of substrates, including plastics and films. These grades provide the extended open time often needed in industrial assembly.

Hot melt and solvent-based pressure-sensitive adhesives, particularly tape and label adhesives.

Elvax 40 is suggested as a minor component in edge banding and bookbinding adhesives based on other Elvax resins.

Coatings: Elvax 40 and 150 are more soluble in common lacquer solvents than other grades of Elvax because of their lower crystallinity and increased polarity. When solubility is an over-riding factor, Elvax 40 is preferred.

Elvax 40 and 150 are suggested for use in—

Lacquers for wood, metal, metal foils, papers and films.

Fabric and decal coatings.

Heat-seal coatings for backing card stock used in blister packaging.

Wood primers and sealers. Combinations of Elvax 40 and nitrocellulose form the basis for wood sealers offering toughness and good intercoat adhesion.

In formulations containing nitrocellulose, Elvax 40 imparts permanent flexibility and improves impact resistance and adhesion. This suggests the effectiveness of Elvax 40 as a permanent plasticizer for other hard, brittle resins.

Elvax 40 and 150 resins are also suggested for use as modifiers in—

Elastomer-based hot melt and hot melt pressure sensitive adhesive formulations.

Natural and synthetic elastomers to improve resistance to ozone cracking.

Caulks and sealants, including those based on asphalt.

Asphalt road repairing compositions.

Traffic paint formulations. Such formulations have superior adhesion to asphalt, abrasion resistance, and flexibility compared with alkyd and chlorinated rubber-based formulations.

Solvent-based and water-based inks.

Elvax 200-Series Resins

Product Name	Inherent Viscosity @ 30°C*	Melt Index (ASTM D1238)	Softening Point (Ring & Ball) (°F)	Tensile Strength (psi)	Percent Elongation at Break	Elastic (Tensile) Modulus (psi)	Hardness Shore A-2 Durometer (ASTM D1706)
210	0.54	400	180	300	700	600	62
220	0.64	150	190	550	850	850	69
240	0.78	43	230	1,050	900	1,110	73
250	0.83	25	260	1,400	950	1,300	75
260	0.94	6	310	2,800	1,000	1,600	80
265	1.01	3	340	3,800	1,000	2,000	83

*0.25 g/100 ml toluene.

Properties Identical for All Grades of the 200 Series:

Vinyl Acetate Content	28% by weight
Ethylene Content	72% by weight
Resdiual Vinyl Acetate Monomer	0.3% maximum
Odor	Slight
Antioxidant (ppm BHT)	750
Specific Gravity @ 73°F	0.951
Elvax 260 & 265	0.955
Bulk Density (lb/ft³)	33
Elvax 260	34
Elvax 265	35
Refractive Index (n_D^{25})	1.485
Elvax 210	1.488

Chemical Type: The Elvax 200-Series are copolymers of ethylene and vinyl acetate, containing 28% vinyl acetate. They range in a melt index of 3.0 to 400.

Key Properties: Recommended for use primarily in hot melt coatings and adhesives. They have excellent compatibility with paraffin wax and other modifying resins and impart improved toughness, flexibility and adhesion to coating and adhesive formulations. In blends with waxes and other modifiers, the 200-series resins are the most broadly useful of all the "Elvax" resins.

Hot Melt Adhesives: High performance, low cost packaging and converting adhesives for porous and nonporous surfaces can be formulated using this series. These adhesives are suitable for—

Corrugated case sealing and tray forming.

Carton assembly and closures.

Manufacture of multiwall bags and weather-resistant fiber drums.

This series is particularly useful in adhesives for frozen food packaging where the exceptional low temperature performance contributed by these resins insures package integrity at freezer temperatures.

Good adhesion to a variety of substrates makes adhesives based on this series useful in other applications, such as—

Bookbinding, collating and industrial assembly.

Kraft laminations, such as reinforced tape and industrial wrap.

Adhesives for high performance carpet tape.

Bonding nonwoven structures.

Pressure sensitive or heat activated labels.

Excellent toughness and adhesion is obtained in adhesives formulated with Elvax 200-series resins for—

Edge and surface veneering for particle board and plywood case stock.

Light duty wood bonding.

Plastic-to-plastic assemblies.

Hot Melt Coatings: This series is suggested for use in modifying waxes and wax-based hot melts for—

Curtain-coated, "rigid-when-wet" corrugated containers used for shipment of iced poultry and produce.

Coating corrugated board for point-of-purchase display stand set-ups.

Heat sealable coatings for frozen food cartons where good adhesion and low temperature seal strength and flexibility are desired.

Heat sealable coatings for flexible packaging materials, such as carton overwraps and soap wraps.

Strippable coatings to protect items such as tools, chrome automobile bumpers and trim from corrosion and abrasion during shipment and storage.

Low viscosity coatings for spray application to preformed items such as cups and tubes.

Elvax 300-Series Resins

Product Name	Inherent Viscosity @ 30°C*	Melt Index (ASTM D1238)	Softening Point (Ring & Ball) (°F)	Tensile Strength (psi)	Percent Elongation at Break	Elastic (Tensile) Modulus (psi)	Hardness Shore A-2 Durometer (ASTM D1706)
310	0.54	400	190	400	700	1,100	70
350	0.84	19	270	1,600	900	1,700	80
360	1.05	2	370	3,500	1,000	2,600	85

*0.25 g/100 ml toluene.

Properties Identical for All Grades of the 300 Series:

Vinyl Acetate Content	25% by weight
Ethylene Content	75% by weight
Residual Vinyl Acetate Monomer	0.3% maximum
Odor	Slight

Antioxidant (ppm BHT)	750
Specific Gravity	0.948
Bulk Density (lb/ft³)	35
Refractive Index (n_D²⁵)	
Elvax 310	1.486
Elvax 350	1.489
Elvax 360	1.491

Key Properties: Elvax 350—General-purpose resin designed for use with microcrystalline waxes to provide good toughness and flexibility at moderate melt viscosity.

Elvax 360—Preferred where maximum toughness and seal strength are desired in blends with microwaxes. Its high molecular weight imparts corresponding high viscosity in hot melt systems.

Elvax 310—Suggested for use in adhesives and coatings where low melt viscosity is essential. Gives less flexible coatings than Elvax 350 or 360 but imparts superior gloss retention.

Combinations of Elvax 310 with Elvax 350 or 360 may be used to obtain intermediate properties.

Chemical Type: Wax-compatible ethylene-vinyl acetate copolymers.

Key Properties (all grades): Developed specifically to enhance the properties of hot melt blends containing microcrystalline waxes. The series shows excellent compatibility with paraffin waxes and with a wide variety of modifying resins and rosins. The slightly lower vinyl acetate content of this series, compared to the 200 series, optimizes compatibility with microcrystalline waxes while maintaining excellent adhesion and barrier properties in blends with these and other petroleum waxes.

Suggested for use in heat-sealable barrier coatings and hot melt adhesives where optimum compatibility with microcrystalline waxes is a major consideration.

Blends of this series with waxes and modifying resins can be formulated to give fiber-or film-tearing bonds between paper substrates and such nonporous packaging materials as aluminum foil, polyethylene and polypropylene films, K cellophane and Mylar polyester films.

Elvax 400-Series Resins

Product Name	Inherent Viscosity @ 30°C*	Melt Index (ASTM D1238)	Softening Point (Ring & Ball) (°F)	Tensile Strength (psi)	Percent Elongation at Break	Elastic (Tensile) Modulus (psi)	Hardness Shore A-2 Durometer (ASTM D1706)
410	0.52	500	190	550	650	2,100	80
420	0.61	150	210	950	700	2,700	84
460	1.02	2.5	390	2,800	850	3,500	90

*0.25 g/100 ml toluene.

Properties Identical for All Grades of the 400 Series:

Vinyl Acetate Content	18% by weight
Ethylene Content	82% by weight
Residual Vinyl Acetate Monomer	0.3% maximum
Odor	Slight
Antioxidant (ppm BHT)	750
Specific Gravity	
Elvax 410	0.934
Elvax 420	0.937
Elvax 460	0.941
Bulk Density (lb/ft³)	36
Elvax 410	35
Refractive Index (n_D²⁵)	
Elvax 410	1.484
Elvax 420	1.492
Elvax 460	1.493

Chemical Type: Ethylene-vinyl acetate copolymers.

Key Properties: Resins in this series are outstanding in their ability to impart high gloss, good gloss retention, scuff resistance, hardness and resistance to blocking in hot melt coatings. When superior flexibility and a low creased water vapor transmission are not required, the "400" series may be more economical than the "200" and "300" series resins.

Elvax 400 series resins are compatible with petroleum waxes and a wide range of tackifying resins. In hot melt adhesives for bonding cellulosic substrates, a good balance of heat resistance and overall adhesive performance can be achieved using formulations based on this series.

Hot Melt Coatings: Suggested for use as wax additives and in hot melts, such as—

High gloss decorative coatings, and scuff-resistant, water-resistant, functional coatings for corrugated board.

Grease-resistant coatings for bakery board.

High-gloss, heat-sealable coatings for folding cartons. This series may be combined with "200" series resins for their greater seal strength while taking advantage of the hardness and gloss retention imparted by the 400 series.

Coatings for rigid containers such as food tubs (nested containers) and composite cans provided the structure is formed before coating.

Low-viscosity, high-gloss coatings for overwraps. Elvax 410 and 420 contribute superior flexibility and adhesion compared with low molecular weight polyethylene.

Hot Melt Adhesives: In hot melt adhesives for sealing porous substrates, performance requirements can often be met at minimum cost using the Elvax 400-series resins.

For example, adhesives based on 400-series resins are suggested for—

Low cost, heat-sealable labels with good seal strength to many packaging materials.

Packaging closures for multiwall bags.

Forming and sealing folding cartons.

Corrugated case sealing and tray forming.

Kraft laminations such as reinforced tape and industrial wrap. Elvax 400 series resins should also be considered in formulating high-viscosity, industrial assembly adhesives for such applications as—

Shoe manufacturing.

Attaching edge bands to particle board and plywood case stock.

Bonding molded plastic articles.

Ethylene-Vinyl Acetate Copolymer Resins

Product Name	Melt Index (ASTM D1238)	Stiffness (ASTM D747) (psi)	Vicat Softening Point (ASTM D1525) (°F)	Tensile Strength (ASTM D638) (psi)	Elongation (ASTM D638) (%)	Hardness Shore A (ASTM D1706)	Compression Set (ASTM D395) (10 Days @ 25°C)
EVA 3120	1.2	14,000	174	2,530	670	97	21

Chemical Type: Contains 7.5% vinyl acetate and 92.5% ethylene.

| EVA 3125 | 0.8 | 10,300 | 174 | 2,640 | 690 | 95 | 26 |

Chemical Type: Contains 9.5% vinyl acetate and 90.5% ethylene.

| EVA 3130 | 2.5 | 8,650 | 166 | 2,650 | 730 | 95 | 52 |

Chemical Type: Contains 12.0% vinyl acetate and 88.0% ethylene.

| EVA 3137 | 0.3 | 8,880 | 176 | 3,250 | 690 | 94 | 57 |

Chemical Type: Contains 12.0% vinyl acetate and 88.0% ethylene.

Product Name	Melt Index (ASTM D1238)	Stiffness (ASTM D747) (psi)	Vicat Softening Point (ASTM D1525) (°F)	Tensile Strength (ASTM D638) (psi)	Elonga-tion (ASTM D638) (%)	Hardness Shore A (ASTM D1706)	Compression Set (ASTM D395) (10 Days @ 25°C)
EVA 3150	2.5	6,780	156	2,620	740	93	26

Chemical Type: Contains 15.0% vinyl acetate and 85.0% ethylene.

| EVA 3154 | 8.0 | 6,200 | 145 | 1,890 | 680 | 93 | 67 |

Chemical Type: Contains 15.0% vinyl acetate and 85.0% ethylene.

| EVA 3165 | 0.7 | 5,450 | 150 | 3,360 | 710 | 92 | 37 |

Chemical Type: Contains 18.0% vinyl acetate and 82.0% ethylene.

| EVA 3168 | 150 | – | – | – | – | – | – |

Chemical Type: Contains 18.0% vinyl acetate and 82.0% ethylene.

| EVA 3170 | 2.5 | 5,530 | 141 | 2,610 | 750 | 91 | 42 |

Chemical Type: Contains 18.0% vinyl acetate and 82.0% ethylene.

| EVA 3172 | 8.0 | 4,550 | 141 | 1,880 | 750 | 92 | 57 |

Chemical Type: Contains 18.0% vinyl acetate and 82.0% ethylene.

| EVA 3175 | 6.0 | 1,780 | 114 | 2,020 | 680 | 85 | 55 |

Chemical Type: Contains 28.0% vinyl acetate and 72.0% ethylene.

| EVA 3180 | 25.0 | 1,320 | 107 | 1,170 | 820 | 82 | 61 |

Chemical Type: Contains 28.0% vinyl acetate and 72.0% ethylene.

| EVA 3184 | 100.0 | – | – | – | – | – | – |

Chemical Type: Contains 28.0% vinyl acetate and 72.0% ethylene.

| EVA 3185 | 43.0 | 890 | 100 | 770 | 960 | 77 | 60 |

Chemical Type: Contains 33.0% vinyl acetate and 67.0% ethylene.

| EVA 3190 | 2.0 | 3,000 | 134 | 2,920 | 660 | 89 | 59 |

Chemical Type: Contains 25.0% vinyl acetate and 75.0% ethylene.

| EVA 3194 | 19.0 | 2,900 | 130 | 1,100 | 750 | 79 | – |

Chemical Type: Contains 25.0% vinyl acetate and 75.0% ethylene.

Chemical Type: DuPont EVA copolymers are members of the polyolefin family and are derived from the basic polyethylene structure. The addition of vinyl acetate to the ethylene chain reduces the polymer's crystallinity, resulting in a thermoplastic with elastomeric-like properties.

Key Properties: EVA copolymers show excellent flexibility and toughness at low temperatures. They are superior to plasticized vinyls and most rubbers in this respect. The higher the vinyl acetate content in the copolymer, the lower the melting point and stiffness. Another outstanding property of EVA copolymers is resistance to environmental stress cracking under continued flexing. In chemical resistance, EVA copolymers appear to be comparable with low density polyethylene. There is, however, a greater tendency for the EVA copolymers to absorb oils or solvents. Also, the EVA copolymers are very susceptible to chlorinated solvents and to aromatic hydrocarbons. EVA copolymers have limited ultraviolet (UV) resistance. However, with the addition of a UV stabilizer to the resin, this property can be improved.

Though classified as thermoplastics, EVA copolymers have many properties similar to elastomers which make them likely candidates for many rubber and plasticized PVA applications.

EVA copolymers are used in a number of applications requiring moderate elastomeric properties. The polar nature of the EVA copolymer makes it an ideal material to be combined with other resins, pigments and fillers. Small quantities are often added to other plastic resins to enhance their toughness and processing characteristics. They are approved for use in contact with food under provisions of Paragraph 121.2570 Code of Federal Regulations.

EVA copolymers with 9.5–12% vinyl acetate have approximately one-half the stiffness of low density polyethylene, but are superior in stress-crack resistance, low-temperature toughness and heat-seal characteristics.

U.S. INDUSTRIAL CHEMICALS CO., 99 Park Ave., New York, NY 10016

Ultrathene Ethylene-Vinyl Acetate Copolymers

Product Name	Vinyl Acetate (%)	Melt Index (ASTM D1238)	Specific Gravity	Tensile at Break (ASTM D638) (psi)	Hardness (Shore A)	Softening Point (Ring & Ball) (°F)
UE 637-04	9	3	0.93	2,000	89	325
UE 631-04	19	2.5	0.94	2,150	87	353
UE 612-04	19	150	0.94	760	85	217
UE 649-04	19	530	0.93	450	78	197
UE 659-04	25	2	0.95	>5,300	83	290
UE 645-04	28	3	0.95	2,760	76	342
UE 634-04	28	6	0.95	2,200	78	298
UE 666-04	28	21	0.95	1,170	74	232
UE 646-04	28	25	0.95	1,100	78	246
UE 636-04	28	43	0.95	1,100	76	231
UE 639-35	28	150	0.95	400	69	210
UE 653-35	28	388	0.94	330	68	188
UE 638-35	31	24	0.95	720	65	233
UE 654-35	33	48	0.95	660	73	232

Chemical Type: Available in grades containing from 9 to 33% by weight in vinyl acetate content. They are divided into three classifications: low vinyl acetate content, up to 19%; intermediate vinyl acetate content, 28–30%; high vinyl acetate content, 30–33%.

Key Properties: All three types can be used for both hot melt adhesives and coatings. In adhesive applications, where an Ultrathene EVA copolymer is the key ingredient, you can tailor formulations with excellent tackiness, adhesiveness, flexibility and stability. Also, these copolymers offer great latitude in formulating high-performance hot melt, solvent-based and pressure-sensitive adhesives. Grades with high vinyl acetate content are recommended for adhesives where strong bonding strength is required.

In hot melt coatings, adding Ultrathene EVA copolymers and various modifiers upgrades the quality of many coatings by enhancing the gloss, clarity, flexibility and strength properties of the coating formulations. These resins generally have up to 28% vinyl acetate content and are compatible with microcrystalline waxes.

In wax-based coatings, an increase in vinyl acetate content improves the following: heat seal strength; creased barrier properties; flexibility; low temperature performance; hot tack.

It reduces the following: wax compatibility; resistance to blocking; gloss retention.

In adhesives, an increase in vinyl acetate content improves the following: adhesion; compatibility; solubility; flexibility; low temperature performance; extends open time.

It reduces the following: temperature resistance.

In general, as vinyl acetate content is increased, copolymers become more elastomeric, exhibit greater tackiness and adhesive properties and are more soluble in organic solvents. They retain a wide compatibility with elastomers, hydrocarbon resins, microwaxes and synthetic resins.

Vynathene Vinyl Acetate-Ethylene Copolymers

Product Name	Vinyl Acetate (%)	Melt Index (ASTM D1238)	Specific Gravity	Tensile at Break (ASTM D638) (psi)	Hardness (Shore A)	Softening Point (Ring & Ball) (°F)
EY 901-25	40	7.5	0.962	1,200	63	295
EY 902-35	40	70	0.962	800	56	232
EY 903-25	45	7.5	0.984	22	35	292
EY 905-25	51	18	0.986	15	32	252
EY 906-25	55	55	1.00	5.5	14	265
EY 907-25	60	3.5	1.02	6.1	14	316

Chemical Type: New and unique series of high vinyl acetate content ranging from 40 to 60% by weight in vinyl acetate content.

Key Properties: They offer the formulator properties heretofore unavailable in a base polymer, especially in tack and weathering characteristics.

Vynathene VAE copolymers are compatible with many modifiers, tackifiers, and both natural and synthetic polymers used in adhesive formulations. Adhesives with good low temperature properties, resiliency and elasticity can be prepared using these resins. Aging characterisitcs are excellent because the basic polymer structure has good resistance to ozone, ultraviolet radiation, and weathering.

Vynathene copolymers can be used alone or in blends with other polymers to produce adhesives for product assembly, packaging, permanent and removable labels and tapes. When formulated into pressure-sensitive systems, they can eliminate environmental and safety problems of solvent evaporation, which can be a problem for user industries.

Sealants, caulks and mastics are other applications for Vynathene resins. The ability of these high vinyl acetate copolymers to accept high filler loadings, while still retaining their tack and flow characteristics, make them unique for these end uses.

In adhesives, a decrease in melt index, improves the following: cohesive strength; increases melt viscosity; flexibility; low temperature performance; heat resistance; hot tack.

It reduces the following: open time.

In adhesives, an increase in vinyl acetate content improves the following: adhesion; compatibility; solubility; flexibility; low temperature performance; extends open time.

It reduces the following: temperature resistance.

In general, as vinyl acetate content is increased, copolymers become more elastomeric, exhibit greater tackiness and adhesive properties and are more soluble in organic solvents. They retain a wide compatibility with elastomers, hydrocarbon resins, microwaxes and synthetic resins.

Section XXIII
Resins–Maleic

FRP, CO., P.O. Box 349, Baxley, GA 31513

Filtrez Maleic Resins

Product Name	Softening Point (Ring & Ball) (°C)	Acid Value (max)	Gardner-Holdt Viscosity*	Gardner Color (max)	Weight per Gallon (lb)	Flash Point (O.C.) (°F)
329	135	30	Q–T	10	9.45	575

Chemical Type: A maleic-modified glycerol ester.

Key Properties: High melting point and exceptionally high alcohol tolerance. Exhibits very good lacquer compatibility and solvent release. Typical uses are: lacquers, varnishes, enamels, hot melt adhesives, furniture finishes and gravure printing inks.

330	140	30	R–V	10	9.45	575

Chemical Type: A maleic-modified glycerol ester.

Key Properties: High melting point, good alcohol tolerance and excellent lacquer compatibility. Lacquers prepared with this material have excellent color retention and adhesion. Typical uses are: lacquers, varnishes, lacquer primers and sealers, tin and furniture finishes, hot melt coatings, enamels and gravure printing inks.

330A	135	39	J–M	10	9.45	575

Chemical Type: A maleic-modified glycerol ester.

Key Properties: High melting point, good alcohol tolerance and excellent lacquer compatibility. Has fast solvent release and excellent color stability. Typical uses are: lacquers, lacquer primers and sealers, varnishes, tin and furniture finishes, baking enamels, hot melt coatings and printing inks.

338	140	30	S–W	10	9.45	575

Chemical Type: A maleic-modified glycerol ester.

Key Properties: High melting point and good alcohol tolerance. Lacquers incorporating this resin have high gloss, hardness and good adhesion. Typical uses are: lacquers, varnishes, lacquer primers and sealers, tin and furniture finishes, hot melt coatings, enamels and gravure inks.

339	135	39	K–O	10	9.45	575

Chemical Type: A maleic-modified glycerol ester.

Key Properties: High melting point and very good alcohol tolerance. Exhibits excellent lacquer compatibility and solvent release. Typical uses are: lacquers, lacquer primers and sealers, varnishes, tin and furniture finishes, baking enamels, hot melt coatings and gravure inks.

345	135	20	W–Z	11	9.4	470

Chemical Type: A maleic-modified penta ester.

Key Properties: Wide solubility range. Typical uses are: heat-set printing inks, varnishes, enamels, machinery finishes, paints and metal primers.

Product Name	Softening Point (Ring & Ball) (°C)	Acid Value (max)	Gardner-Holdt Viscosity*	Gardner Color (max)	Weight per Gallon (lb)	Flash Point (O.C.) (°F)
360	150	19	Z–Z₃	10	9.45	—

Chemical Type: Maleic-pentaerythritol rosin ester with very high melting point and low acid value.

Key Properties: It exhibits light color and excellent color retention. Typical uses are: heat-set inks, gloss letterpress inks and overprint varnishes.

| 379 | 122 | 17 | F–J | 10 | 9.2 | 490 |

Chemical Type: Maleic-modified pentaerythritol ester.

Key Properties: More easily dissolved in all types of drying oils and thinners than the other maleic resins. Resists after-yellowing and after-tack in pale varnishes and white enamels made with tung, oiticica and dehydrated castor oils. Typical uses are: varnishes, enamels, textile print coatings, hardware finishes, hot melt coatings and machinery finishes.

| 3790 | 122 | 16 | D–G | 11 | 9.3 | 490 |

Chemical Type: Maleic-modified penta ester with properties similar to Filtrez 379, but with much greater solubility in low KB solvents.

Key Properties: Typical uses are: varnishes, enamels, textile print coatings, hardware finishes, hot melt coatings, machinery finishes and heat-set printing inks.

| 380 | 145 | 19 | X–Z | 10 | 9.4 | — |

Chemical Type: Maleic-pentaerythritol rosin ester with very high melting point and low acid value.

Key Properties: Exhibits light color and excellent color retention. Typical uses are: heat-set inks, gloss letterpress inks and overprint varnishes.

| 3000 | 130 | 39 | F–K | 10 | — | — |

Chemical Type: Maleic-glycerol rosin ester.

Key Properties: Typical uses are: gravure inks, lacquers and coatings.

| 3007 | 130 | 39 | T–W | 10 | 9.7 | — |

Chemical Type: Modified maleic resin.

Key Properties: Has improved gloss, adhesion, pigment wetting and durability over conventional resins of this type. Compatible with RS nitrocellulose. Typical uses are: C&T type printing inks, lacquers and coatings.

*60% solution in xylene.

Section XXIV

Resins–Miscellaneous

ALLIED CHEMICAL, Fibers and Plastics Co., P.O. Box 2332R, Morristown, NJ 07960

A-C Acrylic Acid-Ethylene Copolymers

Product Name	Softening Point (Ring & Ball) (°C)	Needle Penetration	Specific Gravity	Brookfield Viscosity @ 140°C (cp)	Acid Number (mg KOH/g)
540 + 540A	108	2.0	0.93	500	40
580	102	4.0	0.93	650	75
5120	92	11.5	0.93	650	120

Key Properties: All of the above are emulsifiable.

AMOCO CHEMICALS CORP., 200 East Randolph Drive, Chicago, IL 60601

Amide-Imide Polymers

AI-10* and AI-11 are the two basic grades available.

Product Name	Brookfield Viscosity @ 25°C (cp)	Solids Content (%)	Weight per Gallon (lb)
AI-830	3,250	28	9.2

Chemical Type: Base polymer of AI-10 in 72% N-methyl-2-pyrrolidone.

AI-1130L	1,400	31.5	9.4

Chemical Type: Base polymer of AI-11 in 72% N-methyl-2-pyrrolidone.

AI-Lite	2,500	32.5	9.4

Chemical Type: Proprietary base polymer in 68% N-methyl-2-pyrrolidone.

*AI-10 is also available as a dry powder.

Chemical Type: A family of condensation polymers containing aromatic amide and aromatic imide linkages. AI-Lite is based on a special light resin. All are amber liquids.

Key Properties: Offer an unusual combination of good high temperature, excellent toughness, and good resistance properties. Moderate cost and outstanding performance are the prime reasons for the popularity in high temperature applications. They are designed for continuous service in environments of 250° to 290°C. AI-11 will withstand higher temperatures.

They have superior strength and toughness when compared with other thermally stable polymers such as polyimides and polybenzimidazoles. Have exceptionally useful properties of high flexural strength, hardness, low coefficient of friction and good adhesion. These properties are advantageous when making durable, high performance laminates, adhesives and coatings. Virtually unaffected by most solvents and chemicals when properly cured. Excellent electrical properties. Al-Lite is available for applications requiring light colored enamels with good resistance to yellowing during high heat exposure.

Amoco Resin 18

Product Name	Softening Point (°F)	Gardner-Holdt Viscosity*	Color (TT-O-141B)	Volatile Content (% by wt)	Molecular Weight	Cloud Point (Powers) (°F)
18-210	210	G	2	2.5	685	170
18-240	245	Q	2	2.5	790	212
18-290	290	X	2	2.5	960	251

*60% by wt in toluene.

Physical Properties of All Grades:

Iodine number (Wijs)	nil
Acid number	nil
Saponification number	nil
Appearance	bright and clear
Ash content (ASTM D555)	0.001%
Specific gravity @ 60°/60°F	1.075
Refractive Index @ 20°C	1.61
Volatiles	2.5% max

Chemical Type of All Grades: Poly-α-methylstyrene resin.

Key Properties of All Grades: The following are recommended end uses.

Thermoplastic elastomers used in caulks for hot melt application are reinforced, as the adhesion, tack and flexibility are improved.

Pressure-sensitive adhesives based on thermoplastic elastomers are strengthened and reinforced.

Improved flexibility and increased tack are given.

Can be used to manufacture cost competitive traffic paints which show excellent wear and water resistance. Less frequent repainting means savings of materials, energy and labor.

Glossy overprint varnishes based on Amoco Resin 18 and polystyrene will protect package and magazine cover stocks from smearing, skin oils, stains and abrasion.

BASF CORP., 491 Columbia Ave., Holland, MI 49423

Product Name	Specific Gravity @ 20°C	Saponification Value (max)	Acid Value (approx)
Plastigen G	1.06	10	0

Chemical Type: Plasticizing urea-formaldehyde resin.

Key Properties: Clear, pale yellowish, medium-viscosity resin, which is resistant to aging at temperatures below 100°C. At high temperatures (up to 150°C), it does not yellow. Soluble in practically all conventional solvents, including aliphatic hydrocarbons, but is insoluble in water. Compatible with many coatings raw materials and can be used together with a wide variety of binders. Very high light stability. Increases the flexibility, gloss, gloss retention and resistance to aging of coatings obtained from many binders. Can be used with chlorinated binders in formulations for alkali-resistant coatings, by virtue of its good resistance to hydrolysis. May also be mixed with other nonhydrolyzable plasticizers.

Plastigen G is also suitable for Laroflex MP transparent coatings on concrete and architectural paints. Tin-stamping enamels are formulated from Lutofan 200L and Lutofan 210L and Plastigen G.

CIBA-GEIGY RESINS, Saw Mill River Road, Ardsley, NY 10502

Aracast Hydantoin Resins

Product Name	Brookfield Viscosity (cp)	Color	Epoxy Value	Weight per Gallon (lb)	Flash Point (C.C.) (°C)
XU229	1,150	light	0.63	9.33	353

Chemical Type: Ethyl amyl hydantoin diepoxide, a low viscosity noncrystallizing liquid, recommended for high solids coatings.

XU231	—	white	0.69	—	93 min

Chemical Type: Pentamethylene hydantoin diepoxide, a grindable, nonsintering solid recommended for powder coatings.

XU238	1,400	pale yellow	0.76	—	110 min

Chemical Type: Methyl ethyl hydantoin diepoxide, a water-soluble, noncrystallizing liquid recommended for waterborne coatings.

XB2793	2,500	pale yellow	0.70	10.08	149

Chemical Type: Mostly dimethyl hydantoin diepoxide, a water-dispersible liquid recommended for waterborne coatings/blends with acrylics.

XB2826	5,000	pale yellow	0.62	9.85	113

Chemical Type: Blend of XB2793 and Araldite 6010, with low viscosity, aromatic epoxy character. Recommended for solvent-free coatings, with excellent chemical resistance and mechanical properties.

Chemical Type (of All of The Above): A new generation of epoxies based on the hydantoin ring. This structure is produced from readily available raw materials and a wide range of properties results in the uncured epoxy resins.

Key Properties (of All of The Above): Their advantages are — superior resistance to ultraviolet light, heat stability, outstanding adhesion qualities, exceptional weatherability, wide range of solubility, improved resistance to discoloration at elevated temperatures, good chemical resistance, excellent light stability and high range of compatibility with fillers.

Recommended applications are — waterborne coatings, high solids and solvent-free coatings and powder coatings.

CROWLEY CHEMICAL CO., 261 Madison Ave., New York, NY 10016

Hitac 300

Chemical Type: An amorphous polypropylene/ethylene copolymer.

Physical Properties:

Brookfield viscosity @ 375°F, cp	3,000–15,000
Softening point (Ring and Ball), °F	300
Needle penetration, mm/10	30
Specific gravity	0.86
Weight per gallon @ 375°F, lb	6.1
Glass transition temperature	
°C	–50 to –25
°F	–58 to –13

Key Properties:

Low glass transition temperature
Excellent surface tack
Good oil and grease resistance
Used in caulks and sealants

GAF CORP., Chemical Division, 140 West 51 St., New York, NY 10020

Kolima 35 Resin
Kolima 55 Resin
Kolima 75 Resin

Chemical Type: Vinylpyrrolidone/vinyl acetate copolymer resins in 50% isopropanol solutions.

Key Properties: Used in organic-solvent adhesive systems for general laminating. Modify the properties of other vinyl films. Upgrade inexpensive materials, such as resin esters, to adhesives. Improve adhesion to difficult-to-bond surfaces. Build-in high initial tack with maximum ultimate bond. Produce better gummed coatings for paper and contact adhesives.

THE GOODYEAR TIRE AND RUBBER CO., 1144 East Market St., Akron, OH 44316

Pliolite Resins

Product Name	Specific Gravity	Refractive Index	Softening Point (R&B)	Sward Rocker Hardness (24 hr)	Solvent Requirement (min KB)	Molecular Weight
S-5A	1.05	1.585	160	20	60	163
S-5B	1.05	1.585	155	16	60	132
S-5D	1.05	1.585	155	12	60	114
S-5E	1.05	1.585	135	9	60	71

Chemical Type (of all of the above): Styrene-butadiene.

Product Name	Specific Gravity	Refractive Index	Softening Point (R&B)	Sward Rocker Hardness (24 hr)	Solvent Requirement (min KB)	Molecular Weight
VT	1.026	1.57	160	20	36	152
VT-L	1.026	1.57	135	14	36	78

Chemical Type (of both of the above): Vinyl toluene-butadiene.

Product Name	Specific Gravity	Refractive Index	Softening Point (R&B)	Sward Rocker Hardness (24 hr)	Solvent Requirement (min KB)	Molecular Weight
AC	1.04	1.546	160	46	60	139
AC-L	1.04	1.546	135	40	60	73
HM-L	1.04	1.546	120	22	60	31
BAC	1.06	1.56	160	32	36	128
BAC-L	1.06	1.56	135	29	36	71
BHM-L	1.06	1.56	120	22	36	22

Chemical Type (of all of the above): Styrene-acrylate.

Product Name	Specific Gravity	Refractive Index	Softening Point (R&B)	Sward Rocker Hardness (24 hr)	Solvent Requirement (min KB)	Molecular Weight
VTAC	1.03	1.558	160	32	36	194
VTAC-L	1.03	1.558	135	24	36	83
OMS	1.03	1.527	135	30	26	85.5

Chemical Type (of all of the above): Vinyl toluene-acrylate.

Physical Properties (of all of the above Pliolite resins):

Form	white, friable granules
Solution	clear
Film	colorless
Acid Number	8 maximum

Key Properties (of all of the above Pliolite resins): There are many coating and adhesive systems now being applied from hot melt coating systems. The Pliolite acrylic resins have been found to be very useful as the main binder for coatings of this type (e.g., traffic striping) and as a modifier for adhesives due to their low melt viscosity and extremely good thermal stability.

These resins may also be used in applications such as concrete curing membranes and coatings, various paper coatings, and some conventional pigmented coatings.

NEVILLE CHEMICAL CO., Pittsburgh, PA 15225

Nevtac Resins

Chemical Analysis and Physical Properties:	Nevtac 80	Nevtac 100	Nevtac 115	Nevtac 130	Super Nevtac 99
Softening point (R&B), °C	81	102	114	131	99
Gardner color, 50% in toluene	5	6	7	8	7
Specific gravity @ 25°C	0.95	0.96	0.97	0.98	0.98
Gardner-Holdt viscosity	2.5	4.0	8.0	18.1	3.2
Bubble seconds @ 25°C, 70% toluol	H–I	P	U–V	X–Y	M
Bromine number (ASTM D1159)	26	24	24	22	30
Acid number (ASTM D974), max	0.5	0.5	0.5	0.5	0.5
Molecular weight	1,070	1,215	1,405	1,410	985
Ash content, % max	0.1	0.1	0.1	0.1	0.1
Melt viscosity, cp					
@ 150°C	610	2,270	9,300	30,000	1,650
@ 175°C	175	430	1,230	7,800	315
@ 200°C	65	130	290	1,190	100
Form	solid	flaked/ solid	flaked/ solid	flaked/ solid	flaked

Key Properties:

Nevtac 80   This light colored, low molecular weight aliphatic (or C_5) resin is especially suited for use as a tackifier for natural rubber and cis-polyisoprene. Its 80°C softening point (Ring and Ball) is of great advantage where a plasticizer, modifier, extender or processing aid is required.

It is also compatible with a wide range of aliphatic elastomers and polymers such as butyl rubber, waxes, polyethylene and polypropylene.

Its low molecular weight gives high tack and low solution viscosities, as well as low molten viscosities in hot melt adhesive systems.

Nevtac 100   The popular 100°C softening point and light color of this resin make it particularly useful in hot melt adhesives and coatings.

Its excellent compatibility characteristics with waxes, elastomers, polypropylenes, polyethylenes, ethylene-ethyl acrylates and ethylene-vinyl acetate polymers makes it very useful in a wide range of solvent based pressure sensitive and hot melt adhesives.

Formulators in search of a C_5 resin with exceptional heat and color stability will find that Nevtac 100 offers these properties plus oxidation resistance.

Nevtac 115   Combines many of the fine properties found in the other Nevtac grades, with a higher molecular weight for increased tack and modifying abilities. It has been found extremely useful in pressure sensitive and hot melt adhesive applications where higher viscosity and increased adhesive strength are important.

Nevtac 130   High molecular weight especially developed for formulas that require a higher softening point material than Nevtac 115. Nevtac 130 can offer this extra advantage while still maintaining the excellent compatibility, solubility, exceptional overall performance properties of the other Nevtac resins.

Super Nevtac 99   Latest addition to the Nevtac product line. This grade was primarily designed for use in formulating hot melt pressure sensitive adhesives based on styrene-isoprene block copolymers where high adhesion to paper is required as in the formulation of packaging tapes. Preliminary laboratory work also indicates that it should be very useful in hot melt systems based on high vinyl acetate content EVA resin.

FDA Status: Nevtac resins are cleared substances as defined in many FDA regulations.

PIONEER DIVISION, Witco Chemical Corp., 277 Park Ave., New York, NY 10017

Pioneer Petroleum Resins

	Pioneer 439	Pioneer 442	Pioneer 448
Physical Properties:			
Softening point (Ring and Ball), °F	310–315	305–315	330–340
Needle penetration	0–1	2–4	2–3
Color	black	black	black
Dielectric strength, volts	30,000	30,000	30,000
Solubility, % by wt			
Carbon tetrachloride	99.5	99.5	99.5
Carbon disulfide	99.5	99.5	99.5
Flash point (Cleveland Open Cup), °F	600+	600+	600+

Chemical Analysis (% by wt):

Asphaltenes	56–62
Resins	15–18
Oils	20–28

Typical Physical Properties and Chemical Analysis (All Grades):

Fracture of lump material	conchoidal
Color of mass	black
Hardness, Mohs scale	<4
Luster	bright
Specific gravity @ 60°F	1.11
Mineral matter–ash, % by wt	<0.2
Bitumen soluble in 86° naphtha, %	50–70
Bitumen soluble in ethyl ether, %	93+
Fixed carbon, Conradson method, %	~30
Sulfur, % by wt	trace–2
Saponifiable matter, %	<1
Behavior in petroleum solvents	soluble
Specific heat, Btu/lb	
@ 300°F	0.525
@ 500°F	0.610
Carbon, % by wt	82–85
Hydrogen, % by wt	14–15
Nitrogen, % by wt	0–2
Oxygen, % by wt	trace–3
Thermal coefficient of expansion/°F	0.00035
Thermal conductivity, Btu/hr/ft^2/°F/ft	0.033–0.050
Viscosity @ 450°F	
Saybolt Furol, sec	150–200
Brookfield, cp	300–400

Key Properties: These resins have been used for many years in compounding of rubber goods, brake linings, paints and varnishes, and as substitutes for naturally occurring asphalts such as gilsonite.

REICHHOLD CHEMICALS, INC., RCI Bldg., White Plains, NY 10603

Zinc Resinates

	Zinar	Zitro	Zirex	Zinros
Physical Properties and Chemical Analysis:				
USDA Standards Color	M	N	N	M
Gardner 1963 Color, solid resin*	14	14	14	14
Gardner 1963 Color, flaked resin*	14	14	14	14

(continued)

	Zinar	Zitro	Zirex	Zinros
Capillary tube melting point				
°C	160	132	132	150
°F	320	270	270	302
Softening point (Ring and Ball),				
°C	174	153	153	168
°F	345	307	307	334
Acid number	0**	15	0**	39
Metallic zinc combined, %	5.6	4.9	8.9	2.3
Metallic calcium combined, %	1.8	1.7	0.6	2.6
Ash, %	9.8	8.5	11.8	6.3
Specific gravity @ 25°/25°C	1.150	1.130	1.162	1.111
Flash point (Cleveland Open Cup),				
°F	464	500	487	475
°C	240	260	253	246
Fire point (Cleveland Open Cup),				
°F	482	540	536	525
°C	250	282	280	274
Metallic zinc content	5.6	4.9	8.9	2.3

*In 50% nonvolatile mineral spirits solutions.
**Basic.

Chemical Type (of All Grades): All of these solid zinc resinates are pale, high melting, hard, thermoplastic, friable resins. Zinol (zinc resinate solution) is offered as a mineral spirits solution.

Key Properties:

Zinar Was produced to give a resin with a maximum melting point for use with soft oils, baking alkyds and having fast solvent release for printing inks.

Zitro Was the first solid zinc resinate offered. It has found its greatest use as a general all-purpose zinc resinate.

Zirex Was developed to supply a zinc resinate with maximum zinc content and durability.

Zinros Was created expressly for use in gravure printing inks when a zinc resinate was sought which had a melting point close to Zinar but was considerably lower in viscosity in solution.

Zinol Recommended where the use of zinc resinate in mineral spirits is most convenient. The solvent contained in Zinol complies with Rule 66.

Key Properties (of All Types): Newport has offered a series of zinc resinates for a number of years. Their unique properties, excellent quality and uniformity long ago established them as the standards of the industry.

The zinc resinates comprise a group of resins with sufficiently different physical properties to make them useful in several fields of industry.

The four solid zinc resinates are soluble in all proportions in petroleum solvents, coal tar solvents, terpenes, drying oils, acetates and ethers. They are not soluble in alcohol.

The zinc resinates are compatible with most resins, both natural and synthetic, as well as paraffin wax and practically all types of plasticizers.

Zinc resinates function as pigment wetting and dispersing agents in the process of grinding pigments in vehicles for solvent-based organic protective coatings. This use of zinc resinates will give better pigment color development and will reduce any tendency to flood or float. To realize these benefits, the zinc resinates must be present during the grinding or milling.

None of Newport's resins of this group will react chemically with basic pigments and will not produce livering due to a chemical reaction with the pigments.

Newport's zinc resinates have found increasing use in the rubber adhesive fields. All of the solid zinc resinates are thermoplastic, which readily permits their use in Banbury mixers or rubber mills and other similar equipment.

They are also efficient agents for dispersing and dissolving high molecular weight elastomers and rubbers in oils or solvents. If one of the zinc resinates is milled into rubber and it is subsequently desired to cut this rubber or rubber compound in a solvent, the presence of the zinc resinate in the compound will materially reduce the time required to put the compound into solution. The dispersing power of the zinc resinate will also make it possible to produce solutions of higher solids content for a given viscosity than if zinc resinates were not present.

The dispersing action of zinc resinates plays an important part in incorporating pigments and other resins when milling rubber.

THIOKOL/SPECIALTY CHEMICALS DIVISION, 930 Lower Ferry Road, Trenton, NJ 08650

LP Liquid Polysulfide Polymers

Physical Properties and Chemical Analysis:	LP-31	LP-2	LP-32	LP-12	LP-3	LP-33
Color (MPQC-29-A), max	150	100	100	70	50	30
Viscosity @ 25°C, poises	950–1,550	410–525	410–525	410–525	9.4–14.4	15–20
Moisture content, max %	0.22	0.25	0.25	0.22	0.10	0.10
Mercaptan content, max %	1.5	2.00	2.00	2.00	7.7	6.5
Molecular weight, avg	8,000	4,000	4,000	4,000	1,000	1,000
Refractive index	1.5728	–	1.5689	–	1.5649	–
Pour point, °F	50	45	45	45	-15	-10
Flash point (PMCC), °F	437	406	414	406	345	367
Crosslinking agent, %	0.5	2.0	0.5	0.2	2.0	0.5
Specific gravity @ 77°F	1.31	1.29	1.29	1.29	1.27	1.27
Viscosity, poises						
@ 40°F	7,400	3,800	3,800	3,800	90	165
@150°F	140	65	65	65	1.5	2.1
Low temperature flexibility (G10,000 psi), °F	-65	-65	-65	-65	-65	-65

Chemical Type: Polysulfides are polymers of bis(ethyleneoxy)methane containing disulfide linkages. The reactive terminal groups used for curing are mercaptans (-SH). The general structure is: $HS(C_2H_4-O-CH_2-O-C_2H_4SS)_xC_2H_4-O-CH_2-O-C_2H_4SH$.

Applications:

	LP-31	LP-2	LP-32	LP-12	LP-3	LP-33
Aircraft sealants	X	X	X	–	–	–
Automotive sealants	–	X	X	–	–	–
Building sealants	X	X	X	X	–	–
Flow type sealants	–	X	X	X	–	–
Heavy construction sealants	X	X	X	–	–	–
Insulating glass sealants	–	X	X	–	–	–
Marine sealants	X	X	X	–	–	–
Dental molding compounds	–	X	–	–	–	–
Epoxy modifiers	–	–	–	–	X	X
Fluid membranes	–	X	X	–	–	–
Concrete coatings	–	–	–	–	X	X
Intumescent coatings	–	–	–	–	X	X
Electrical potting	–	X	X	–	X	X
Leather impregnation	–	X	–	–	X	–
Propellant binders	–	–	–	–	X	X

Applications listed here (specified by an X) typify the many usages for the polysulfide base sealants which have evolved over the past 30 years.

Key Properties (All Grades): Sealants based on LP polysulfide liquid polymers have found wide acceptance all over the world, in industries such as construction, automotive, marine, aircraft and insulating glass.

Construction — With the advent of the modern curtain wall and high rise structures came the need for a sealant which would adhere tenaciously to almost any surface and provide a weatherproof seal against leakage. Polysulfide-base sealants provided the construction industry with such a material, thereby offering a solution to the problem of reducing maintenance costs.

Marine — The resistance to the corrosive effects of sea water, the permanent adhesion to a number of surfaces, and continued flexibility are properties which have given polysulfide-base sealants a range of uses in the marine industry. Lapstrake seams and joints, centerboard wells, stern and keel joints, portholes, deck hardware fittings and deck seams are just some of the many areas where polysulfide-base compounds can be used on marine craft as a sealant, adhesive or glazing compound.

Aircraft — Since the 1940s, sealants based on polysulfide polymers have been used on virtually every commercial and military aircraft. With excellent resistance to fuels, these compounds have been specified for sealing integral fuel tanks and numerous other areas on aircraft of all sizes.

Insulating Glass — Only a few compounds can meet the rigid requirements placed on insulating glass sealants. Polysulfide-base sealants, because of their excellent adhesion and high resistance to UV, are used in the manufacture of insulating glass units. They are also used in glazing applications, specifically windows, patio doors, and refrigerator windows.

Other Applications — There are many other kinds of polysulfide-base sealant applications. Some of these include: cold molding, potting, formed-in-place gaskets, trailer waterproofing, flexible molds, auto glass and body sealing, filleting and mold sealing, membrane coating, gas main sealing, concrete bonding, plastic tooling, wire and cable sealing, anticorrosion coating, leather impregnation, printing rollers, propellant binders, vibration damping and dental molding compounds.

Key Properties of Cured Compositions:

Stress-Strain Properties — Liquid polysulfide polymers can be cured to an elastomeric rubber with moderate stress-strain properties. These properties can be widely varied.

Oil, Solvent and Chemical Resistance — Cured liquid polysulfide-based compositions display excellent resistance to a wide range of oils and solvents.

Electrical Properties — Cured liquid polysulfide polymers have good electrical properties which can be improved through compounding.

Aging and Weathering Resistance — Cured liquid polysulfide polymer-based compositions display excellent resistance to aging, ozone, oxidation, sunlight and weathering.

Service Temperature Range — The service temperature range for cured compositions based on polysulfide polymer is –65° to +250°F.

Adhesion Properties — Excellent adhesion to most common construction materials, coupled with good flexibility, is obtained with properly compounded polysulfide-based compositions.

Selection of Curing Agents — The final selection of a curing agent is based on its overall performance with respect to a number of requirements, including: cost, stability, controllable cure rate, heat stability of cured composition, elastomeric properties, toxicity, etc.

Toxicology: LP liquid polysulfide polymers are relatively nontoxic.

UPJOHN POLYMER CHEMICALS, Box 685, LaPorte, TX 77571

PAPI Polymeric Isocyanates

Chemical Type: A dark brown medium viscosity polymethylene polyphenylisocyanate of lower reactivity and an average functionality of 2.7, in brown liquid form.

Chemical Analysis and Physical Properties (All Grades):

Boiling point @ 5 mm Hg	392°F (200°C)
Vapor pressure, mm Hg	2×10^{-4}
Vapor density, Air: 1	8.6
Specific gravity @ 25°C	1.2
Volatile, % by vol	nil
Flash point (ASTM D93, Closed Cup)	>400°F (205°C)

PAPI

Key Properties: The reduced reactivity of PAPI may be useful in applications where better liquid flow and low exotherm are desired such as for continuous panel, discontinuous and continuous block foam. Specialty moldings.

PAPI 50

Chemical Type: 50% solution of PAPI in monodichlorobenzene.

Key Properties: The primary application is as a bonding agent for promoting the adhesion of synthetic fabrics to elastomers, vinyl or metal.

Section XXV
Resins–Natural

O.G. INNES CORP., 10 East 40th St., New York, NY 10016

Product Name	Melting Point (°F)	Acid No.	Saponification No.	Iodine No.	Moisture Content (%)	Gardner-Holdt Viscosity*
Alphacopal	235	125	180	110	1	H-I

*50% solids in 95% denatured alcohol.

Chemical Type: A processed Manila gum, in purified and powdered form. Completely soluble in alcohols. Usable in amine solutions.

Key Properties: This purified resin maintains the characteristics of Manila resin and is recommended for spirit varnishes, printing inks, sanding sealers, floor and wall coatings, primers and undercoats, paper coatings, lacquers, adhesives, liquid polishes, etc.

Product Name	Refractive Index	Softening Point (°F)	Melting Point (°F)	Specific Gravity	Acid No.	Saponification No.	Iodine No.
Batu East India	1.538	245	345	1.03	24	42	76

Chemical Type: High melting point Damar-type resins, available in the following gradings:
Batu – bold scraped, bold unscraped, nubs and chips, dust.

Key Properties: Has adhesion, toughness, flexibility and color retention. It is scuff and fume resistant. When cut cold, yields a flat. As heat is increased, if cooked in oils, the flat may be reduced until a high temperature provides sturdy, high gloss vehicles. This resin has fast solvent release, is extremely adhesive and shows resistance to water, greases and wide temperature changes. Its water resistance and adhesion under tough wearing conditions particularly recommend it for undercoats on metals, traffic marking paints and similar outlets.

It has good compatibility with a wide range of oils, resins, waxes, cellulose derivatives, chlorinated rubber, stearic acid, asphalt and pitches. Batu is used in the adhesive field for cements of all types. Soluble in aryl hydrocarbons and hydrogenated aliphatic hydrocarbons. Insoluble in alcohols, ketones, esters, ethers and terpenes.

Note: FDA approved.

Black East India	1.541	247	324	1.04	23	34	81

Chemical Type: High melting point Damar-type resins, available in the following gradings:
Black – bold scraped, bold unscraped, nubs and chips.

Key Properties: Has adhesion and flexibility and is particularly interesting in metal undercoats. It has brine and scuff resistance. Although it is a black color, it yields an opaque type varnish and dries to a surprisingly attractive color. It may be used in a broad range of outlets, including floor varnishes and porch and deck enamels. An interesting ingredient in undercoats for ship-bottom paints and automotive and can coatings where flexibility is needed. The East Indias cover broad usages for wood, metal and cement surfaces, inks, stains, sizings, adhesives, gasket compounds, outdoor as well as inside finishes, undercoats and tough-wearing vehicles.

Soluble in aryl hydrocarbons and halogenated aliphatic hydrocarbons. Insoluble in alcohols, ketones, esters and terpenes.

Note: FDA approved.

Product Name	Refractive Index	Softening Point .... (°F)	Melting Point	Specific Gravity	Acid No.	Saponification No.	Iodine No.
Congo Gum	1.541	208	345	1.06	103	128	129

Chemical Type: Available in the following gradings:

Water White	Pale Nubs
Ivory	Amber Nubs
Pale Bold AA	Small Nubs
Pale Amber BB	Chips
Dark Amber Bold CC	Dust
Amber Sorts DD	

Key Properties: Congo Gum is not soluble in any regular solvent. It must be fused to render it oil-soluble and compatible with a broad range of other products.

Product Name	Gardner-Holdt Viscosity (50% Xylol)	Melting Point (°F)	Specific Gravity	Acid No.	Weight per Gallon (lb)
#871 Processed Congo	A–B	196	1.08	73	8.83

Chemical Type: Pure processed Congo.

Key Properties: Yields tough gloss varnishes for a broad range of outlets including furniture, floors, porch and deck vehicles, wrinkle finishes, undercoats of all types, flexible metal finishes, insulating varnishes, can container or food contract coatings and artists' materials where the unusually superior color retention of Congo is of prime importance. It also has good brine resistance. Congo may be finely fused for use in amine or caustic solutions in conjunction with waxes and similar outlets.

Congo may be processed as pure fused material, modified or esterified to meet broad specifications.

Note: FDA approved.

Product Name	Refractive Index	Softening Point .... (°F)	Melting Point	Specific Gravity	Acid No.	Saponification No.
Damars	1.537	160	224	1.05	26	35

Chemical Type: General all-around varnish resin, available in the following gradings:

Singapore Damar	Batavia Damar		Siam Damar
No. 1	Standard A/E Mixed	Nubs C	
No. 2	Standard A/D Mixed	Chips D	
No. 3	Standard A/C Mixed	Grain E	
Seeds	Bold A	Seeds F	
Dust	Medium B	Dust	

Key Properties: Damars provide unique depth of gloss, superior color and gloss retention, adhesion and flexibility and resistance to greases, cold checking and scuffing. A general all-around varnish resin for cold cuts, blends or cooked oil vehicles. Its fume resistance and binding value recommend Damar for white enamels and pastels. Recommended for adhesives and inks. A toner resin with miscibility and versatility. Soluble in aryl and aliphatic hydrocarbons and terpenes. Insoluble in alcohols and esters.

Note: FDA approved.

Manilas	1.539	187	252	1.07	125	165

Chemical Type: Alcohol-soluble resins (Iodine No. of 119), available in the following gradings:

Loba A Bold	DBB Chips	Philippine Pale Bold
Loba B Nubs	CNE Nubs	Philippine Pale Nubs
Loba C Small Nubs	DK Chips	Philippine Pale Sorts

Loba D Chips	WS Extra Pale Soft Sorts	Philippine Pale Chips
CBB Nubs	MA Amber Soft Sorts	Philippine Seeds and Dust
Loba Seeds and Dust		

Key Properties: These alcohol-soluble resins are broadly used for paper and packaging coatings of all types, label varnishes, wallpaper finishes, floor finishes, wall sizes, primers, lacquers, adhesives and cements.

Manila gum may be rendered water-soluble in amine or caustic solutions for inclusion in liquid floor polishes and similar materials.

Note: FDA approved.

Product Name	Refractive Index	Softening Point .... (°F)	Melting Point	Specific Gravity	Acid No.	Saponification No.	Iodine No.
Mastic	–	131	168	1.05	60	76	103

Chemical Type: Soft spirit-soluble resin of ancient usage.

Key Properties: Compatible with a wide range of vegetable oils, cellulose derivatives and both nitrocellulose and ethylcellulose. Readily miscible with asphalts and pitches on heating. Can be used in spirit varnish lacquers and adhesives, printing inks and specialty purposes, particularly in the artists' field.

| **Sandarac** | 1.545 | 239 | 289 | 1.07 | 136 | 151 | 127 |

Chemical Type: One of the oldest known resins in the artist and coating fields.

Key Properties: Gives a hard, white spirit varnish with excellent color retention, stability, adhesion and flexibility. Used in spirit varnishes, lacquers, dental cements and specialty products.

| **Pale East India** | 1.541 | 245 | 299 | 1.05 | 24 | 38 | 80 |

Chemical Type: High melting point Damar resins, available in the following gradings:
Pale – Bold, Nubs, Chips, Dust

Key Properties: Have adhesion, flexibility, toughness and color retention. Scuff and fume resistant. May be controlled for specially attractive sheen by adjustment of solvents and thinners. Soluble in aryl hydrocarbons and hydrogenated aliphatic hydrocarbons. Insoluble in alcohols, ketones, esters, ethers and terpenes.

Note: FDA approved.

| **Pontianaks** | 1.540 | 243 | 307 | 1.08 | 116 | 165 | – |

Chemical Type: Alcohol-soluble resins, available in the following gradings:
Genuine Bold Scraped, Genuine Split Chips, Genuine Nubs, Genuine Chips

Key Properties: These alcohol-soluble resins are broadly used for paper and packaging coatings of all types, label varnishes, wallpaper finishes, floor finishes, wall sizes, primers, lacquers, adhesives and cements.

The Pontianak resins are similar to the alcohol-soluble Manilas, but have higher melting points. Therefore, they are used in more exacting specifications, particularly in the specialty coating and adhesive fields. While the Manilas are not generally cooked, Pontianak has been popular in certain types of cooked oil vehicles.

Note: FDA approved.

Product Name	Melting Point (°C)	Acid No.	Viscosity * @ 32°C	Specific Gravity
Processed Elemi PEI	85	25	A	0.960

*50% in mineral spirits.

Chemical Type: Plasticizing resin.

Key Properties: Used to promote adhesion and flexibility. Has good level and flow properties and excellent water resistance. Used in lacquers, adhesives and cements, wax compositions, printing inks, textile and paper coatings.

Note: FDA approved.

	Softening Point	Melting Point	Specific Gravity	Acid No.	Saponification No.	Iodine No.
Product Name	. . . . (°F).					
Yacca Gum (Red Gum Accroides)	167	246	1.34	130	70	200

Chemical Type: A very low-cost resin with unusual characteristics.

Key Properties: This alcohol-soluble resin does not "run" in a molten state, but is heat reactive. It sets at a temperature between 160° and 225°C and becomes completely insoluble. Yacca is highly resistant to oils and has excellent insulation qualities. The attractive red color has not been bleached or removed from the resin and is recommended as a dye ingredient for lacquers and plastic products. Soluble in alcohol and alkali solutions. This resin is attractive because of its very low cost, but it is inclined to be brittle and not always soluble according to usual rules.

Note: FDA approved.

S. WINTERBOURNE & CO., INC., P.O. Box 316-E, Rahway, NJ 07065

Manila Copal Resins

Product Name	Melting Point (°C)	Specific Gravity	Acid No.	Iodine No.
Loba C Manila Nubs	119	1.075	140	122
DBB Pale Chips	125	1.065	139	113
Bold Pale Philippine	129	1.07	109	123
Philippine Chips	130	1.08	113	116

Key Properties: Compatible with phenolics, ester gum, maleic-rosin, waxes, ethyl and spirit-soluble nitrocellulose. Soluble in alcohols and ketones. Recommended for oil and spirit varnishes, primer-sealers, lacquers, japans, adhesives, shellac substitutes, floor polishes and coatings.

Pontianak Resins

Pontianak Nubs	153	1.075	115	106
Pontianak Chips	146	1.075	120	125

Key Properties: Compatible with phenolics, rosin, ester gum, maleic-rosin, alkyds, ethyl and spirit-soluble nitrocellulose. Soluble in alcohols and ketones. Recommended for spirit varnish and shellac substitutes.

Damar Resins

No. 1 Singapore Damar	115	1.045	17	113
No. 2 Singapore Damar	114	1.05	32	104
Singapore Damar Dust				
No. 1 Siam Damar	114	1.05	32	84

Key Properties: Compatible with phenolics, alkyds, rosin, ester gum, maleic-rosin, paracoumarone, waxes, chlorinated rubber, ethyl and nitrocellulose. Soluble in aryl and aliphatic hydrocarbons, terpenes. Recommended for wood and metal lacquers, inks and enamels.

East India Resins

Pale East India Nubs	158	1.04	22	80
Pale East India Chips	144	1.045	27	80
Bold Black Scraped	162	1.04	20	83
Black Unscraped	159	1.04	25	81
Bold Scraped Batu	174	1.025	19	81

Key Properties: Compatible with phenolics, alkyds, rosin, ester gum, maleic-rosin, paracoumarone, waxes, chlorinated rubber. Soluble in aryl hydrocarbons and hydrogenated aliphatic hydrocarbons.

Elemi Resin

Product Name	Melting Point (°C)	Specific Gravity	Acid No.	Iodine No.
No. 1 Elemi	plastic	1.05	27.5	118

Key Properties: Compatible with phenolics, rosin, ester gum, maleic-rosin, waxes, paracoumarone, chlorinated rubber, ethylcellulose and nitrocellulose. Soluble in aryl hydrocarbons and esters. Recommended as plasticizer, for metal lacquers for adhesion, inks, cements and adhesives.

Gum Accroides (Yacca)

	Melting Point (°C)	Specific Gravity	Acid No.	Iodine No.
Regular	119	1.34	123	200
Powdered (Extra Fine)	134	–	133	196

Key Properties: Compatible with maleic-rosin, paracoumarone, cellulose acetate, nitrocellulose. Soluble in alcohols. Recommended for flexographic inks, paper coatings, spirit varnishes, building materials and fireworks.

Section XXVI
Resins–Phenolic

ASHLAND CHEMICALS, P.O. Box 2219, Columbus, OH 43216

Arofene Heat Reactive Alkyl Phenolics

Product Name	Softening Point (Ring & Ball) (°C)	Color (Gardner) (max)	Specific Gravity	Methylol Content (%)
612	100	4	1.115	13

Key Properties: Very fast drying. Short open tack time. High heat resistance. Employed as a blending resin for higher heat resistance. Recommended for nitrile and neoprene.

669	112	4	1.100	13

Key Properties: General purpose. Good specific adhesion and strength properties. Recommended for nitrile and neoprene.

6403	96	4	1.115	15

Key Properties: Fast solvent release and rapid strength development. High heat resistance. Resistant to phasing. Recommended for nitrile and neoprene.

6690	95	4	1.083	12

Key Properties: Exhibits very long open tack times. Promotes good strength and adhesion. Recommended for nitrile and neoprene.

8723	110	4	1.100	14.5

Key Properties: Produces high performance adhesives. Resistant to phasing. Recommended for nitrile and neoprene.

Note: All of the above are in crushed form.

Arofene Thermosetting One-Step Phenolics

Product Name	Softening Point (Dennis Bar) (°C)	Color (Gardner) (max)	Specific Gravity	Hot Plate Cure @ 150°C
24780	88	11	1.213	107.5

Key Properties: Light color. Stain resistant. Excellent chemical resistance. Recommended for nitrile and urethanes.

85678	85	12	1.235	65 @ 160°C

Key Properties: Good compatibility. Excellent metal adhesion and flexibility. Recommended for nitrile.

Note: Both of the above are in crushed form.

Arofene Thermosetting Two-Step Phenolics

Product Name	Softening Point (Dennis Bar) (°C)	Hot Plate Cure @ 150°C
877	88	75

Key Properties: High heat resistance and strength properties. Recommended for nitriles.

6781	98	50

Key Properties: Good compatibility characteristics. Improves strength and specific adhesion. Recommended for nitriles and natural rubber.

Note: Both of the above are in powder form.

Arotap Thermosetting Liquid Phenolics (Water Soluble)

Product Name	Brookfield Viscosity (cp)	Solids Content (%)	pH	Specific Gravity	Hot Plate Cure @ 150°C
352-W-71	375	71	7.7	1.245	90

Key Properties: Primarily used with vinyl acetate latex to improve heat and chemical resistance.

Arofene 744-W-55	300	55	9.2	1.190	22.5

Key Properties: Self-emulsifying. Good heat and solvent resistance. Recommended for SBR.

Arotap 8095-W-50	100	50	8.8	1.170	40

Key Properties: Provides good adhesion, strength and chemical resistance to latex blends. Recommended for SBR.

Arotap Thermosetting Liquid Phenolics (Solvent Based)

Product Name	Brookfield Viscosity (cp)	Solids Content (%)	pH	Specific Gravity	Hot Plate Cure @ 150°C
Arofene 536-E-56	325	56	8.1	1.065	47.5

Key Properties: Imparts high chemical and heat resistance. In ethanol solvent. Recommended for nitrile and urethane.

Arofene 986-A1-50	600	50	7.2	1.095	80

Key Properties: Improves heat and chemical resistance. Promotes high specific adhesion. Good flexibility. In Cellosolve acetate solvent. Recommended for nitrile.

Water Soluble Phenolic Resins

Product Name	Brookfield Viscosity (cp)	Solids Content (%)	Specific Gravity @ 25°C	Hot Plate Cure @ 150°C (sec)	pH (As Is)	Water Tolerance (%)
Arofene 201-W-51	28	51.5	1.195	100	7.6	2,400

Key Properties: Recommended as a binder for fiber glass and mineral wool type insulation.

Arotap 352-W-71	375	71	1.245	90	7.7	1,000

Key Properties: Recommended as an impregnant for honeycomb and paper laminates, and as a modifier for polyvinyl acetate emulsions and latex systems.

Arotap 670-W-40	150	42	1.180	10 max	12.3	2,400

Key Properties: Recommended for hardboard manufacture by the wet process. Hot-press bonding of wood.

Product Name	Brookfield Viscosity (cp)	Solids Content (%)	Specific Gravity @ 25°C	Hot Plate Cure @ 150°C (sec)	pH (As Is)	Water Tolerance (%)
Arotap 744-W-55	300	55	1.190	28	9.2	250

Key Properties: Water soluble phenolic for wet end addition. Impregnant for use in honeycomb and paper saturations.

Arofene 760-W-70	388	70	1.205	120	8.1	250

Key Properties: Recommended as modifier for polyvinyl acetate emulsions to improve strength properties and water resistance.

Arofene 8095-W-50	100	50	1.170	40	8.8	100

Key Properties: Recommended as a water emulsifiable phenolic for wet end addition.

Arofene 8828-W-80	3,500	80	1.230	—	7.4	40

Key Properties: Recommended as high solids and high viscosity for use in corrosion resistant cements.

Arofene 9580-W-72	5,000	72	1.228	—	9.9	1,500

Key Properties: Recommended as binder for use in cork bonding applications.

Key Properties: This line of water soluble phenolic resins offers a wide variety of performance characteristics. These resins are employed in applications where high heat and chemical resistance is desired from phenolics exhibiting high water tolerances. Applications would include insulation binders, impregnating resins for specialty paper saturations, modifying resins for latex-based adhesives and binders for numerous wood and paper composites. These resins impart high heat resistance, coupled with good water and chemical resistance in these applications.

Key Properties of All Arofene Phenolic Resins: Imparts improved heat resistance, specific adhesion, strength, tack and chemical resistance to a wide range of adhesive systems. The comprehensive line includes oil soluble, thermosetting one-step and two-step resins and water and alcohol soluble liquid resins. Use of these resins imparts improved performance characteristics to a wide range of natural and synthetic elastomer based adhesives.

DUREZ DIVISION, Hooker Chemicals & Plastics Corp., North Tonawanda, NY 14120

Durez Resins

Product Name	Contraction Point (°C)	Cure Time @ 165°C (sec)	Through 200 Mesh (% max)
Durez 175	52	40	10

Chemical Type: Thermosetting, phenol-formaldehyde resin of the one-step type, supplied in crushed form. Does not contain any fillers or lubricants.

Key Properties: Intended for many applications wherein a one-step type of phenolic resin is intended. Being a one-step type, it possesses thermosetting characteristics which cannot be modified to any appreciable extent by the addition of hexamethylene-tetramine or other accelerators.

Useful in adhesives.

Product Name	Melting Point (Capillary Tube) (°C)	Softening Point (Ring & Ball) (°C)	Acid Number	Color (Gardner)	Specific Gravity
Durez 225	156	175	58	12	1.09

Chemical Type: Oil-soluble, wax-compatible, high-melting, thermoplastic, terpene phenolic lump resin.

Product Name	Melting Point (Capillary Tube) (°C)	Softening Point (Ring & Ball) (°C)	Acid Number	Color (Gardner)	Specific Gravity
Durez 240	168	186	60	11	1.09

Other Properties:

Color (Rosin Scale)	M–H
Refractive Index at 25°C	1.565
Gardner-Holdt Viscosity in—	
Toluol at 60% Solids	Z_3-Z_4
Mineral Spirits at 50% Solids	Z_3-Z_6

Chemical Type: Terpene phenolic, oil-soluble, very high-melting, high-viscosity, thermoplastic resin.

Key Properties: Soluble in many solvents.

Product Name	Contraction Point (°C)	Cure Time @ 165°C (sec)	Cure Time @ 150°C (sec)	I.P. Flow (mm)	Through 200 Mesh (%)
Durez 7031A	80	21	37	19	1

Chemical Type: Finely pulverized thermosetting phenol-formaldehyde resin of the two-step type.

Key Properties: It has a short flow, fast cure and a medium set. Soluble in alcohols and ketones. Used to reduce the nerviness and impart reinforcement to NBR stocks used in tape manufacture.

Product Name	Melting Point (Capillary Tube) (°C)	Melting Point (Ring & Ball) (°C)	Acid Number	Color (Rosin Scale)	Color (Gardner Scale) (max)	Specific Gravity @ 25°C
Durez 9633	142	161	18	N–H	12	1.09

Other Physical Properties:

Weight per Gallon	9.10 pounds
Refractive Index at 25°C	1.553
Gardner-Holdt Viscosity in—	
Toluol at 60% Resin	1
Mineral Spirits at 50% Resin	1

Chemical Type: Terpene phenolic, oil-soluble, high-melting, low-viscosity, thermoplastic resin.

Key Properties: Soluble in many solvents.

Product Name	Solids Content @ 135°C (%)	Brookfield Viscosity @ 25°C (cp)	Cure Time @ 165°C (sec)	Cure Time @ 150°C (sec)	Water Tolerance (%)	Specific Gravity @ 25°C
Durez 10694	80	4,250	48	74	45	1,246

Chemical Type: Thermosetting, phenol-formaldehyde, liquid resin of the one-step type.

Key Properties: Slow to medium cure and a rigid set. The use of 10 phr concentrations of this resin in polysulfide sealant formulations improves adhesion of these sealants to non-porous substrates such as glass, steel and aluminum. It may also be used to seal joints in concrete where staining is not a problem.

Product Name	Melting Point (Capillary Tube) (°C)	Softening Point (Ring & Ball) (°C)	Acid Number	Color (Gardner)	Specific Gravity @ 25°C	Weight per Gallon (lb)
Durez 12603	133	152	65	10	1.09	9.1

Chemical Type: Thermoplastic, oil-soluble, high melting terpene phenolic resin, supplied in flake form.

Key Properties: Recommended for use in neoprene rubber solvent and latex cements. Has high melting point and low viscosity in most solvent solutions.

One of the principal applications of this resin is as a component in neoprene solvent adhesives. The resin has good compatibility with all types of neoprene and has good solubility in the solvents normally used in neoprene adhesive formulations.

The following properties are improved—

Specific adhesion.

Initial tack and tack retention.

Generally soluble in the solvents used.

Bond strength will increase with age.

Negates undesirable effects of hydrated calcium silicate.

Will produce an increase in the heat resistance.

Reduces acid tendering characteristics of the adhesive.

Improves aging stability.

Reinforces and tackifies.

Product Name	Contraction Point (°C)	I.P. Flow (0.2 g) (mm)	Cured Specific Gravity (With 8% Hexa)	% Retained on 10 Mesh Screen	% Retained on 60 Mesh Screen	Specific Gravity (Uncured Resin)
Durez 12686	60	140	1.18	10	60	1.18

Chemical Type: Cashew nut shell modified novolak phenolic resin supplied in ground form.

Key Properties: Recommended as a processing aid and reinforcing resin for NBR compounds and adhesives. Requires the addition of hexamethylene tetramine to make it thermosetting and to impart maximum reinforcement and heat resistance in synthetic elastomers, especially NBR compounds and adhesives.

Product Name	Contraction Point (°C)	I.P. Flow (0.2 g) (mm)	Cured Specific Gravity (With 8% Hexa)	Cured @ 165°C (sec)	Retained on 200 Mesh Screen (% max)	Specific Gravity (Uncured Resin)
Durez 12687	60	75	1.18	23	1.0	1.20

Chemical Type: Finely pulverized, thermosetting phenolic resin of the two-step type.

Key Properties: Has a long flow, medium cure and a rigid set.

Durez Resins 12686, 12687, 12707

Chemical Type: The same base resin is used in each. The difference between them is in the hexamethylenetetramine content.

	% Hexamethylenetetramine
Durez 12687	8
Durez 12707	3
Durez 12686	None

Key Properties: Intended for use in rubber compounding.

Durez 12687 Resin—Cures rapidly to a hard, inflexible, heat resisting product. When cured, it is insoluble in oils, acids, alkalies and organic solvents.

Durez 12707 Resin—Due to its lower hexamethylenetetramine content, cures to a semi-flexible product with a tendency to soften at elevated temperatures. Its cure speed is slower than that of 12687. Hexamethylenetetramine can be increased to produce the same properties as obtained with 12687.

Durez 12686 Resin—Will not cure unless hexamethylenetetramine is added.

The 12707 and 12686 modifications of 12687 Resin are made available for use in special recipes and to eliminate scorching tendencies in Banbury mixing.

Product Name	Contraction Point (°C)	Cure @ 165°C (sec)	I.P. Flow (mm)	Retained on 200 Mesh (% max)	Cured Specific Gravity
Durez 12987	63	24	72	1	1.20

Chemical Type: Finely pulverized, thermosetting phenolic resin of the two-step type.

Key Properties: Recommended for use in the manufacture of NBR solvent cements and is especially suited for applications involving rubber to metal bonding.

Toxicity: Some of the above are slightly hazardous. Refer to the manufacturer's Material Safety Data Sheets for specific details.

REICHHOLD CHEMICALS, INC., RCI Bldg., White Plains, NY 10603

Nirez 2000 Series Terpene Phenol Resins

Grades: The following five resins currently comprise this series: **Nirez 2019; Nirez V-2040; Nirez V-2150; Nirez 2092WG;** and **Nirez 2092KM.**

Chemical Types: Nirez 2019 and Nirez V-2040, the original resins in this series, have been joined by a new high melting homolog designated Nirez V-2150. Its development was prompted by requests for a higher melting point resin than Nirez V-2040. These three resins are brilliant, pale, hard, friable, thermoplastic resins.

Key Properties (all of the above): They are soluble in a variety of solvents and are compatible with many types of resins, oils and polymers. They will all show a very slight acid number, but they do not react with basic pigments and fillers.

Nirez 2092WG and Nirez 2092KM have principal application in neoprene cements.

Super-Beckacite Pure Phenolics

Product Name	Softening Point (Ring & Ball) (°C)	Color (USDA) (max)	Specific Gravity	Weight per Gallon (lb)
24-001	91	X	1.09	9.08

Chemical Type: Oil compatible-reactive.

24-003	91	E	1.09	9.08

Chemical Type: Oil compatible-reactive.

24-011	130*	X	1.05	8.75

Chemical Type: Oil compatible-nonreactive.

24-020	105	K	1.04	8.66

Chemical Type: Oil compatible-nonreactive.

24-021	163	I	1.07	8.91

Chemical Type: Oil compatible-nonreactive.

24-024	150	X	1.08	9.00

Chemical Type: Oil compatible-reactive cold blend.

24-025	96*	X	1.22	10.16

Chemical Type: Oil compatible-nonreactive.

24-050	95	X	1.09	9.08

Chemical Type: Oil compatible-reactive.

Product Name	Softening Point (Ring & Ball) (°C)	Color (USDA) (max)	Specific Gravity	Weight per Gallon (lb)
24-051	103	X	1.09	9.08
Chemical Type: Oil compatible-reactive.				
24-052	103	X	1.09	9.08
Chemical Type: Oil compatible-reactive.				
24-054	103	X	1.09	9.08
Chemical Type: Oil compatible-reactive.				

*Navy Ring & Ball Method.

Varcum Pure Phenolics

Product Name	Softening Point (Ring & Ball) (°C)	Color (USDA) (max)	Specific Gravity	Weight per Gallon (lb)
29-000	110	Light Amber	1.06	8.83
Chemical Type: Oil compatible-nonreactive.				
29-001	110	Light Amber	1.065	8.87
Chemical Type: Oil compatible-nonreactive.				
29-002	103	Amber	1.21	10.08
Chemical Type: Oil compatible-nonreactive.				
29-004	140	Amber	1.055	8.80
Chemical Type: Oil compatible-nonreactive.				
29-005	143	Light Amber	1.10	9.17
Chemical Type: Oil compatible-nonreactive.				
29-008	135	Light Amber	1.055	8.80
Chemical Type: Oil compatible-nonreactive.				
29-009	113	Amber	1.045	8.71
Chemical Type: Oil compatible-nonreactive.				
29-033	93	Light Yellow to Light Amber	1.10	9.17
Chemical Type: Oil compatible-reactive.				
29-400	90	Clear Light Amber	1.175	9.79
Chemical Type: Oil compatible-reactive.				
29-401	98	Clear Light Yellow	1.115	9.29
Chemical Type: Oil compatible-reactive.				
29-402	95	Water White to Light Yellow	1.10	9.17
Chemical Type: Oil compatible-reactive.				
29-424	100	Clear Light Yellow	1.115	9.29
Chemical Type: Oil compatible-reactive.				

Chemical Type (all of the above): Substituted phenols, not pure phenol, must be used to produce phenol-formaldehyde resins for use in chemical coatings. The reaction of substituted phenols and formaldehyde to form linear polymers takes place in the presence of either an acid catalyst or an alkali catalyst.

Acid-catalyzed phenolics have no terminal CH_2OH groups. Therefore, although these are soluble in oils, they do not react with oils. The alkali-catalyzed phenolics have terminal CH_2OH groups which react with oils and release water. Care must be taken when using oil-reactive phenolic resins in a varnish cook because when the water is released it will usually boil out with considerable foaming.

Key Properties (all of the above): Because they have good antioxidant properties and weather resistance, phenolics give good service in chemical coatings. They are best where lightness of color is not critical, for phenolics do have a tendency to yellow on exposure. Spar varnishes, marine primers, floor varnishes and enamels, adhesives, hot-melt coatings, gelation inhibitors, and chemical resistant varnishes are all formulated using phenolics for the resin base.

Varcum Phenolic Resins for Use in Metal Coatings Meeting F.D.A. Requirements

Product Name	Form	Brookfield Viscosity (cp)	Softening Point (Capillary) (°C)	Hot Plate Cure @ 150°C (sec)	Specific Gravity	Solids Content (%)
29-101	Solid-Lump	225	85	113	1.155	—

Key Properties: Used in phenolic epoxy formulations where relatively high percentages of phenolic may be used for cost savings.

| 29-104 | Solid-Lump | 225 | 70 | 90 | 1.230 | — |

Key Properties: Meat Inspection Division of U.S.D.A. sanctioned. General purpose resin for can coatings.

| 29-105 | Solid-Crushed | 313 | 58 | 68 | 1.265 | — |

Key Properties: M.I.D. of U.S.D.A. sanctioned. Metal coatings requiring maximum chemical resistance. Low flexibility.

| 29-106 | Solid-Crushed | 120 | 80 | >4 min @ 185°C | 1.195 | — |

Key Properties: Low melt point version of Varcum 29-107 for improved compatibility with tung oil.

Chemical Type (all of the above): Phenolic resin types, some of which fall under Section 121.2514 Subsection IV of the F.D.A. Regulations.

Key Properties (all of the above): The Varcums are included in a class of materials which may be used for coatings constituents under the terms of the above Section. Therefore, they can be used in coatings for food packaging.

An increasingly important application for phenolic resins is in metal coatings, such as for drums and cans. Resins for this application are thermosetting one-step resins which can be cured to coatings which protect the metal surface from corrosion and the contents from contamination. These coatings require baking temperatures of 300° to 400°F.

Differing requirements, as to flexibility, chemical resistance and rate of cure make it desirable to have available a wide range of resins. Varcum offers one of the most complete lines available. Resins are offered both in the solid and solution form. Variations due to the properties of the resins can be further extended by combining the resins with any of a number of compatible modifiers.

| 29-107 | Solid-Crushed | 225 | 95 | >4 min @ 185°C | 1.200 | — |

Key Properties: M.I.D. of U.S.D.A. sanctioned. Slow cure modifying resin, good vinyl compatibility. Improves vinyl adhesion.

| 29-108 | Solid-Crushed | 138 | 70 | 150 | 1.215 | — |

Key Properties: M.I.D. of U.S.D.A. sanctioned. Bisphenol resin for use involving critical taste considerations.

Product Name	Form	Brookfield Viscosity (cp)	Softening Point (Capillary) (°C)	Hot Plate Cure @ 150°C (sec)	Specific Gravity	Solids Content (%)
29-109	Solid-Lump	900	75	65	1.230	–

Key Properties: Excellent chemical resistance. Limited flexibility. Used in drum and tank linings.

| 29-112 | Solid-Crushed | 225 | 70 | 85 | 1.230 | – |

Key Properties: M.I.D. of U.S.D.A. sanctioned. General purpose resin for can coatings.

| 29-150 | Solution | 8,500 | – | 50 | 1.076 | 65 |

Key Properties: General purpose resin for containers. Compatible with epoxies and polyvinyl butyral. Contains butyl and isopropyl alcohols.

| 29-153 | Solution | 13,000 | – | 100 | 1.080 | 68 |

Key Properties: M.I.D. of U.S.D.A. sanctioned. Excellent for post forming operation. Contains butanol and diacetone alcohol.

| 29-170 | Liquid | 5,000 | – | – | 1.155 | 86 |

Key Properties: Coreactant with epoxies, vinyls or alkyds in metal coatings. Alkali resistant. Low percentages of phosphoric acid recommended to promote cure. Liquid phenolic resin free of added solvents.

General Purpose Industrial Coating Resins

Product Name	Form	Brookfield Viscosity (cp)	Softening Point (Capillary) (°C)	Hot Plate Cure @ 150°C (sec)	Specific Gravity	Solids Content (%)
29-152	Solution	800	–	7.5	1.030	54

Key Properties: Recommended for hardware finishes with moderate flexibility and compatibility. Has fast cure and is highly chemical resistant.

| 29-154 | Solution | 650 | – | 25 | 1.065 | 56.5 |

Key Properties: Has high chemical resistance.

| 29-164 | Solution | 750 | – | 50 | 1.020 | 52 |

Key Properties: Has good chemical resistance and compatibility with epoxy resins.

| 29-172 | Solution | 325 | – | 30 | 1.033 | 53 |

Key Properties: Used in tank car linings. Outstanding acid chemical resistance.

| 29-462 | Solution | 400 | – | 40 | 1.065 | 57 |

Key Properties: Used with polyvinyl butyral resins in Western Pine Association & Primer Knot Sealer formulations.

SCHENECTADY CHEMICALS, INC., P.O. Box 1046, Schenectady, NY 12301

Schenectady Phenolic Resins

Product Name	Form	Melting Point (Capillary Tube) (°F)	Color (U.S.D.A. Standard)	Specific Gravity	Methylol Content
SP-102	Lump	153	D–F	1.10	–
SP-103	Lump	153	X or lighter	1.10	–
SP-126	Lump	155	I–N	1.10	–
SP-134	Lump or Crushed	155	X or lighter	1.10	–

Product Name	Form	Melting Point (Capillary Tubes) (°F)	Color (U.S.D.A. Standard)	Specific Gravity	Methylol Content
SP-144	Lump	145	X	—	18
SP-154	Lump or Crushed	175	X or lighter	1.10	—
SP-155	Lump	165	X or lighter	—	10
SP-174	Lump	300	Amber	1.10	—
SP-553	Flake	115*	10**	1.00	—
SP-560	Flake	150*	10***	1.10	—
SP-8014	Powdered	160	—	1.25	—
FRJ-551	Lump	185	X or lighter	1.10	—

*Ring & Ball.
**Gardner-Hellige (50% in Toluol).
***Gardner-Hellige (60% in Toluol).

SP-102

Chemical Type: Oil-soluble, heat-reactive, phenolic resin based on a para-substituted alkyl phenol.

Key Properties: Its widest use is in the formulation of neoprene-based contact cements, both general-purpose and heat-activating. It is also used in the formulation of adhesives with NBR, SBR, natural and reclaimed rubber.

SP-103

Chemical Type: Oil-soluble, heat-reactive, phenolic resin based on a para-substituted alkyl phenol.

Key Properties: Its widest use is in the formulation of general-purpose and heat-activating neoprene cements. The outstanding feature of this resin is its extremely light color. It is also used in the formulation of adhesives with NBR, SBR, natural and reclaimed rubbers.

In the specific case of neoprene contact cements, SP-103 resin extends tack life, increases the specific adhesion of the film to metal and glass and increases the cohesive strength of the adhesive film itself.

SP-126

Chemical Type: Oil-soluble, heat-reactive, phenolic resin.

Key Properties: The proper incorporation of this resin into neoprene contact cements yields formulations with excellent static and dynamic strength properties, at both room and elevated temperatures.

Neoprene cements utilizing SP-126 Resin are extensively used in the shoe, automotive, furniture, aircraft and building industries. It is compatible with both neoprene and nitrile rubbers in all properties.

SP-134

Chemical Type: Oil-soluble, heat-reactive, phenolic resin.

Key Properties: Developed primarily for the formulators of neoprene solvent-type adhesives desiring outstanding qualities of lightness in color, high heat resistance and optimum cohesive strength. Recommended for the shoe industry, automotive industry, building industry and furniture industry.

SP-144

Chemical Type: Oil-soluble, heat-reactive, alkyl phenolic resin.

Key Properties: Developed for neoprene contact adhesives, both general-purpose and heat-activated.

SP-154

Chemical Type: Oil-soluble, heat-reactive, phenolic resin.

Key Properties: Developed specifically for formulating non-phasing neoprene solvent-type contact cements. In addition, it has excellent qualities of high heat resistance, light color and high cohesive strength.

Since neoprene adhesives based on SP-154 resin are phase resistant, the applications available are considerably broader than with the previous unstable neoprene adhesives. The new areas include the show, automotive, aircraft, furniture, building and numerous other industries. The added features of high heat resistance, high cohesive strength and light color further increase the number of potential uses.

SP-155

Chemical Type: Oil-soluble, heat-reactive, alkyl phenolic resin.

Key Properties: Developed for neoprene contact adhesives. In this application, SP-155 affords heat resistance intermediate between adhesives formulated with SP-144 and SP-154. Adhesives formulated with SP-155 are phase resistant.

SP-174

Chemical Type: Oil-soluble, heat-reactive, phenolic resin.

Key Properties: Designed primarily for neoprene solvent-type contact adhesives, where high heat-resistance is required for peel and/or shear adhesion. The relatively short open time enables this resin to be used in adhesives designed for highly automated production applications such as the automotive industry.

SP-553

Chemical Type: Oil-soluble, thermoplastic, terpene phenolic resin.

Key Properties: SP-553 is used to increase the tack retention of both neoprene and SBR adhesives. In the case of neoprene adhesives, the use of excessive amounts of this resin will cause a considerable reduction in the cohesive strength of the adhesive film. SP-553 will function as an effective tackifier for neoprene cements even in the presence of magnesium oxide where it will not react to form a non-tacky product.

SP-560

Chemical Type: Thermoplastic, oil-soluble, terpene phenol resin.

Key Properties: Used extensively as the base resin or as a modifying resin in the formulation of neoprene cements. Neoprene cements are directly formulated from simple mixtures of neoprene and SP-560 Resin. These cements display good strength and tack retention properties. However, they are somewhat deficient in high-temperature performance. To overcome this deficiency, SP-560 may be blended in any proportion with heat-reactive resins.

SP-8014

Chemical Type: One-step, thermosetting, phenolic resin.

Key Properties: Used in the formulation of nitrile rubber (NBR) adhesives with exceptional shelf-life. This resin is also used in acrylic and urethane adhesives. Completely compatible with all NBR rubbers. Adhesives formed from such combinations have good heat resistance, excellent oil and vinyl plasticizer resistance and high bond strengths. In general, these adhesives have a short tack range and, therefore, bonds must be completed while the adhesive is still wet. They are used to bond NBR rubber, metal and vinyl plastics.

FRJ-551

Chemical Type: Heat-reactive, alkyl phenolic resin.

Key Properties: Designed primarily for use in the formulation of Neoprene AF-based contact adhesives. FRJ-551 imparts low initial viscosity and excellent viscosity-stability to adhesives of this type. In adhesives based on Neoprene AC or AD, FRJ-551 gives good bond strengths, both at room and elevated temperatures.

Section XXVII

Resins–Polyamide

AZS CHEMICAL CO., 762 Marietta Blvd., N.W., Atlanta, GA 30318

Azamide Polyamide Resins

Product Name	Softening Point (°C)	Gardner Color	Acid Value (max)	Amine Value (max)	Brookfield Viscosity @ 160°F (cp)	Specific Gravity
11	120	9	4	4	350	0.98

Chemical Type: Alcohol soluble.

Key Properties: Designed for use in flexographic and rotogravure inks, overprint varnishes, metal lacquers, heat seal coatings and hot melt adhesives. Excellent adhesion to paper, foil and plastic films with good resistance to water, grease and detergents.

30	110	9 max	4	4	2,400	0.98

Chemical Type: Thermoplastic. Produced by the condensation reaction of dimerized fatty acids and polyamines.

Key Properties: Finds extensive application in heat seal adhesives and in formulating flexographic inks due to its excellent adhesive quality, flexibility and sharp melting point. Generally classified as a cosolvent resin.

40	110	9 max	4	4	1,500	0.98

Chemical Type: Thermoplastic.

Key Properties: Finds widespread use in flexographic ink, hot melt adhesives, heat seal coatings and overprint coatings.

Product Name	Solids Content (%)	Gardner Color	Amine Value	Brookfield Viscosity (cp)	Flash Point (°C)	Specific Gravity
90	100	8	90	950 @ 150°C	295	0.97

Chemical Type: High equivalent weight dimer acid based polymeric polyamide resin. Reactive, semisolid.

Key Properties: Solutions are used to formulate epoxy coatings with excellent chemical and corrosion resistant properties. Can also be used in hot melt resin systems as a flexibilizer and modifier to improve properties of heat seal, overprint coating and adhesive formulations. (This resin is also available in various solvent-based variations.)

215	100	8	238	100 @ 150°C	295	0.97

Chemical Type: Intermediate equivalent weight dimer acid based polyamide resin.

Key Properties: Used with epoxy resins to formulate coatings and adhesives. Ratio used is non-critical and the proportion can be varied to achieve varying properties of the cured formulation. Increasing the amount of Azamide 215 will increase flexibility while a decrease in the ratio of Azamide 215 will increase hardness. Imparts roughness and resiliency to epoxy adhesives. Can be used as a corrosion inhibitor, asphalt additive or plasticizer for other resin systems. (This resin is also available in various solvent-based variations.)

Product Name	Solids Content (%)	Gardner Color	Amine Value	Brookfield Viscosity (cp)	Flash Point (°C)	Specific Gravity
300DM	100	16	400	900 @ 25°C	155	0.96

Chemical Type: Low cost, modified fatty imidazoline-amino amine.

Key Properties: Can be used as a coreactant for both liquid and solution type epoxy resins. Increasing the quantity of Azamide 300DM will increase flexibility, while decreasing the amount will increase hardness. Polyamines, tertiary-amines, amine adducts, bisphenol A, resorcinol and phenol accelerators have been used successfully to speed cure mixtures of Azamide 300DM and epoxy resins in thin films. The quantity of accelerator can be increased or decreased to meet specific curing requirements.

325	100	8	345	9,600 @ 40°C	265	0.97

Chemical Type: Low equivalent weight, dimer acid based polyamide resin.

Key Properties: The ratio of Azamide 325 to liquid epoxy resin is not critical and can be varied. Decreasing the concentration will produce harder, more brittle systems while increasing the concentration will increase flexibility. The use of accelerators, such as phenol, bisphenol A, resorcinol, polyamines, polyamine adducts, tertiary amines and other accelerators will decrease curing time and shorten pot life. Sand, wood flour, crushed walnut shells, silica flour, aluminum powder, glass fibers, asbestos and other fillers can be used. The handling properties of the uncured formulation can be modified with thixotropes. Azamide 325 can be used also as a corrosion inhibitor, asphalt additive or as a plasticizer for other resin systems.

340	100	8	375	3,600 @ 40°C	200	0.97

Chemical Type: Low viscosity dimer acid based imidazoline type polyamide resin.

Key Properties: Ideal curing agent for flooring systems, low viscosity adhesives or high solids coatings. The ratio of Azamide 340 to liquid epoxy resin is not critical and can be varied to achieve various degrees of flexibility and/or hardness. Accelerators, such as tertiary amines, bisphenol A, phenol, resorcinol, polyamines, polyamine adducts, other accelerators and/or heat can be used to speed up the cure. Sand, aluminum powder, alumina, silica flour, and a variety of other fillers can be used in these systems. Thixotropes can be used to impart thixotropy. Azamide 340 can be used also as a corrosion inhibitor, asphalt additive or as a plasticizer for other resin systems.

360	100	7	413	330 @ 25°C	175	0.94

Chemical Type: A fatty imidazoline-amino amide. Can be used as a coreactant for both liquid and solution type epoxy resins.

Key Properties: Increase the quantity of Azamide 360 to increase flexibility and decrease the amount to increase hardness. Accelerators such as polyamines, tertiary amines, amine adducts, bisphenol A, resorcinol and phenol can be used to meet specific curing requirements. Azamide 360 can also be used as a corrosion inhibitor, asphalt additive or as a plasticizer for other resin systems.

390	100	9 max	435	T–V*	>450°F	0.97

*Gardner-Holdt.

Chemical Type: A very low viscosity dimer acid based polyamide type epoxy resin hardener.

Key Properties: Ideal curing agent for epoxy flooring, casting, adhesive and laminating applications. The ratio of Azamide 390 to liquid epoxy resin is not critical and can be varied to change the physical properties of the finished product. The reaction can be speeded up through the use of the various accelerations normally used in epoxy resins.

430-XB-60	60	9	123	1,150	95°F	0.95

Chemical Type: Polyamine adduct, 60% in n-butanol-xylene.

Key Properties: Used with epoxy resins to formulate coatings and adhesives. The ratio to epoxy resins is noncritical and the proportion can be varied to achieve different properties of the cured formulation. Increasing the amount of Azamide will increase flexibility while a decrease in the ratio will increase hardness.

Product Name	Solids Content (%)	Gardner Color	Amine Value	Brookfield Viscosity (cp)	Flash Point (°C)	Specific Gravity
450	100	8	550	625	140	0.98

Chemical Type: Modified fatty imidazoline-amino amide.

Key Properties: Typically used as a coreactant for liquid epoxy resins. The ratio of Azamide 450 to liquid epoxy resin is relatively noncritical and the proportions can be varied to change the physical properties of the system. Azamide 450 will produce minimum amine blush when used with liquid epoxy resins and will cure well under conditions of high atmospheric humidity.

Blends of Azamide 450 with Azamide 360 can be used with liquid epoxy resin to develop low viscosity 100% solids systems with varying pot life and cure time.

| 600 | – | 10 max | 600 | 1,450 | >200°F | 0.98 |

Chemical Type: Fatty amine-amide coreactant for epoxy resins.

Key Properties: Normal mix is 30 parts by weight of Azamide 600 with 100 parts epoxy resin. Recommended for coatings requiring good chemical and water resistance. May also be used in adhesive and laminating applications.

| 680B75 | 75 | 10 max | 247.5 | 2,250 | 115°F | 0.97 |

Chemical Type: Reactive polyamine adduct solution, 75% solids solution in butanol.

Key Properties: Recommended as a hardener for epoxy resin coating systems. Both solid epoxy resin solutions and liquid epoxy resins may be cured with it. Shows excellent compatibility with epoxy resins and the resulting coatings will exhibit good solvent and chemical resistance properties.

| 900 | 100 | 1.5 | 1,025 | 500 | 140 | 1.01 |

Chemical Type: Light colored modified polyamine safety hardener for use with liquid epoxy resins. May be used to cure these resins both at room and elevated temperatures.

Key Properties: Very low order of toxicity compared to typical polyamine curing agents. Suggested ratio is 22-28 parts of Azamide 900 to 100 parts of liquid epoxy resin. Typical end uses are in epoxy formulations for flooring, laminating, casting and other applications where physical contact is likely. Mixes well with all resins and modifiers at room temperature. Cured formulations will have low viscosity and good wetting characteristics. They will also have good electrical characteristics and will give low shrinkage and good dimensional properties. Various accelerators such as bisphenol A, resorcinol and phenol have been used to successfully speed cure mixtures.

HENKEL CORP., 425 Broad Hollow Road, Melville, Long Island, NY 11746

Versamid Polyamide Resins

Product Name	Amine Value	Gardner Color (max)	Brookfield Viscosity (cp)	Resinous (%)	Specific Gravity @ 25°C	Flash Point (°C)
100	90	9	950 @ 150°C	100	0.97	319
100-C60	54	9	950 @ 40°C	60	0.95	42
100-CX60	54	8	1,250 @ 40°C	60	0.94	28
100-IT60	54	9	700 @ 40°C	60	0.90	27
100-X65	58.5	9	2,500 @ 40°C	65	0.94	28
115	238	9	3,450 @ 75°C	100	0.97	295
115-X70	167	9	550 @ 40°C	70	0.94	27

(continued)

Product Name	Amine Value	Gardner Color (max)	Brookfield Viscosity (cp)	Resinous (%)	Specific Gravity @ 25°C	Flash Point (°C)
115-173	175	9	1,150 @ 40°C	73.2	0.94	20
125	345	9	800 @ 75°C	100	0.97	265
140	385	9	350 @ 75°C	100	0.97	185
150	385	8	150 @ 75°C	100	0.96	182
Waterpoxy	180	9	5,750 @ 25°C	65.5	0.97	39
280-B75	250	10	1,750	75	0.96	38

Solvent Code:

B	=	n-butanol	R	=	cellosolve
C	=	cellosolve	T	=	toluene
H	=	water	W	=	C$_8$ or greater aromatic hydrocarbon
I	=	isopropanol	X	=	xylene

Chemical Type: Condensation products of polyamines with dimer acids or fatty acids. Many isomers of dimer fatty acids are possible. Versamids are polyamide resins. Genamids are amidoamine resins.

Key Properties: Versamid and Genamid resins are epoxy coreactants which have been industry standards for many years. When they are blended with appropriate epoxy resins they will cause a reaction between the amine groups of the coreactant resin and the epoxide groups of the epoxy resin. This results in a nonreversible chemical reaction forming a solid, crosslinked, high molecular weight polymer. Because these coreactants are manufactured from fatty acids, the cured epoxy blends exhibit a high degree of internal plasticization and high impact strength. In addition, the high polarity of these coreactants contributes excellent substrate wetting, adhesion and cohesion.

Considering these properties, these coreactant resins have found wide use in the coatings and adhesives industries.

Recommended Usage:

	. . . . Versamid					
	100	115	125	140	150	5201-HR65
Coatings (general)	x	x	x	x	x	x
Maintenance	x	x	x	x	x	x
Primers	x	x	x	x	x	x
Enamels	x	x	x	x	x	x
High Build/High Solids	–	–	–	x	x	x
Vinyl Modified	x	x	–	–	–	–
Coal Tar Modified	–	–	x	x	x	–
Masonry	x	x	–	–	–	x
Paper	x	–	–	–	–	–
Flexible	x	x	–	–	–	–
Grouts	–	–	–	x	x	x
Toppings	–	–	–	x	x	x
Water Systems	–	–	x	–	–	x
Adhesives (general)	–	x	x	x	x	x
Laminations	–	–	x	x	x	x
Castings	–	–	–	x	x	–
Pottings	–	–	–	x	x	–
Encapsulating	–	–	–	x	x	–

Note: Versamid 100, 115 and 125 are also supplied in various solvents for ease of formulation.

	. . . . Genamid			
	250	747	2000	5701-H65
Adhesives (general)	x	x	x	x
Laminations	x	x	x	x
Castings	x	x	x	–
Pottings	x	x	x	–
Encapsulating	x	x	x	–

R.T. VANDERBILT CO., INC., 30 Winfield St., Norwalk, CT 06855

Van-Amid Polyamide Resins

Product Name	Solids Content (%)	Gardner Color (Liquid or Molten)	Amine Value	Viscosity @ 75°C (poises)	Specific Gravity
300	100	7	90	11.5 (@ 150°C)	0.97

Key Properties: Also available in three solvent-based solutions. One of these is exempt from air-pollution regulations.

315	100	7	238	34.5	0.97

Key Properties: Also available in two solvent-based solutions. One of these is exempt from air-pollution regulations.

325	100	7	345	8	0.97
340	100	7	375	4	0.97

Chemical Type: Condensation products of dimerized fatty acids with various polyamines.

Key Properties: Van-Amid 300, 315, 325 and 340 represent a complete line of chemically reactive polyamide resins. This series was the first group of polyamide resins commercially available with improved color and a narrow color specification range. Manufactured to assure batch-to-batch uniform quality. Specifically used in conjunction with epoxy resins for protective and decorative coatings, laminatings, castings, adhesives and potting and encapsulating compounds.

Product Name	Solids Content (%)	Gardner Color (max.)	Amine Value	Viscosity @ 160°C (poises)	Softening Point (°C)	Specific Gravity
3030	100	12	1.5	24	110	0.98
3040	100	12	1.3	15	110	0.98

Chemical Type: Van-Amid 3030 and 3040 are solid polyamide resins derived from the reaction of dimer acids and linear diamines. They are thermoplastic in nature and exhibit characteristically sharp melting points.

Key Properties: These resins are designed for use in the formulation of flexographic and rotogravure printing inks, overprint varnishes, lacquers and hot melts. They are primarily useful in inks employing a co-solvent system, and provide many useful properties there. They are soluble in a wide range of solvent combinations. They are manufactured under the maximum possible control to conform to specifications.

Section XXVIII
Resins–Polyethylene

ALLIED CHEMICAL, Fibers and Plastics Co., P.O. Box 2332R, Morristown, NJ 07960

A-C Oxidized Polyethylene Homopolymers

Product Name	Softening Point (Ring & Ball) (°C)	Needle Penetration	Specific Gravity	Viscosity (Brookfield @ 140°C) (cp)	Acid Number (mg KOH/g)
629 & 629A	104	5.5	0.93	200	16
655	107	2.5	0.93	210	16
656	100	9.0	0.92	185	15
680	110	1.5	0.94	250	16
690	111	1.2	0.95	250	16
316 & 316A	140	<0.5	0.98	30,000 @ 149°C	16
392	138	<0.5	0.99	9,000 @ 149°C	28
395	134	<0.5	1.00	2,400 @ 149°C	41

A-C Polyethylene Homopolymers

1702	85	90.0	0.88	40	Nil
617 & 617A	102	7.0	0.91	180	Nil
6 & 6A	106	4.0	0.92	350	Nil
7 & 7A	107	2.5	0.92	350	Nil
8 & 8A	116	1.0	0.93	400	Nil
9 & 9A	117	0.5	0.94	450	Nil
712	108	3.5	0.90	1500	Nil
715	109	2.5	0.92	4000	Nil
725	108	3.5	0.92	1400	Nil
735	110	2.5	0.92	6000	Nil

Polymist Polyethylene Fine Powders

Product Name	Specific Gravity	Needle Penetration	Softening Point (ASTM E28) (°C)	Acid Number	Particle Size (μ)	Particle Size Range (μ)
A-12	0.99	0.5	140	30	12	2-24

Chemical Type: Modified high density polymer type.

B-6	0.96	0.5	128	Nil	6	2-12

Chemical Type: High density polymer type.

Product Name	Specific Gravity	Needle Penetration	Softening Point (ASTM E28) (°C)	Acid Number	Particle Size (μ)	Particle Size Range (μ)
B-12	0.96	0.5	128	Nil	12	2-24

Chemical Type: High density polymer type.

Product Name	Specific Gravity	Needle Penetration	Softening Point (ASTM E28) (°C)	Acid Number	Particle Size (μ)	Particle Size Range (μ)
B-21	0.96	0.5	128	Nil	21	3-35

Chemical Type: High density polymer type.

BARECO DIVISION, 6910 East 14th St., P.O. Drawer K, Tulsa OK 74112

Polywax Polyethylenes

Product Name	Melting Point (ASTM D127) (°F)	Needle Penetration (@ 77°F)	Viscosity (ASTM D2669) (@ 300°F)	Color Saybolt (ASTM D156)	Acid No.	Sapon. No.
500	187	7.0	3.0	5.0	–	–
655	215	3.0	6.0	5.0	–	–
1000	235	1.0	11.0	5.0	–	–
2000	257	0.5	50.0	5.0	–	–
E-730	200	6.0	10.0	0.5*	30	50
E-2018	242	1.0	38.0	0.5*	18	34
E-2020	242	1.0	38.0	0.5*	21	38

*ASTM D1500

DUPONT CO., Wilmington, DE 19898

Product Name	Specific Gravity	Bulk Density (lb/ft³)	Melt Index (ASTM D1238)	Softening Point (Vicat) (°F)	Tensile Strength (psi)	Percent Elongation at Break	Elastic Tensile Modulus (psi)
Elvax 1310	0.922	30	2.1	201	1750	600	35,000

Chemical Type: High molecular weight, low density polyethylene homopolymer in colorless pellet form.

Cloud point in fully refined paraffin wax, AMP 146°F:

5% in Wax	190°F (88°C)
10% in Wax	198°F (92°C)

Key Properties: Suitable for use in modified wax coatings, printing inks and hot melt adhesives. It improves hardness, gloss and scruff resistance of wax coatings without significantly affecting water vapor barrier properties.

In applications requiring higher seal strength or improved grease barrier properties, use of an "Elvax" ethylene/vinyl acetate copolymer is suggested.

Approved for food contact use under F.D.A. Food Additive Regulation 177.1521.

U.S. INDUSTRIAL CHEMICALS CO., 99 Park Ave., New York, NY 10016

Petrothene Low Density Polyethylene Resins

Product Name	Melt Index (ASTM D1238)	Specific Gravity	Tensile @ Break (ASTM D638) (psi)	Hardness (Shore A)	Softening Point (Ring & Ball) (°C)
NA 592	1.8	0.918	2,020	>95	379
NA 593	22	0.915	1,440	>95	267
NA 594	70	0.914	1,100	>95	248
NA 596	150	0.913	990	94	238
NA 597	250	0.927	1,520	95	252
NA 595	425	0.910	980	94	216

Chemical Type: Low molecular weight polyethylene (PE) homopolymers.

Key Properties: Useful in low cost systems where they impart strength, gloss and flexibility to a variety of adhesive and coating applications. They are compatible with a broad range of waxes and modifiers and have flow characteristics which are suitable for many end uses.

In wax-based coatings, a decrease in melt index improves the following:

Heat seal strength.
Flexibility.
Creased barrier properties.
Low temperature performance.
Hot tack.

Melt viscosity is increased and gloss retention is reduced.

In adhesives, a decrease in melt index improves the following:

Cohesive strength.
Melt viscosity.
Flexibility.
Low temperature performance.
Heat resistance.
Hot tack.

Open time is reduced.

Resins—Polypropylene

CROWLEY CHEMICAL CO. 261 Madison Ave., New York, NY 10016

Polymer C

Chemical Type: Amorphous polypropylene of low softening point and low viscosity. Waxlike, noncrystalline, slightly tacky solid.

Physical Properties:

Specific Gravity @ 60°F	0.865
Color	Off-white
Softening Point (Ring and Ball)	250°F
Viscosity, Thermacel @ 300°F	240 cp
Viscosity, Thermacel @ 350°F	130 cp
Heavy Metals	25 ppm
Fire Point	535°F

Key Properties: Polymer C, or combinations of it with microcrystalline wax, provides excellent lamination bonds for glassine, kraft, paperboard, film and foil to paper combinations.

Polymer C can be used as a base for proprietary formulations, since it is compatible with commonly used additives, such as resins, rosins, EVA, etc.

Polymer C improved MVTR at bends and creases when blended with wax on paper, canvas or other membranes.

In combination with oils and other ingredients in the preparation of lubricants and anti-rust compounds.

As a wax extender in coatings for glassine and greaseproof papers and wax-impregnated corrugated boxes.

Polypol 19

Chemical Type: Low molecular weight amorphous polypropylene. Soft, noncrystalline, slightly tacky polymer. Off-white color.

Physical Properties:

Specific Gravity @ 60°F	0.8544
SSU Viscosity @ 210°F	3150
Molecular Weight	2000
Fire Point	460°F
Metals Content	25 ppm
Softening Point	68°F

Key Properties: Caulking & Sealing Compounds—Replaces higher viscosity grades of polybutenes. Its viscosity curve at lower temperatures reduces cold flow vs. polybutenes.

Rubber—Tackifier for EPDM. Provides improved tear strength and good mill tack. Also of value in Neoprene formulations.

Other Uses.

Polypropylenes

Product Name	Softening Point (Ring & Ball) (°F)	Viscosity (Brookfield) (cp)	Viscosity Temp. (°F)	Specific Gravity @ 60°F	Metals Content	Fire Point (°F)
Polytac R-500	300	1500	375	0.866	25 ppm	525
Polytac R-1000	300	4000	375	0.866	25 ppm	525

Chemical Type: Amorphous (atactic) polypropylene, a fully saturated polymer, available in two grades.

Key Properties: Compatible with a wide range of organic materials. Widely used, either by itself, or formulated, in the general area of hot melt adhesives, sealants, mastics, potting compounds, etc., where advantage can be taken of its relative low cost, inertness, low odor, etc.

Polymer C	250	240	300	0.865	25 ppm	535
	–	130	350	–	–	–

Chemical Type: Amorphous polypropylene of low softening point and low viscosity.

Key Properties: Polymer C, or combinations with microcrystalline wax, provide excellent lamination bonds. Can be used as a base for formulations with commonly used additives, such as resins, rosins, EVA, etc. Valuable as a wax extender in coatings. Polymer C improved MVTR at bends and creases when blended with wax.

		(SSU)				
Polypol 19	68	3150	210	0.8544	25	460

Chemical Type: Low molecular weight amorphous polypropylene.

Key Properties: Replaces higher viscosity grades of polybutenes in caulking and sealing compounds. Its viscosity curve at lower temperatures reduces cold flow vs. polybutenes.

Hitac 300	300	9000	375	0.870	25	450

Chemical Type: Amorphous polypropylene and ethylene copolymer available in slab form.

Key Properties: Has excellent oil and grease resistance in caulking and sealing compounds. Compatible with EPDM, butyl, polybutenes and similar sealant compounds. Has excellent surface tack in adhesives. Compatible with hydrocarbon resins, EVA and rosins.

F.D.A. Approval: All grades of Crowley amorphous polypropylene are approved for food-packaging adhesives under F.D.A. Regulation 121.2520.

Polytac

Chemical Type: Amorphous (atactic) polypropylene is a fully saturated polymer available in two grades based on Brookfield Thermacel Viscosity @ 375°F.

Grade R-500	500-2,500 cp
Grade R-1000	2,500-5,500 cp

Physical Properties:

Specific Gravity @ 60°F	0.866
Color	Cream
Softening Point (Ring and Ball)	290°-310°F
Metals Content	25 ppm
Appearance	Solid
Fire Point	525°F

Key Properties: Compatible with a wide range of organic materials, including the following:

Petroleum hydrocarbon resins.

Hydrogenated rosin and tall oil rosin and rosin esters.

Low molecular weight polyethylene.

Butyl and similar elastomers and TPR elastomers.

Most waxes and mineral oils and polybutenes.

Asphalt, nonoxidized.

Ethylene-vinyl acetate and ethylene-ethyl acrylate.

Key Properties: Polytac is widely used by itself, or formulated, in the general area of hot melt adhesives, sealants, mastics, potting compounds, etc. where advantage can be taken of its relatively low cost, its inertness, its low odor, etc. Some uses are:

Adhesive Applications: As a base in hot melt formulations, combined with resins, rosins, EVA, etc.

Hot Melt Lamination Applications:

In multi-wall bags, kraft-to-kraft or kraft-to-burlap.

Fiberglass reinforced kraft tape.

Fiberglass reinforced kraft carpet wrap, lumber wrap and steel wrap.

Kraft-to-film, film-to-foil and kraft-to-polyethylene.

Paperboard/kraft laminate. Barrier for moisture, oil and grease. Prevents delamination at elevated temperatures.

Carpet Applications:

Lamination of secondary backing to latex tie-coated carpet at exit end of drying oven to obtain faster oven through-put.

To improve tuft lock in automotive carpet, as a tiecoat prior to sintering polyethylene.

It can be formulated to provide an inexpensive backing replacement for vinyl and latex for modular carpet tiles.

Lamination of jute in automotive carpet underlay to provide heat barrier.

Sealant Applications:

Architectural sealants, particularly when modified with polybutenes. The combined product has good flexibility and tack at low temperatures. Eliminates cold flow in compounded sealants.

In tape caulks, combined with butyl, polybutenes and inert filler.

In automotive and appliance sealants where its low odor is of importance. It is compatible with most elastomers. It can be compounded with up to 75% inert fillers.

Polytac is an excellent replacement for asphalt when nonstaining is a prerequisite.

Acoustical Applications: In auto and mobile home sound deadening materials, where after high filler loading, it can be contour shaped.

Rubber Applications: As a processing aid in tread rubber to improve extrusion and mold flow and tackifier for EPDM.

Section XXX
Resins–Polyterpene

ARIZONA CHEMICAL CO., Wayne, NJ 07470

Zonarez Polyterpene Resins

Product Name	Softening Point (Ring & Ball, °C)	Gardner Color	Specific Gravity
7010	10	2	0.97
7025	25	2	0.97
7040	40	2	0.97
7055	55	2	0.98
7070	70	3	0.98
7085	85	3	0.97
7100	100	3	0.97
7115	115	3	0.99
7125	125	3	0.99

Physical Constants of All Grades:

Appearance	clear and bright
Acid Number	less than 1
Saponification Number	less than 1
Ash content	less than 0.005%
Toluene insolubles	less than 0.005%
Chlorine content	negative to Beilstein Test less than 50 ppm

B-10	10	2+	0.92
B-25	25	2+	0.93
B-40	40	2+	0.94
B-55	55	3	0.96
B-70	70	3	0.96
B-85	85	3	0.99
B-100	100	3	0.99
B-115	115	3	0.98
B-125	125	3	0.97

Physical Constants of All Grades:

Appearance	clear and bright
Acid Number	less than 1
Saponification Number	less than 1
Ash content	less than 0.005%
Toluene insolubles	less than 0.005%
Chlorine content	negative to Beilstein Test less than 50 ppm

Chemical Type: Thermoplastic polymers which are produced by the polymerization of terpene hydrocarbons consisting primarily of beta-pinene and dipentene. The Zonarez B resin series is produced by the polymerization of beta-pinene. The Zonarez 7000 resin series is based primarily upon dipentene. The primary difference between the series is molecular structure which results from basic differences in the terpene monomers used for their production. They are differentiated mainly by molecular weight, molecular weight distribution and polymer structure.

Key Properties: Characterized by outstanding quality, uniformity and FDA approval. Bright, clear and pale colored, low molecular weight polymers which impart high levels of tack and adhesion to many elastomeric and polymeric materials. Excellent stability in adhesive systems whereby variations in color, viscosity, holding power and tack are minimized. They are neutral in nature and have good resistance to attack by acids, alkalis, salts and water and exhibit excellent compatibility with numerous organic solvents, rubbers and elastomers, drying oils, polyethylenes, ethylene copolymers, waxes, esters, terpenes, rosin and rosin derivatives.

Recommended for use for pressure-sensitive adhesives, rubber cements, solvent-based adhesives, emulsion adhesives, hot melt adhesives and coatings, can sealants, caulking and general sealant compounds, paints, concrete waterproofing agents, varnishes and many other uses.

FDA Approval: Zonarez B and 7000 series of resins are approved for use in many applications covered by FDA regulations.

CROSBY CHEMICALS, INC., 600 Whitney Bldg., New Orleans, LA 70130

Crosby Terpene Resins

Product Name	Color	Softening Point (Ring & Ball, °C)
Croturez B-85	WW	85
Croturez B-100	WW	100
Croturez B-115	WW	115

Chemical Type: High quality pale terpene resins produced by the polymerization of terpene hydrocarbons consisting of beta-pinene.

Key Properties: Characterized by low acid number, pale color and low ash content. The Croturez resins have very good resistance to alkalies and acids. They also have excellent compatibility with solvents and resins which are commonly used in adhesives. Croturez resins are generally used as tackifiers in adhesive systems.

Croturez B-115 does have FDA approval to use in chewing gum.

GOODYEAR CHEMICALS, 1144 East Market St., Akron, OH 44316

Wingtack Plus Resin

Chemical Type: A synthetic polyterpene resin.

Chemical Analysis and Physical Properties:

Appearance	light yellow flake
Gardner Color, 50% in toluene	4
Softening Point, R&B	94°C
Ash Content	0.1% max
Acid Number	1.0 max
Specific Gravity	0.93
Heat Stability, 5 hr @ 350°F	
Volatiles	3.2%
Skinning	0.0%
Aged Color	8.5

Key Properties: A new, modified resin developed to help improve the properties and economics of adhesives based on SIS block polymers. Wingtack Plus can be used at higher than normal resin levels in SIS systems with no loss of tack, peel and shear adhesion. Since you can increase resin levels by 10 to 15%, significant cost savings can be realized. Other advantages include good heat stability, low volatility and unusually light color.

Compatible with SIS, natural rubber, Natsyn, EPDM and butyl. Soluble in aromatic solvents, aliphatic solvents, chlorinated solvents and esters.

Regulatory: Wingtack 95 may be used in accordance with the following regulations: 121.2514, 121.2526, 121.2571, 121.2550, 121.2577, 121.2520 and 121.2562.

Wingtack 10 Liquid Tackifying Resin

Chemical Type: Unique liquid synthetic polyterpene resin.

Chemical Analysis and Physical Properties:

Appearance	light amber liquid
Softening point (R&B)	10°–15°C
Gardner color, 50% in toluene	4
Specific gravity @ 25°C	0.880
Brookfield viscosity	20–40,000 cp
Molecular weight	490
Iodine number	100
Acid number	0.18
pH	7

Key Properties: Developed for plasticizing and tackifying applications. It is very similar chemically to Wingtack 95 and 115, yet is a completely polymeric liquid at room temperature. The resin contains no solvents, oil additives, or chlorides.

Compatible with a wide variety of elastomers and hot melt polymers. Compared to other plasticizing materials, the liquid resin exhibits relatively low volatility. It can be used in hot melt compounds to lower molten viscosity with little effect on softening point, or blended with other Wingtack resins to achieve a lower softening point.

In pressure-sensitive adhesive applications, Wingtack 10 resin can be used effectively to increase tack adhesion, modify shear and peel strength and lower viscosity and softening point. It is recommended to improve adhesion in low temperature tapes, surgical tape and electrical tapes.

When used as a plasticizing agent in hot melt, pressure-sensitive adhesives, Wingtack 10 increases tack adhesion and helps prevent "stringing" on hot melt applicators and air bubble formation on roller applicators. These processing advantages, which reduce possibilities of degradation, are gained by lowering viscosity, not by increasing temperature. Hot melt, pressure-sensitive adhesive applications for Wingtack 10 include self-stick floor tiles and various types of tapes and labels.

The liquid resin is also suggested for use as an aid in rubber milling to ease processing.

Wingtack 95 Tackifying Resin

Chemical Type: A unique synthetic polyterpene resin.

Chemical Analysis and Physical Properties:

Appearance	light yellow flake
Gardner color, 50% in toluene	4
Softening point, R&B	95°–105°C
Ash content	0.10% max
Acid number	1.0 max
Iodine number	30
Specific gravity	0.93

Key Properties: Recommended for use in –

Hot Melt Coatings: Proven to be an excellent modifying resin in these applications. In hot melt coatings, Wingtack has a number of distinct qualities—high gloss, heat sealability, a minimum water vapor transmission rate and viscosity stability.

Hot Melt Adhesives: When used in hot melt adhesives, exhibits good hot tack and wettability, excellent shear and peel adhesion, low volatility and viscosity stability under prolonged heat.

Pressure-Sensitive Adhesives: Include a number of application areas such as surgical tape, masking tape, electrical tape and various labels. In all of these uses, the performance of the adhesive is dependent upon four factors: wetting ability (quick stick), adhesion, cohesive strength, and good aging strength without loss of strength or color. As a replacement for natural polyterpenes, Wingtack 95 possesses these qualities, both in solvent-based and hot melt systems and offers an economic advantage over the natural polymers.

Other applications.

Regulatory: Wingtack 95, recognized as a synthetic polyterpene, meets FDA regulations for use in food contact applications where natural polyterpenes are allowed: 121.2514, 121.2520, 121.2526, 121.2550, 121.2562, 121.2571, 121.2577.

Wingtack 115 Tackifying Resin

Chemical Type: A new synthetic polyterpene resin with a high softening point.

Chemical Analysis and Physical Properties:

Appearance	light yellow flake
Softening point, R&B	115°–120°C
Gardner color (50%)	
Initial	7 max
Aged	12
Gardner viscosity (70%)	R–S
Solubility*	–5°C
Volatility, aged**	0.5% max
Ash content	0.1% max
Specific gravity	0.94

*The temperature at which a precipitate is formed in a 20% solution of Wingtack 115 resin in 50/50 MEK and toluene.
**Aged 5 hours at 350°F.

Key Properties: Recommended for use in pressure-sensitive adhesives. Especially suited to pressure-sensitive applications such as cellophane tape, masking tape, industrial tape, surgical tape and various types of labels.

Compared to natural polyterpenes, Wingtack 115 Tackifying Resin has exceptionally high shear strength after aging and good 180° peel strength. It offers an economic advantage over the more expensive 115° polyterpene resins. The low specific gravity of Wingtack 115 resin (0.93 average) gives more volume per pound than competitive resins with higher specific gravities.

Hot Melt Adhesive Applications: Can also be used to advantage in hot melt adhesives. Its high softening point and wide compatibility with a variety of elastomers, polyolefins, waxes and oils make it a good choice for higher temperature hot melt adhesive applications. The resin has good tack and wettability, low volatility and viscosity stability under prolonged heat.

REICHHOLD CHEMICALS, INC., RCI Bldg., White Plains, NY 10603

Nirez Series 1000 Polyterpene Resins

Physical Properties:	1010 (36-500)	1040 (36-501)	1085 (36-503)	Nirez 1100 (36-504)	1115 (36-505)	1125 (36-506)	1135 (36-507)
Softening point, R&B	—	37-43	82-88	97-103	112-118	122-128	132-138
Gardner color, 1933, 50% N.V. in toluene, max	4	4	4	4	4	4	4
Gardner-Holdt Viscosity @ 25°C	Z7	—	—	—	—	—	—
Capillary tube melting point, °C	—	—	65	80	95	105	115
Flash point, C.O.C.							
°F	240*	334	382	409	445	467	528
°C	116	168	194	209	229	242	276
Fire point, C.O.C.							
°F	360	374	398	428	>500	535	580
°C	182	190	203	220	>260	279	304

*Pensky-Martens Closed Cup.

Properties (All Grades):

Chlorine, Beilstein Test	negative
Appearance	light in color
Acid number	less than 1
Saponification number	less than 1

Chemical Type: The seven resins currently comprising this series are all pale, nonyellowing, thermoplastic, polyterpene hydrocarbon resins.

Key Properties: A broad range of softening points is presented from which to select. Chemical inertness is a prime feature of all of these resins. They are neutral and unsaponifiable, displaying excellent resistance to acids, alkalies and saline solutions. They remain color stable upon exposure to elevated temperatures.

They are completely soluble in a wide variety of low cost mineral oils, petroleum solvents, chlorinated hydrocarbons, long chain alcohols and other industrial solvents. These resins exhibit excellent compatibilities with many materials. The following are suggested uses:

Paper Coatings: With polyethylene for grease- and oil-resistant, moisture-vapor-resistant, heat-sealing coatings.

Rubber Compounding: As softeners and tackifiers for pale rubber compounds.

Adhesive Manufacture: For masking and binding tapes, and heat-sensitive, solvent-sensitive and pressure-sensitive adhesives including those used for automotive and electrical applications. Pale colored adhesives for floor tile.

Architectural: Moisture-proof concrete coatings for curing and sealing purposes.

Nirez T-4115 (Resin)

Chemical Type: Pale, hard, thermoplastic, friable, polyterpene resin produced from beta-pinene.

Physical Properties:

Gardner Color, 50% non-volatile in toluene	3–4
Softening point, R&B	116°C (241°F)
Melting point, capillary tube	96°C (205°F)
Acid number	0.2 max
Saponification number	1 max
Ash content	0.1% max
Specific gravity @ 25°/25°C	0.995

Key Properties: A prime feature is chemical inertness. Neutral and unsaponifiable and exhibits excellent resistance to acids, alkalies and saline solutions. Soluble in low cost mineral oils, petroleum solvents, long chain alcohols, chlorinated hydrocarbons and other industrial solvents. Compatible with polyolefins, ethylene-vinyl acetate copolymers, and various waxes. Compatible, also, with most rubbers and other terpene resins.

The principal application for Nirez T-4115 is the production of high-quality pressure-sensitive adhesives, particularly in combination with natural rubber. It also has application with polyethylene for grease- and oil-resistant, moisture-vapor-resistant, heat-sealing coatings; with wax, polyethylene, and ethylene-vinyl acetate copolymers for moisture-vapor-resistant, heat-sealing coatings.

Toxicity: Threshold Limit Value (TLV) unknown. Refer to manufacturer's Material Safety Data Sheet.

Section XXXI
Resins–Powder Coating

HERCULES, INC., 425 Broad Hollow Road, Melville, Long Island, NY 11746

Hercotuf Powder Coatings

Chemical Type: Fine particle size of polypropylene that can be formulated into a powder coating.

Physical Properties

Maximum Particle Size (100%)	<115 mesh
Melting Point	333°F
Specific Gravity	0.906
Refractive Index	1.5
Impact Resistance (Gardner, Inch-Pound)	>84
Hardness (Sward)	20
Lap Shear Bond Strength	3,500 psi
Flexibility	Good
Abrasion Resistance	Excellent
Chemical Resistance	Excellent
UV Resistance	Poor
Initial Gloss	Fair
Dielectric Constant (Molded-Resin Specimens @ 10_{kc}, 100_{kc}, 1000_{kc})	2.22–2.26
Dissipation Factor (Molded-Resin Specimens @ 10_{kc}, 100_{kc}, 1000_{kc})	0.0004–0.0009

Key Properties: Hercotuf Powder Coatings are based on one of the most versatile polymers in use in the plastics industry today. These coatings have the inherent advantages of polypropylene while overcoming the poor adhesion to metal characteristic of this inert polymer. These important advantages are offered:

Low density	Excellent dielectric properties
High melting point	High degree of chemical inertness
High tensile strength	Thermoplasticity
High abrasion resistance	Excellent adhesion to metal

Hercotuf Powder Coatings have these unique features:

(1) Self-priming: A single coat fuses to give excellent adhesion to a variety of metal substrates.

(2) Chemically inert: Polypropylene, like most crystalline polyolefins, has excellent resistance to water, greases, many common solvents, including esters, ketones, aromatics, aliphatics, and many weak and strong alkalies and acids.

(3) Thermoplastic structural adhesive: Since the polymer is thermoplastic, coated metal parts can be heat-fused at any time during or after the coating operation to form structural bonds.

(4) Impact- and abrasion-resistant: The coating from this relatively high molecular weight thermoplastic has excellent impact resistance and shows exceptional resistance to abrasion.

Hercotuf formulations can be applied by electrostatic fluidized bed, electrostatic gun and conventional fluidized bed. These coatings show excellent fluidization and give continuous films at thickness as low as 2½ mils. They will fuse to a continuous film at 400°F in 3 to 4 minutes once the substrate is up to temperature. A cold-water quench after removal from the heat source is recommended for optimum properties. Easy to apply.

Powder coatings are easily applied in uniform thickness without sagging. There is less chance of pinholes or bubbles. Excellent edge protection and wraparound are obtained with these coating techniques. Thus, rejects are minimized. Plant fire hazards can be reduced since powder coatings do not require volatile carriers. More favorable insurance rates should result. Collecting and recycling equipment permits almost total material use. Since air from powder reclaim units can be exhausted through filters back into the plant, fuel bills for air replacement are eliminated.

Hercotuf Powder Coatings are new to the market and are being evaluated for a variety of end uses. The properties inherent in polypropylene suggest their use for any metal component that needs protection from physical and chemical attack. Applications for these functional coatings could include appliance parts, such as dishwasher racks, refrigerator racks, washer tubs; metal containers such as pails and drums; pipe and electrical applications; and industrial equipment components. An important aspect is that Hercotuf powder coatings are available to comply with FDA regulations governing food contact surfaces and food packaging materials.

Section XXXII
Resins–Proprietary Composition

EXXON CHEMICALS, 3020 West Market St., Akron OH 44313

Escorez Resins

Product Name	Softening Point (°C)	Color (Gardner)	Melt Viscosity @180°C (cp)	Acid Number (mg KOH/g) (max)	Molecular Weight Average	Specific Gravity @18°C
1102	100	8	650	1	4800	0.97

Chemical Type: Hydrocarbon resin containing linear, branched and cyclic structures of an aliphatic nature, with a wide range of compatibility.

Key Properties: Recommended for use in pressure-sensitive adhesives, based on natural rubber and SIS block copolymer, rubber compounding, waterproofing treatments and carpet backing compounds.

Product Name	Softening Point (°C)	Color (Gardner)	Melt Viscosity @180°C (cp)	Acid Number (mg KOH/g) (max)	Molecular Weight Average	Specific Gravity @18°C
1304	100	6	410	1	1450	0.97
1310	94	6	250	1	1152	0.96
1315	118	7	1760	1	2210	0.97

Chemical Type (1300 Series): Hydrocarbon resins containing linear, branched and cyclic structures of an aliphatic nature. A range of pale-colored, aliphatic resins developed specifically for adhesives and hot melts.

Key Properties (1300 Series): Recommended for use in natural rubbers, SIS block copolymer and polyisoprene pressure sensitive adhesives, EVA-based hot-melt adhesives, hot melt road marking compounds, hot melt coatings and wax blends.

Product Name	Softening Point (°C)	Color (Gardner)	Melt Viscosity @180°C (cp)	Acid Number (mg KOH/g) (max)	Molecular Weight Average	Specific Gravity @18°C
2101	92	8	130	1	925	1.02

Chemical Type: An aromatic-aliphatic hydrocarbon resin with excellent stability and adhesive properties.

Key Properties: Recommended for use in natural rubber, random SBR, SIS block copolymer and SBS block copolymer pressure sensitive adhesives and EVA based hot melt adhesives.

Product Name	Softening Point (°C)	Color (Gardner)	Melt Viscosity @180°C (cp)	Acid Number (mg KOH/g) (max)	Molecular Weight Average	Specific Gravity @18°C
5380	85	1	1	1	100	1.10
5300	105	1	1	1	130	1.10
5320	125	1	1	1	350	1.10

Chemical Type (5300 Series): Premium quality water-white hydrogenated hydrocarbon resins with exceptional tackifying properties and outstanding thermal stability.

Key Properties: Recommended for use in natural rubber, random SBR, SIS block copolymer and SBS block copolymer pressure sensitive adhesives, hot melt road marking compounds, hot melt coatings and wax blends, tackifiers for EP and butyl rubbers and sizing agents for textiles.

HERCULES, INC., 910 Market St., Wilmington, DE 19899

Hercoprime Resins G-35 and A-35

Chemical Type: Proprietary composition crystalline polymer resin polypropylene-to-metal bonding agents. Hercoprime G-35 is a high-functionality, low molecular weight resin. Hercoprime A-35 is a low-functionality, high molecular weight resin.

Physical Properties (Both Grades):

Particle Size (Average)	35–40 μ
Particle Size (Maximum)	100% <115 Mesh
Specific Gravity	0.9

Key Properties: Hercoprime resins make it possible to achieve outstanding polypropylene-to-metal bonds. The Hercoprime resins may be blended with Hercotuf resins to form self-priming powder coatings. These coatings demonstrate outstanding adhesion to steel, aluminum and many other metals while retaining virtually all the properties which have made polypropylene a major factor in the plastics industry.

The Hercoprime resins may also be used as adhesives to form polypropylene-to-metal laminations and sandwich constructions. An extremely thin film of Hercoprime, applied by conventional techniques from a dilute organosol, provides polypropylene-to-steel bonds which reach high T-peel bond strength and high lap shear bonds, with suitable metal pretreatment.

The higher functionality of Hercoprime G-35 enables the formulator to use lower concentrations to achieve adhesion. However, Hercoprime A-35, by virtue of its higher molecular weight, provides maximum bond strengths and, in the case of powder coatings, maximum impact resistance. Both are insoluble in all common solvents at ambient temperatures.

LAWTER CHEMICALS, INC., 990 Skokie Blvd., Northbrook, IL 60062

Petro-Rez Resins

Product Name	Softening Point (Ring & Ball) (°C)	Color (Gardner) (50% in Toluol)	Acid Value (max)	Iodine Number (ASTM D1959)	Molecular Weight (Number Average)	Specific Gravity @25°C
100	100	10+	1	35	1250	1.07

Key Properties: Truly the rubber compounder's workhorse. Product of an unvarying raw material stream and closely controlled manufacturing technique. It is a consistent and dependable performer. Mixing characteristics, color and odor, as well as those peripheral qualities essential to its uniqueness do not change from batch to batch. Reliable low reactivity and ability to hold color and viscosity under heat. BHT is present to insure stability.

102	100	10+	1	35	1250	1.07

Key Properties: A new development which differs from Petro-Rez 100 to the extent that it is polymerized in a manner that yields a very narrow molecular weight range. Unique characteristics may be noted, such as broader compatibilities and solubilities, which result from the elimination of higher molecular weight components.

103	100	11+	1	90	1170	1.07

Key Properties: The most versatile resin, with applications in rubber, adhesives and printing inks. Somewhat darker in color and slightly more reactive than 100, 101 and 102. It is an effective and economical alternative to other resins of similar softening point.

Chemical Type (Of All): Petroleum derived hydrocarbon resins-thermoplastic, highly aromatic and nonreactive. Ash content: <0.1%

Key Properties (Of All):

Tackify hot melts	Resist water, acid and alkali
Control hot melt viscosity	Broad compatibility range
Wet pigments	Nonstaining
Tackify rubber systems	Heat stable
Reinforce block copolymers	Light stable

NEVILLE CHEMICAL CO., Pittsburgh, PA 15225

Nevex 100 Resin

Chemical Type: Proprietary composition light-colored thermoplastic resin.

Physical Properties:

Specific Gravity @ 25°/15.6°C	1.12
Weight per Gallon	9.3 pounds
Softening Point (Ring and Ball)	99°C
Color (Neville)	1
Color (Gardner-50% in Toluene)	12
Viscosity (Gardner-70% in toluene, Bubble-Seconds)	2.65 (J-K)
Viscosity (Gardner-70% in Mineral Spirits, Bubble-Seconds)	22.0 (Z)
Acid Number	Nil
Ash	Trace
Refractive Index @ 25°C	1.620

Key Properties: Primarily designed for use in hot melt coatings and adhesives that are based upon ethylene-vinyl acetate copolymer resins.

It has been determined that these hot melt systems containing this resin show improvement in the areas of:

Ease of application at reduced temperatures.
Resistance to alkali, grease and oil.
Gloss and gloss retention.
Moisture vapor transmission.
Adhesion.
Lower sealing temperature.
Scuff resistance.

Has many other applications in the coatings field. Compatible with a wide variety of alkyd resins and other polymers. Nevex 100 Resin extender of systems based on these polymers leads, in some cases, to improved economics, and in others, improved performance.

Has been found to be extremely compatible with many alkyd resins and other polymeric film formers. Its compatibility in these areas is much greater than other hard resins now available. When used as a replacement for a portion of the alkyd or polymer solids in a finished coating, Nevex 100 contributes:

Better gloss and gloss retention.
Increased water and caustic resistance.
Hardness.
Faster dry.

F.D.A. Status: Nevex 100 Resin is an improved substance under many of the regulations governing indirect food additives.

REICHHOLD CHEMICALS, INC., RCI Bldg., White Plains, NY 10603

Nirez 3098 LM Resin

Chemical Type: Proprietary composition pale, hard friable resin.

Physical Properties:

Color (U.S.D.A. Standards)	WG max
Capillary Tube Melting Point	57°-60°C
Softening Point (Ring & Ball)	79°C
Acid Number	148
Weight per Gallon @ 25°C	8.9 lb
Saponification Number	154

Key Properties: Has application in the adhesives field.

Processed specially to impart excellent compatibility and stability to it for use with natural rubber. It is soluble in all of the commonly used aliphatic and aromatic hydrocarbon solvents, and is compatible with a wide variety of resins, waxes, elastomers and polyolefins.

Designed primarily for use as a tackifier for natural rubber adhesives.

Toxicity: Threshold Limit Value (T.L.V.): Unknown. Please refer to manufacturer's Material Safety Data Sheet.

VIRGINIA CHEMICALS, INC., 3340 West Norfolk Road, Portsmouth, VA 23703

Virset A-125 Resins

Chemical Type: Proprietary composition coating insolubilizer, in water-white liquid form.

Chemical Analysis and Physical Properties:

Nonvolatile Content (P_2O_5 Dehydration Method)	55% min
Weight per Gallon @ 25°C	10.0 lb
Brookfield Viscosity @ 25°C	100 cp
Solubility in Water	Infinite
pH (Phenol Red Indicator Method)	7.2-7.6
Odor of Formaldehyde	Slight

Key Properties: Polyvinyl Alcohol: Virset A-125 provides a low-cost means for reducing the water sensitivity of polyvinyl alcohol based adhesives systems. It imparts good wet-rub resistance to coating colors, anti-blocking properties to sizings and moisture resistance to adhesives.

Butadiene-Styrene Latex: Virset A-125 is used to improve the early development of wet-rub resistance in paper and paperboard coatings formulated with butadiene-styrene binders.

Casein and Soya Protein: Virset A-125 gives rapid wet-rub resistance development in low-solids paper coating formulations based upon casein and soya protein binders. If a serious problem is encountered with viscosity, it is recommended that Virset 656-4 be used as an alternate.

F.D.A. Status: Ingredients approved.

Section XXXIII
Resins–Radiation Coating

CARGILL, INC., Chemical Products Division, P.O. Box 9300, Minneapolis, MN 55440

Cargill Radiation Curable Resins

Product Name	% Oligomer	Brookfield Viscosity at 20°C (cp)	Color (Gardner) (max)	% NCO (max)	Weight per Gallon (lb)	Flash Point (C.C.) (°F)
XP-1511	75	12,150	2	0.1	8.70	168

Chemical Type: Acrylate functional, aliphatic urethane oligomer solution in 2-ethylhexyl acrylate.

XP-1512	75	19,500	2	0.1	9.01	>220

Chemical Type: Acrylate functional, aliphatic urethane acrylate solution in 1,6-hexanediol diacrylate.

XP-1521	75	19,500	2	0.1	8.87	158

Chemical Type: Acrylate functional, aliphatic urethane oligomer solution in 2-ethylhexyl acrylate.

XP-1522	70	16,200	2	0.1	9.17	203

Chemical Type: Acrylate functional, aliphatic urethane oligomer solution in 1,6-hexanediol diacrylate.

Key Properties: All of the above are suitable for use in U.V. and E.B. curable coatings and inks. They comprise the second generation of radiation curable resins from Cargill. These new resins offer improved stability and handling ease compared with previous products. They also offer, due to processing improvements, greatly reduced levels of free isocyanate and hydroxyl acrylate.

XP-1511 and XP-1512 both contain the same oligomer.

XP-1521 and XP-1522 both contain a different oligomer.

These oligomers may be supplied in different diluents and/or diluent blends.

THIOKOL CORP., 930 Lower Ferry Road, P.O. Box 8296, Trenton, NJ 08650

Uvithane Urethane Oligomers

Chemical Analysis and Physical Properties:

	782	783	788	893
Appearance at Room Temperature	Low Melting Solid	Viscous Liquid	Low Melting Solid	Viscous Liquid
Color-APHA (max)	50	50	50	110
Odor	Mild	Mild	Mild	Mild
Weight per Gallon (lb) @ 77°F	10.2	10.5	9.9	10.0

	782	783	788	893
Viscosity (Poises) @ 120°F	1,200	1,300	3,350	1,450
Viscosity (Poises) @ 160°F	275	190	525	130
Viscosity (Poises) @ 180°F	125	80	150	55
Unsaturation (Equivalent/100 g)	0.045	0.19	0.16	0.165
Isocyanate Content (% max)	0.3	0.3	0.2	0.2

Key Properties: Uvithane urethane oligomers are abrasion-resistant, acrylated urethane materials which may be cured by either ultra-violet or electron-beam irradiation as well as thermal processes.

Each oligomer cures to a specific type of film:

Uvithane 782—Soft and extensible.

Uvithane 783—Tough and flexible.

Uvithane 788—Nonyellowing, tough and flexible.

Uvithane 893—Nonyellowing, tough and flexible.

Applications: Uvithane urethane oligomers display versatility in many radiation-curable applications. Whether cured using ultra-violet (UV) or electron beam (EB) application, the resulting coatings, inks and adhesives require lesser amounts of energy to manufacture and reduce pollution levels to a minimum. The table below lists some of the many application areas for these products.

Radiation Curable Applications:

	Adhesives	Metals	Textiles	Plastics	Wood	Auto Parts
			...Coatings for...			
Uvithane 782	x	x	x	x	—	x
Uvithane 783	x	x	x	x	x	x
Uvithane 788	x	—	x	x	x	x
Uvithane 893	x	—	x	x	x	x

In applications involving Uvithane oligomers, monofunctional acrylic monomer Reactomer RC-20 (ethoxy-ethoxyethyl acrylate) has been found to be a highly desirable reactive diluent. Thiokol also manufactures polyfunctional monomers which are suitable for radiation-cured inks and coatings.

Toxicity: The oligomers do not require warning labels under the Federal Hazardous Substances Act based on tests for eye irritation, acute oral toxicity and dermal toxicity. However, it is advisable to avoid skin contact or inhalation of vapor.

Section XXXIV

Resins–Urethane

CHEMICAL COMPONENTS, INC., 20 Deforest Ave., East Hanover, NJ 07936

Waterbase Urethanes

Product Name	Viscosity (Brookfield) (cp)	Percent Solids Content	pH
SI-822	250	35	10.0

Chemical Type: Aqueous dispersion of a urethane polymer.

Key Properties: It yields a soft high gloss flexible film when cured by evaporation. The film has excellent abrasion and outstanding resistance to solvents and water. Application can be by rotogravure or knife over roll. It can also be sprayed if diluted to a lower viscosity with water. Can be cured at ambient temperatures or 1 to 2 minutes at 220°F.

Product Name	Viscosity (Brookfield) (cp)	Percent Solids Content	pH
SI-810	250	35	9.0

Chemical Type: Aqueous dispersion of an aliphatic urethane polymer.

Key Properties: It yields a hard high gloss flexible film when cured by evaporation. The film has excellent abrasion and outstanding resistance to solvents and water. Application can be by rotogravure or knife over roll. It can also be sprayed if diluted to a lower viscosity with water. Can be cured at ambient temperatures or 1 to 2 minutes at 220°F.

Section XXXV

Resins—Vinyl Chloride

FIRESTONE PLASTICS CO., P.O. Box 699, Pottstown, PA 19464

FPC 450

Chemical Type: Vinyl chloride type resin of intermediate molecular weight which may be used in solution resin applications. Powder form.

Physical Properties

Specific Gravity	1.36
Gallons per Solid Pound	0.088
Relative Viscosity (1% in Cyclohexanone)	1.85

Key Properties: Compatibility with vinyl plasticizers, pigments and stabilizers allows compounding latitude in specific applications, such as strippable coatings, printing inks and protective coatings. Film properties are the following:

Transparent.
Air drying—thermoplastic.
High tensile and elongation.
Strippable from nonporous substrates.
Adhesion to vinyl surfaces; molecular weight high enough to allow application over plasticized surfaces.
Adhesion to porous substrates: paper, cloth, etc.
Resistant to moisture, grease, chemicals.

FDA Regulations: Suitable for use under subpart F of the following FDA regulations:

121.2514: Container Coatings.
121.2520: Adhesives.
121.2526: Paper Coatings for Aqueous Fatty Foods.
121.2550: Closures.
121.2571: Paper Coatings for Dry Foods.

FPC 454

Chemical Type: Very low molecular weight vinyl chloride copolymer resin. Powder form.

Physical Properties

Specific Gravity	1.36
Gallons per Solid Pound	0.088
Relative Viscosity (1% in Cyclohexanone)	1.45

Key Properties: Designed specifically for solution application. Will adhere to primers made with FPC 470 and to porous substrates, such as paper and cloth. Excellent solubility with resultant high solids in vinyl type solvents. Film properties are the following:

Transparent, odorless, tasteless.
Air drying—thermoplastic.
Resistant to water, moisture vapor, chemicals, oils, greases, etc.
Nonflammable.
Adhesion to porous surfaces: paper, cloth, etc.
Adhesion to "primed" metal surfaces.

FDA Regulations: Suitable for use as described under several FDA regulations (21 CFR) for use in contact with food:

> 175.105 (Subpart B): Adhesives.
> 175.300 (Subpart C): Resinous and polymeric coatings.
> 176.170 (Subpart B): Components of paper and paperboard in contact with aqueous and fatty foods.
> 176.180 (Subpart B): Components of paper and paperboard in contact with dry food.
> 177.1210 (Subpart B): Closures with sealing gaskets for food containers.

FPC 470

Chemical Type: Vinyl chloride terpolymer solution resin. Powder form.

Physical Properties

Specific Gravity	1.31
Gallons per Solid Pound	0.091
Relative Viscosity (1% in Cyclohexanone)	1.35

Key Properties: Versatile solution resin with ketone solubility, toluene and xylene tolerance, alkyd and extender resin compatibility and excellent adhesion. Can be used in many new applications which were previously considered impractical for vinyl chloride resins. The usual qualities attributed to polyvinyl chloride resins are retained.

FDA Regulations: Suitable for use under Subpart F of the following FDA regulations:

> 121.2514: Container Coatings.
> 121.2520: Adhesives.
> 121.2526: Paper Coatings for Aqueous or Fatty Foods.
> 121.2550: Closures.
> 121.2571: Paper Coatings for Dry Foods.

FPC 471

Chemical Type: Vinyl chloride copolymer resin. Powder form.

Physical Properties

Specific Gravity	1.31
Gallons per Solid Pound	0.091
Relative Viscosity (1% in Cyclohexanone)	1.45

Key Properties: Excellent for coating applications requiring high vinyl resin solids. The resulting film, which may be either rigid or flexible, has good resistance to water, moisture vapor, chemicals and weather. Porous surfaces such as paper can be coated with excellent results. However, a primer is needed where adhesion to metal is a requirement. Film properties are the following:

> Transparent.
> Air drying—thermoplastic.
> Weather resistant.
> Resistant to water, moisture vapor, chemicals, grease, etc.
> Adhesion to porous surfaces: paper, cloth, etc.
> Adhesion to "primed" metal surfaces (primers based on FPC 470).
> Excellent light stability.

FDA Regulations: Suitable for use under Subpart F of the following FDA regulations:

> 121.2514: Container coatings.
> 121.2520: Adhesives.
> 121.2526: Paper Coatings for Aqueous or Fatty Foods.
> 121.2550: Closures.
> 121.2571: Paper Coatings for Dry Foods.

FPC 481

Chemical Type: Vinyl chloride copolymer resin of high molecular weight. Powder form.

Physical Properties

Specific Gravity	1.35
Gallons per Solid Pound	0.088
Relative Viscosity (1% in Cyclohexanone)	1.92

Key Properties: Developed primarily for solution applications. Compatibility with other raw materials allows considerable latitude for formulating printing inks, strippable coatings, protective coatings and decorative coatings. Coatings based on FPC 481 exhibit high tensile strength and may be applied by using the common lacquer techniques. Film properties are the following:

Transparent.
Air drying—thermoplastic.
High tensile and elongation.
Strippable from nonporous substrates.
Adhesion to vinyl surfaces. Molecular weight high enough to allow application over plasticized surfaces.
Adhesion to porous substrates: paper, cloth, etc.
Resistant to water, moisture, grease, chemicals, etc.

FPC 497

Chemical Type: Medium molecular weight vinyl chloride copolymer resin. Powder form.

Physical Properties

Specific Gravity	1.36
Gallons per Solid Pound	0.088
Relative Viscosity (1% in Cyclohexanone)	1.55

Key Properties: Developed especially for solution application. FPC 497 will not adhere to nonporous substrates unless an adhesive type resin such as FPC 470 is blended into the solution. FPC 497 will adhere to the above based primers and porous substrates such as paper and cloth. Film properties are the following:

Air drying, thermoplastic and transparent.
Weather resistant.
Resistant to water, moisture vapor, chemicals, grease, etc.
Adhesion to porous surfaces: paper, cloth, etc.
Adhesion to "primed" metal surfaces such as primers based on FPC 470.

FDA Regulations: Suitable for use under Subpart F of the following FDA regulations:

175.300: Container Coatings.
175.105: Adhesives.
176.170: Paper Coatings for Aqueous or Fatty Foods.
177.1210: Closures.
176.180: Paper Coatings for Dry Foods.

FPC XR-2313

Chemical Type: Vinyl solution terpolymer.

Physical Properties

Relative Viscosity	1.57
Solution Viscosity (Resin 20, MEK 40, Toluene 40)	124 cp
Solution Color	White to Off-White
Solution Clarity	Good
Solution Seeds	0.0 mil
Appearance (Glass Slide)	Excellent
Butanol Tolerance	Good
Blush Resistance (Cure 10'-300°F)	Excellent
Aliphatic Tolerance (Mineral Spirits & VM&P Naphtha)	Good
Photovolt 60° Gloss	
Clear Over Black Card	87
White	41
Heat Stability (Time to Yellow at 149°C)	>165 minutes

Key Properties: The OH portion of the resin increases solubility, compatibility and crosslinking possibilities. Adhesion is as follows:

Wood: Good
Wash Primer: Good
Shellac: Excellent
Urea-Formaldehyde: Excellent
Nitrocellulose: Excellent

Section XXXVI
Resin Emulsions and Dispersions

SCHENECTADY CHEMICALS, INC., P.O. Box 1046, Schenectady, NY 12301

Schenectady Modified Phenolic Resin Dispersions

Product Name	% Solids	pH @ 20% Solids
HRJ-790	50	2.75

Chemical Type: Modified phenolic resin dispersion, in light tan to white color in water medium.

Key Properties: When used with neoprene and acrylic latices, HRJ-790 improves bonding range and heat and moisture resistance of contact adhesive formulations.

Product Name	% Solids	pH @ 20% Solids
HRJ-587	54	2.75

Chemical Type: Modified alkyl phenolic resin dispersion, in light tan to white color in water medium.

Key Properties: When used with neoprene and acrylic latices, HRJ-587 improves bonding range and heat and moisture resistance of contact adhesive formulations.

Section XXXVII

Resin Esters

ARIZONA CHEMICAL CO., Wayne, NJ 07470

Zonarez Resin Esters

Product Name	Softening Point (Ring & Ball) (°C)	Color (USDA)	Acid No.	Specific Gravity @ 25°C
55	52	M	8	—

Chemical Type: Glycerol derived resin ester based on Arizona DR-24 disproportionated tall oil rosin, a thermoplastic, amber-colored resin.

Key Properties: Exhibits exceptional stability to oxidative attack. This ester should find extensive use as a tackifier for difficult to tackify elastomers, such as SBR. The lower softening point of this resin will also find use where good flexibility and tackification are required at lower temperatures. Can be readily emulsified for use in tackifying both natural and synthetic latices. Also, recommended for use in hot melt, pressure sensitive adhesives and contact cements. Meets certain applications governed by F.D.A. Regulations.

75	78	M	8	—

Chemical Type: Glycerol derived ester based on Arizona DR-22 disproportionated tall oil rosin, a thermoplastic, amber-colored resin.

Key Properties: Exhibits exceptional stability to oxidative attack. Specifically designed for use in hot melt and other applications where elevated temperature stability is a problem. Very attractive for use in applications where long term stability, under ambient conditions, is critical. Can be readily emulsified for use in tackifying natural, SBR and neoprene latices. Also, recommended for use in hot melt adhesives and coatings, pressure sensitive adhesives and contact cements.

85	83	WW-WG	7	1.02

Chemical Type: Rosin ester based on the unique Actinol R Type 3A Tall Oil rosin, a glycerol resin ester, available in either solid or flake form.

Key Properties: Has a low acid number, high softening point, excellent color, good heat stability and low cost. Exhibits excellent compatibility with solvents, elastomers, rubbers, polymers, waxes and other tackifying resins used to compound adhesives. Finds widespread use in chewing gums, hot melt adhesives, hot melt coatings, mastic adhesives, contact cements and pressure sensitive adhesives.

100	100	G	11	—

Chemical Type: Amber colored, high melting, pentaerythritol ester of tall oil rosin.

Key Properties: May be used in a large number of adhesive applications where good economics and high performance are the prime considerations. Exhibits excellent compatibility with solvents, elastomers, rubbers, polymers, waxes and other tackifying resins which are normally used in compounding adhesives. Can be readily emulsified to permit use in SBR latex, natural rubber and neoprene adhesives. May be used in the preparation of hot melt adhesives, mastic adhesives, contact cements and pressure sensitive adhesives.

Zonester 55 Resin Ester

Chemical Type: Glycerol derived ester based on Arizona DR-24 disproportionated tall oil rosin. Thermoplastic, amber-colored resin.

Physical Properties:

Softening Point (Ring & Ball)	52°C	Color (USDA Rosin Scale)	M
Acid Number	8		

Key Properties: This ester will exhibit exceptional stability to oxidative attack, since it is derived from a disproportionated rosin base. Additionally, being derived from a soft disproportionated rosin, this ester should find extensive use as a tackifier for difficult to tackify elastomers, such as SBR. The lower softening point of this resin will also find use where good flexibility and tackification are required at lower temperatures.

Recommended for use in hot melt and pressure sensitive adhesives, contact cements and latex adhesives.

FDA Regulations: Tackifying resin which can be used in certain applications governed by FDA regulations. Acceptable for use in adhesives, sealing gaskets and coatings for packaging food products.

Zonester 75 Resin Ester

Chemical Type: Glycerol derived ester based on Arizona DR-22 disproportionated tall oil rosin. Thermoplastic, amber-colored resin.

Physical Properties:

Softening Point (Ring & Ball)	78°C	Color (USDA Rosin Scale)	M
Acid Number	8		

Key Properties: Being derived from a disproportionated rosin base, this ester will exhibit exceptional stability to oxidative attack. Zonester 75 ester has been specifically designed for use in hot melt and other applications where elevated temperature stability is a problem. This resin has also been shown to be very attractive for use in those applications where long term stability, under ambient conditions, is critical.

Zonester 75 resin is recommended for use in hot melt adhesives and coatings, pressure sensitive adhesives, contact cements and latex adhesives.

Zonester 85 Resin Ester

Chemical Type: A glycerol ester which is first in Arizona's series of rosin esters based on the unique Actinol R Type 3A tall oil rosin.

Physical Properties:

Color (USDA Rosin Scale)	WW-WG	Acid Number	7.0
Softening Point (Ring & Ball)	83	Specific Gravity @ 25°/25°C	1.02

Key Properties: Finds widespread use in chewing gums, hot melt adhesives, hot melt coatings, mastic adhesives, contact cements and pressure sensitive adhesives. Possesses a low acid number, high softening point, excellent color, good heat stability and low cost. This resin exhibits excellent compatibility with solvents, elastomers, rubbers, polymers, waxes and other tackifying resins normally used to compound adhesives. Zonester 85 is available in either solid or flake form.

FDA Regulations: Zonester 85 is approved for use under most FDA regulations covering food and food related applications.

Zonester 100 Resin Ester

Chemical Type: Amber colored, high melting, pentaerythritol ester of tall oil rosin.

Physical Properties:

Softening Point (Ring & Ball)	100°C	Color (USDA Rosin Scale)	G
Acid Number	11		

Key Properties: May be used in a large number of adhesive applications where good economics and high performance are the prime considerations. May be used in the preparation of hot melt adhesives, mastic adhesives, contact cements and pressure sensitive adhesives.

This resin exhibits excellent compatibility with solvents, elastomers, rubbers, polymers, waxes and other tackifying resins which are normally used in compounding adhesives. Emulsification of Zonester 100 permits its use in SBR latex, natural rubber and neoprene adhesives.

FDA Approval: Zonester 100 is approved for use under certain applications by the FDA for use in adhesives, sealing gaskets and coatings.

FRP CO., P.O. Box 349, Baxley, GA 31513

Product Name	Softening Point (Ring & Ball) (°C)	Acid Value (max)	Viscosity (Gardner-Holdt)	Color (Gardner) (max)	Weight per Gallon (lb)	Flash Point (O.C.) (°F)
Isoester B	80	10.5	R–V	10	9.0	435

Chemical Type: An ester of isomerized resin.

Key Properties: Offers good resistance to oxidation, light and crystallization. Typical uses are: water-dispersed adhesives, pressure-sensitive adhesives, heat-sensitive adhesives, embossing inks and lacquers.

Section XXXVIII
Rosins

CROSBY CHEMICALS, INC., 600 Whitney Bldg., New Orleans, LA 70130

Crosby Resins

Product Name	Acid Number	Drop Melt (°C)
Resin 721	160	77
Valros	163	94
Polros A	150	107
Crosdim	145	150

Chemical Type: Polros A is a high melt polymerized tall oil rosin. By distillation, Polros A can be separated into pure high melt dimerized rosin called Crosdim and a low melt pale rosin named Resin 721, which is a partially polymerized rosin.

Key Properties: Polros A is recommended for use in those products which require greater hardness and higher viscosities than can be obtained with natural rosin. This polymerized rosin is noncrystalline in nature and is very resistant to oxidation. Polvos A has many features not found in regular wood, gum or tall oil rosin.

These rosins are used extensively to manufacture resins for use in inks, coatings and adhesives. The polymerized rosins have much better color stability properties than rosin. These rosins have a very wide range of compatibility and solubility properties.

NATRO CHEM, INC., P.O. Box 1205, Savannah, GA 31402

Galex Disproportionated Rosin

Chemical Type: Stable nonoxidizing rosin consisting principally of dehydroabietic acid. The stability is brought about by an intramolecular rearrangement and dehydrogenation of abietic acid, the major constituent of rosin, to dehydroabietic acid which is a more stable type of rosin acid containing the benzenoid nucleus.

Key Properties: Because of its relative immunity to oxygen absorption, its light color, its low softening point and excellent tack retention, Galex is recommended for rubber-based pressure-sensitive adhesives.

In addition to the above, Galex is finding an ever-increasing use in adhesives of the water-insoluble type, as an extender for natural and synthetic resins, in the manufacture of rubber cement and many other uses.

Chemical Analysis and Physical Properties:

	NXD	G-75
Gardner color (50% solution)	7	7
Acid number	158	158
Softening point (Ring & Ball), °C	70	70
Specific gravity @ 15.5°C	1.05	1.05

(continued)

	NXD	G-75
Benzene, insoluble, %	nil	<0.01
Ash content, %	0.06	0.06
Flash point (C.O.C.), °C	210	210
Fire point (C.O.C.), °C	240	240

REICHHOLD CHEMICALS, INC., RCI Bldg., White Plains, NY 10603

Poly-Tac Adhesive Resins

Chemical Type: Zincated modified rosins.

Chemical Analysis and Physical Properties:

	Poly-Tac 85	Poly-Tac 100
RCI product code	36–626	36–627
Softening point (Ring & Ball), °C	83–87	95–100
Gardner color (1963), 50% non-volatile in mineral spirits, max	7	8
Melting point (capillary tube), °C	63	73
Ash content, %	12.8	13.5
USDA color	N	N
Acid number, calculated	–17	–18
Specific gravity @ 25°/25°C	1.158	1.172

Key Properties: Both of these resins are pale colored, hard, friable products which differ from each other in melting point. They are based on rosin modified to impart superior stability and are exceptionally stable to heat and oxidation. The Poly-Tac resins are soluble in aliphatic and aromatic hydrocarbon solvents. They are not soluble in the alcohols. They are compatible with a wide variety of other resins, copolymers and elastomers.

Recommended end uses are as tackifying resins or as extending resins in pressure-sensitive adhesives, contact cements, hot-melt adhesives, hot-melt coatings, sealants and mastics.

Toxicity: Threshold Limit Value (TLV) is unknown. Please refer to manufacturer's Material Safety Data Sheet.

Section XXXIX
Silicas (Silicon Dioxide)

<u>BASF CORP., 491 Columbia Ave., Holland, MI 49423</u>

Silica Gel B Wind-Sifted

Chemical Type: Finely dispersed silica produced by a special technique.

Key Properties: Primarily intended for sealers and fillers, but can also be used in road-marking paints and high-build systems. It is not recommended for flat paints.

<u>DAVISON CHEMICAL DIVISION, W.R. Grace & Co., P.O. Box 2117, Baltimore, MD 21203</u>

Product Name	Silica Content (Ignited Basis) (%)	Color	Density Centrifuge (g/cc)	pH (5% Slurry in Water)	Particle Size (μ)	Oil Absorption (lb/100 lb)
Syloid 244	99.3	white	0.15	7.6	4.0	300
Syloid 378	96.6	white	0.26	2.5	4.0	200

Chemical Type: Ultrafine amorphous silica gel.

Key Properties: Can increase bond strength markedly when added to adhesive formulations. These strengths were improved in the following types of adhesives:

> Neoprene-type adhesive in a solvent solution.
> Polyvinyl alcohol acetate thermoplastic.
> Epoxy resin base.
> Synthetic rubber base.
> Thermosetting polyester base.

Other uses are:

> Syloid 244, in latex bases.
> Syloid 244, in pressure-sensitive adhesives.
> Syloid AL-1 for prevention of blushing in cements in high humidity regions.

<u>DEGUSSA CORP., Pigments Division, Route 46 at Hollister Road, Teterboro, NJ 07608</u>

Aerosil Silicas

Chemical Type (All Grades—Standard and Special): Extremely pure silicon dioxide of amorphous structure. Its remarkable thixotropic filler action is a function of the silanol groups present on its surface in optimal density and their propensity to form hydrogen bonds.

Produced from silicon tetrachloride in a flame hydrolysis process with oxygen-hydrogen gas. This process produces a highly dispersed silica of great purity with controlled particle size and chemical constitution offering a wide field of special properties.

Physical Properties (All Grades—Standard and Special):

Number of Particles	At 200 ± 25 m^2/g, 1 g of Aerosil 200 comprises approximately 3×10^{17} individual particles; these would extend for 6 million km, if lined up.
X-Ray Structure	amorphous
Specific Gravity	2.2 g/cm^3
Solubility	Insoluble in all solvents and liquids except hydrofluoric acid and very concentrated caustic solutions.
Silanol Groups	Approximate density 3 silanol groups (SiOH) per nm^2 surface area.
Refractive Index	1.45
Electrical Charge	negative
Specific Electrical Resistivity	The electrical conductivity of Aerosil is very poor and qualifies it in effect as an insulator.
Thermal Conductivity @ 0°C	0.022 kcal/m-hr-°C

Key Properties (All Grades—Standard and Special):

Very large specific surface (up to 380 m^2/g)
Hydrophilic and hydrophobic grades
Extremely fine particle size
Very high chemical purity
Amorphous x-ray structure—no hazard of silicosis
Refractive index of 1.45 transparency in media with similar refractive index.
Efficient thickener in nonpolar and low polar media
Absorbs water up to 40% of its own weight without losing status of a powder.

Toxicity: Aerosil is compatible with the skin and medical evidence is available to the effect that it is harmless when administered orally. It does not cause silicosis and thus does not constitute an industrial health hazard.

GREFCO, INC., Minerals Division, 3450 Wilshire Blvd., Los Angeles, CA 90010

Dicalite Diatomaceous Silica Diatomite Filler

Product Name	Brightness (G.E.)	Ignition Loss (Dry Basis) (%)	Loose Weight (lb/ft^3)	Retained on 325 Mesh (%, max)	pH	Oil Absorption (lb/100 lb)	Hegman Fineness of Grind
White Filler	93	0.2	8.0	0.07 (400 mesh)	9.5	128	5½
395	92	0.2	8.0	0.07	9.7	128	5
305	92	0.2	8.0	0.15	9.7	140	4

Key Properties (of All of the Above): Recommended for use as a flatting varnish and lacquer, polish (fine), plastic antiblocking agent.

WB-5	92	0.2	8.0	0.80	9.5	140	3½

Key Properties: Recommended for use in paint and general filler use.

L-5	91	0.2	8.0	3.5	9.5	145	3

Key Properties: Recommended for use in paint, match heads and cleansers.

L-10	88	0.2	8.0	15.0	9.5	165	0

Key Properties: Recommended for use in paint, general filler use and cleansers.

SP-5	88	0.5	11.0	6.0 (150 mesh)	9.7	155	—

Key Properties: Recommended for use in polish (coarse) and cleansers.

Physical Characteristics (of All of the Above Grades):

Color	white
Moisture	0.5% maximum
Specific gravity	2.33
Bulking value	19.4 lb/gal
Tamped weight	22.0 lb/ft^3
L-10	23.0 lb/ft^3
SP-5	24.0 lb/ft^3
Porosity	85.0
SP-5	83.5

Average Particle Size Distribution (Sedimentation Method), % by wt:

Microns	White Filler	395	305	WB-5	L-5	L-10	SP-5
Over 40	0.0	0.0	0.0	0.5	1.0	3.0	3.0
20-40	0.5	0.5	1.0	1.5	2.5	8.0	27.0
10-20	3.5	2.5	4.0	4.0	9.5	16.0	26.5
6-10	9.0	8.0	10.0	14.0	15.0	18.0	23.5
3-6	32.0	38.0	40.0	42.0	40.0	33.0	16.0
<3	55.0	51.0	45.0	38.0	32.0	22.0	4.0

Physical Characteristics:

	PS	SF-5	103	SA-3
Color	light pink	deep pink	cream white	cream white
Moisture, max %	0.5	0.5	6.0	6.0
Ignition loss, dry basis, %	0.5	0.5	5.0	5.0
Specific gravity	2.25	2.25	1.98	2.00
Bulking value, wt/gal	18.7	18.7	16.5	16.7
Loose weight, lb/ft^3	8.0	9.0	8.5	7.5
Tamped weight, lb/ft^3	22.0	24.0	18.0	21.0
Porosity	84.5	83.0	85.5	83.0
pH	7.0	6.7	7.5	8.0
Oil absorption, lb/100 lb	163	160	150	175
Maximum % retained on screen noted	1.0 (on 325)	7.0 (on 150)	0.3 (on 400)	3.0 (on 325)
Surface area by air permeability, m^2/g	2.5	1.8	4.5	3.7
Brightness, G.E.	—	—	76	76
Hegman fineness of grind	3	—	6	2

Average Particle Size Distribution (Sedimentation Method), % by wt:

Microns				
Over 40	0.0	4.0	trace	1.0
20-40	2.0	8.0	2.0	1.0
10-20	4.0	15.0	1.0	11.0
6-10	9.0	17.0	3.0	14.0
3-6	30.0	43.0	16.0	23.0
<3	55.0	13.0	78.0	50.0

Key Properties:

PS   Recommended for use in match heads, polish (medium coarse), general filler use and insulation.

SF-5   Recommended for use as a ground filler and in coarse polishes.

103   Recommended for use in ultrafine polishes.

SA-3   Recommended for use in medium polishes, paints, catalyst carriers, and for general filler uses.

Chemical Type (All Grades): Composed of silica, which is essentially inert. They are divided into the following three classes:

Natural — Buff colored. Manufactured from selected crude ore that is dried, milled and air classified to produce various particle size products.

Calcined — Light pink colored. These start with the finished natural product, are calcined at high temperatures in a rotary kiln and then again are milled and air classified.

Flux Calcined – White colored. They begin with the finished natural product, are calcined in a rotary kiln in the presence of a fluxing agent, then are milled and air classified. Calcination reduces the surface area, changes the color and renders trace minerals less soluble.

Key Properties (All Grades): Recommended for use as a paint extender and flatting agent. Economical, as it replaces portions of expensive prime pigment with no loss in hiding power. Other advantages are improved brushing and leveling characteristics and improved adhesion and durability. This is the most effective flatting agent for semi-gloss and flat inside and outside enamels and paints and flat varnishes, and lacquers. The gloss is easily controlled over a wide range down to dead flat.

Can be used to good advantage in countless high viscosity applications, such as caulking compounds, adhesives, dopes, pastes and wood fillers.

Can be used in many other applications.

ILLINOIS MINERALS CO., 2035 Washington Ave., Cairo, IL 62914

Silica Air Floated Grades

Physical Properties:

	200	250	54	1160	1240	0
Oil absorption						
Gardner-Coleman	29–31	29–31	29–31	29–31	29–31	29–31
Spatula rub-out	23–25	23–25	23–25	23–25	23–25	23–25
Apparent density (Scott Volumeter), lb/ft³	32–34	31–33	30–32	29–31	28–30	27–29
Particle size (Fisher Sub-Sieve), μ	—	—	3.00	2.95	2.82	2.35
Mean particle size, μ	—	—	—	7.00	6.40	5.30
Specific surface area, cm²/g	—	—	7,547	7,673	8,029	9,635
Brightness (G.E.), %	85	86	86.5	87.5	86.0	87.75
Particle size distribution (Tyler screen), % by wt						
Through 200 mesh	90–95	96–99	99.58	99.98	99.99	99.99
Through 325 mesh	75–79	80–87	90–95	96–98	98–99.4	99.5+
Through 400 mesh	—	—	84.58	92.65	95.5	98.92
Specifications for shipment						
Through mesh	200	200	325	325	325	325
Percent passing	90–95	96–99	90–95	96–98	98–99.4	99.5+
Moisture, max %	0.25	0.25	0.25	0.25	0.25	0.25

Silica Micronized Grades

Physical Properties:

	Imsil A-25	Imsil A-15	Imsil A-10	Imsil A-108
Oil absorption				
Gardner-Coleman	29–31	29–31	29–31	29–31
Spatula rub-out	23–25	23–25	23–25	23–25
Apparent density (Scott Volumeter), lb/ft³	24–26	22–24	21–23	20–22
Hegman fineness of grind	5	6	7	7+
Particle size (Fisher Sub-Sieve), μ	1.97	1.82	1.55	1.12
Mean particle size, μ	4.30	2.85	2.20	1.80
Specific surface area, cm²/g	11,493	12,440	14,607	20,216
Brightness (G.E.), %	90.5	89.5	88.5	88.0
Particle size diameter distribution, % by wt				
Below 40 μ	99.0	100.0	100.0	100.0

(continued)

	Imsil A-25	Imsil A-15	Imsil A-10	Imsil A-108
Below 20 μ	96.0	100.0	100.0	100.0
Below 15 μ	90.0	99.0	100.0	100.0
Below 10 μ	77.0	96.0	99.0	100.0
Below 5.0 μ	51.0	70.0	76.0	96.0
Specifications for shipment				
Particle size	–400 mesh	15 μ	10 μ	8 μ
Percent	99.9+	99.0	99.0	99.0
Moisture, max %	0.25	0.25	0.25	1.0

Chemical Analysis (of Air Floated and Micronized Grades), % by wt:

Silica (SiO_2)	99.5	Calcium oxide (CaO)	0.15
Ferric oxide (Fe_2O_3)	0.025	Magnesium oxide (MgO)	0.008
Titanium dioxide (TiO_2)	0.005	Loss on ignition	0.30
Aluminum oxide (Al_2O_3)	0.009		

Physical Properties (of Air Floated and Micronized Grades):

Specific gravity	2.65	Refractive Index	1.545
Weight per solid gallon, lb	22.07	Hardness, Mohs scale	6.5
Bulking value	0.04531	Melting point, °F	3100
pH value	7	°C	1722
Specific resistance, ohms	25,700		

Key Properties (of Air Floated and Micronized Grades):

Neutral pH	Not hygroscopic
Low in soluble salts	Inert
High purity	High G.E. brightness
High loading	Excellent suspension
Soft silica	Excellent dispersion
Low moisture content	Soluble only in hydrofluoric
Controlled oil absorption	acid
Good electrical resistance	Low cost

Surface Treated Silicas

The 1240 and the Imsil grades are also treated with the following silane and titanate organo-functional modifiers to make them suitable for special applications. The following are the standard surface treatments:

Surface Treatments	Nomenclature	End-Use Polymers
H	gamma-aminopropyltriethoxysilane	phenolic, melamine, polybenzimidazole, polyimide, nylon, polycarbonate, urethane, polybutyleneterephthalate
P	gamma-methacryloxypropyltrimethoxysilane	diallyl phthalate (DAP) thermosetting polyester, crosslinked polyethylene, and peroxide cured polymers of EPM, EPDM and silicone rubber
E	gamma-glycidoxypropyltrimethoxysilane	cycloaliphatic epoxide, epichlorohydrin, epoxy, polyvinyl chloride, phenolic urethane
S	gamma-mercaptopropyltrimethoxysilane	polybutadiene, urethane, epichlorohydrin, EPDM (sulfur cured), polyvinyl chloride, neoprene, polybutadiene, polyisoprene, SBR
K	di(dioctylpyrophosphate)ethylene titanate	polypropylene, polyethylene, other olefin resin systems
V	. . . . see below	

The V treatment is a silicone fluid similar to dimethylpolysiloxane, chemically altered to make coupling possible with SiO_2. Even though the silicone fluid is a highly reactive material, the compound is easily blended and cured to produce an inert and nonreactive hydrophobic treated silica as the end product.

The hydrophobic silica finds use as a free-flow agent in all types of dry powders and as an extender pigment where low moisture content is desired—protective coating primers, plastic fibers of all types, especially in high temperature molding.

Key Properties and Chemical Reaction: The wide variety of surface treatments has the effect of expanding the array of amorphous silica applications enormously. This means that one of the available products will meet almost any filler and/or extender need ever required.

The majority of surface treatments listed are composed of an organo-functional group (such as amino, mercapto, epoxy, etc.) and a hydrolyzable alkoxy group. One end of the chain hydrolyzes with the surface of the silica leaving the organo-functional group attached but exposed and unreacted (to react later when the pigment particle is added to the resin). Therefore, comparing a surface treated filled system to an unreacted filled system, at constant pigment volume concentration and using the appropriate surface treatment, one can expect the use of treated silica plastics to increase the mechanical strength and electrical properties of the total system. Other performance improvements, such as weatherability, chemical resistance, salt spray and corrosion resistance, and reduced wicking from vapor transmission can be expected.

Surface modifiers also improve processing of filled plastics by reducing viscosity, making wet-out faster and by facilitating dispersion. Potential applications for surface treated silicas include both thermoplastic and thermosetting resins for molded engineering plastics, electrical grade encapsulants, casting compounds, adhesives and coatings.

PPG INDUSTRIES, INC., One Gateway Center, Pittsburgh, PA 15222

Hi-Sil Silicas

Grades and Physical Properties:

	T-600	422
Median agglomerate size, μ	3.5	—
Ultimate particle size, μ	0.021	0.12
Surface area, m^2/g	150	45.0
Oil absorption, lb/100 lb	150	100
pH, 5% solution @ 25°C	6.9	9.0
Total ignition loss, %	—	11.5
Loss at 105°C, %	6.3	5.5
Equilibrium moisture content, %		
at 50% relative humidity	7.2	—
at 70% relative humidity	10	—
at 90% relative humidity	20	—
Refractive Index	1.455	1.45
Bulk density, lb/ft^3	3	—
Specific gravity	2.1	2.08
Weight per gallon, lb	17.5	17.33
Bulking value, gal/100 lb	5.8	5.8
Wet sieve residue, 325 mesh, %	0.002	—
Brightness, Hunter	—	96+

Chemical Analysis:

	T-600	422
Silica (SiO$_2$),		
Anhydrous basis, %	97.5	—
As shipped, %	87.1	—

Chemical Type of Hi-Sil 422: Amorphous, fine particle, precipitated, hydrated silica pigments; a precipitated reaction product of a continuous chemical process, in white powder form, which contains no residue of mineral impurities that impart color or lower brightness.

Key Properties of Hi-Sil 422: Designed specifically as an extender pigment for use in paints, with minimum paint variations from batch to batch. By its spacing action it improves the optical efficiency of titanium dioxide. As a partial replacement for titanium dioxide, it reduces raw material costs, yet maintains or improves opacity and brightness. Alternatively, as a partial replacement for coarse extenders, it significantly improves paint quality.

Hi-Sil 422 is easy to wet and disperse and is a stir-in pigment. It offers better suspension properties than most pigments and has a lower dispersant demand than other fine-particle extenders.

Because of its excellent suspension properties, it performs well over a wide range of loadings. It also imparts excellent package stability to latex paints.

Hi-Sil 422 provides the following improvements:

> Hiding power and tint efficiency
> Increased dry hiding
> Improved brightness
> High stain resistance, washability and enamel holdout
> More uniform sheen
> pH stability
> Eases flow and leveling problems
> High film integrity of exterior wood paints

Almost any coating containing titanium dioxide, except high gloss finishes, can be improved in hiding power and efficiency with Hi-Sil 422.

Chemical Type of Hi-Sil T-600: Classed as wet-process, hydrated silicas because they are produced by a chemical reaction in a water solution, from which they are precipitated as ultra-fine, spherical particles. The particles have an average diameter of 0.021 micron and tend to agglomerate in a loose structure that looks like a grape cluster. Fully hydrated silica with the surface saturated with silanol groups.

Key Properties of Hi-Sil T-600: Highly efficient, low-cost thickener designed to increase viscosity and provide thixotropic action for a variety of liquids.

Prevents the sagging of paints, sealants and other materials on vertical surfaces. It also keeps coarse particles in uniform suspension in a container of paint or other material during storage. Stabilizes emulsions of immiscible liquids, acting to prevent phase separation.

Being relatively inert and having a neutral pH, it can be cost effective in most applications where thickening, thixotropic, antisag or antisettling agents are needed. Since it has a Mohs hardness of zero, it will not abrade spray tips or equipment.

Primary uses occur in areas where inorganic thickeners are customarily used. Also, where organic thickeners such as gums, starches, cellulosics or soaps create difficulties like fermentation, poor heat stability or lack of thixotropic action.

Products that could be enhanced include all types of paints, coatings and many other applications.

Toxicity (of Both Grades): A time-weighted average (TWA) maximum dust exposure limit of 20 million particles per cubic foot is currently listed in OSHA standard, 29 CFR 1.910.1000, Table Z-3 for amorphous silicas. A TWA of 15 milligrams per cubic meter is the OSHA standard for nuisance dusts and is an absolute maximum for any dusty material.

Persons exposed to dust concentrations exceeding the OSHA standards should wear a NIOSH/MSHA approved mechanical filter type respirator.

TAMMSCO, INC., North Front St., Tamms, IL 62988

	Silver Bond B	S Micron Silica	Gold Bond Silica	00 Smoke Silica
Physical Properties:				
Specific gravity	—	2.65	—	—
Weight per solid gallon, lb	—	22.07	—	—
One pound bulk in gallons	—	0.0453	—	—
Coarseness, μ	—	30	48	44
S.S.D.	2.51	2.38	3.01	2.51

(continued)

	Silver Bond B	S Micron Silica	Gold Bond Silica	00 Smoke Silica
Dispersion parameter	538	395	464	421
Oil absorption (ASTM D1483-60)	24	28–30	27–29	27–29
Hegman fineness of grind	—	5–6	3–3½	4½–5
pH value	7.0	6.7	6.85	6.7
Refractive index	—	1.54	—	—
Moisture limit, %	0.50	0.25	0.25	0.25
Mohs hardness	—	6.5	—	—
Average particle size, μ	—	5.5	8.5	7
Percent through 325 mesh	99.00	99.99	98.5	99.5
Weight per cubic foot, lb	60.00	—	—	—
Melting point, °F	3100	—	—	—

Silica Products

Physical Properties:

	Velveteen R	1A Rouge	Ruff Buff	Neosil A	Neosil XV
Specific gravity	—	—	2.65	—	—
Weight per solid gallon, lb	—	—	22.05	—	—
One pound bulk in gallons	0.0453	0.04356	0.04536	0.0453	0.0453
Coarseness, μ	56	5	—	10	<15
S.S.D.	3.10	0.72	—	1.60	1.85
Dispersion parameter	5.13	228	—	210	260
Oil absorption (ASTM D1483-60)	26–28	30–31	26–28	29.5–31.5	28.5–30.5
Hegman fineness of grind	3–4	7.5	—	7+	6–7
pH value	—	—	6.7	—	—
Refractive index	—	—	1.54	—	—
Moisture limit, %	0.25	—	0.50	0.25	0.25
Mohs hardness	—	—	6.5	—	—
Average particle size, μ	12	<1	—	1	<15
Percent through 325 mesh	—	100	25–30	100	100

Chemical Analysis (Typical of All Grades):

	Percent by Weight
Silica (SiO_2)	99.0
Alumina (Al_2O_3)	0.18
Calcium oxide (CaO)	0.04
Ferric oxide (Fe_2O_3)	0.02
Magnesium oxide (MgO)	0.01
Undetermined	0.08
Loss on ignition	0.19

Section XL
Stabilizers

ARGUS CHEMICAL, 633 Court St., Brooklyn, NY 11231

Mark 292 and Mark X Organotin Stabilizers

Chemical Type: Alkyl tin mercaptide stabilizers.

Physical Properties:

	Mark 292	Mark X
Appearance	. . . Clear pale yellow liquid	
Specific Gravity @ 25°C	1.127	1.075
Refractive Index @ 25°C	1.5052	1.4910

Key Properties: Either will provide polyvinyl chloride compositions with outstanding heat stability and crystal clarity.

Mark 292 represents the most advanced type of general-purpose "thioglycolate" and, for all practical purposes, supersedes Mark X. The qualities of Mark 292 also apply to Mark X. For processing plasticized vinyl compounds where an organotin stabilizer is indicated, Mark 292 is an ideal choice. Its high efficiency allows the use of minimum concentrations to provide excellent heat stability.

Compounds made from Mark 292 will not water blush.

In most cases, solutions for metal coating are prepared from polymers containing free-acid groups, and the use of the standard organo metallic stabilizers generally results in precipitation or gellation. Mark 292, however, will not produce this adverse effect.

Mark 462, Mark 462A, Mark 462B, Mark 462C

Chemical Type: Liquid Ba/Cd/Zn Stabilizers.
Liquid Ba/Cd Stabilizers.

Physical Properties:

	Mark 462	Mark 462A	Mark 462B	Mark 462C
Appearance	. . . . Amber Liquid			
Specific Gravity @ 25°C	1.033	0.988	0.993	0.984
Refractive Index @ 25°C	1.501	1.486	1.484	1.481
Zinc Level	None	Low	Medium	High

Key Properties: Highly compatible and efficient liquids recommended for use in plasticized and semi-rigid vinyl compounds where clarity is the prime consideration.

Section XLI
Surfactants/Surface Active Agents

AMERICAN CYANAMID CO., Wayne, NJ 07470

Aerosol Alkylaryl Sulfonate Surfactant

Product Name	..... Solubility ..... Water at 25°C (g/100 ml)	Organic Solvents	Biodegrad- ability	Non- Phytoxic Range	Surface Tension in Water (dynes/cm) (min)	CMC (approx.)
OS	>20	Partially	Slowly	–	37	0.65

Chemical Type: Sodium diisopropyl naphthalene sulfonate, an anionic surfactant in 75% active powder form.

Key Properties: Recommended for dispersing pigments and colors in paints and plastics. Emulsifier for emulsion polymerization and additive to latexes. For agricultural wettable and dispersible powders. Metal cleaning, bottle washing, paint stripper, brick and tile cleaning, pickling, acid etching agents and hard surface cleaners. Emulsion stabilizer. For dispersing, solubilizing, wetting, emulsifying. EPA approved under 180.1001c. Meets FDA 175.105.

Aerosol Amphoteric Surfactants

A-30	Infinite	Insoluble	Partially	–	35	0.06

Chemical Type: Cocoamidopropyl betaine, an amphoteric surfactant in 35% liquid form. (U.S. Patent 3,225,074).

Key Properties: Foaming agent for latexes. Excellent foaming agent for shampoos in combination with anionics. Nonirritating and gentle to skin. Imparts good manageability to hair.

Aerosol Cationic Surfactants

C-61	Forms Dispersions	Soluble when alcohols are present	Partially	–	34	0.001

Chemical Type: Alkylamine-guanidine polyoxyethanol, a cationic surfactant in 70% liquid paste form.

Key Properties: Antistat and dispersant for pigments and fillers. Cationic dispersing and fixing agent for pigments, dyes and fillers. Emulsifying agent used alone or with Aerosol OT surfactant. Softening agent for textiles.

Aerosol Disodium Mono Ester Sulfosuccinate Surfactants

A-268	Infinite	Insoluble	–	–	28	0.1

Chemical Type: Disodium isodecyl sulfosuccinate, an anionic surfactant in 50% liquid form.

Key Properties: Recommended for emulsion and suspension polymerization of polyvinyl chloride and polyvinylidene chloride. Excellent heat stability of polymer is unique feature. Foaming agent for foamed coatings. Recommended for emulsifying and solubilizing. Good tolerance to Fe, Al, Sb and Mg cations. FDA petition submitted. EPA exempt 180.1001(d).

Product Name	Solubility Water at 25°C (g/100 ml)	Organic Solvents	Biodegradability	Non-Phytoxic Range	Surface Tension in Water (dynes/cm) (min)	CMC (approx.)
A-102	Infinite	Insoluble	–	–	29	0.11

Chemical Type: Disodium ethoxylated alcohol half ester of sulfosuccinic acid surfactant, containing nonionic and anionic groups in 31% liquid form.

Key Properties: Recommended for use in vinyl acetate and acrylate based latexes, especially cross-linkable types. May be used with N-methylolacrylamide without detracting from cross-linking properties. Gives intermediate particle size (0.15-0.20 μ). Effective as post stabilizer. Tasteless in polymers. Solubilizing, foaming, dispersing, emulsifying, lime-soap dispersing, fluidizing, cleaning, reducing surface tension. High electrolyte compatibility. Nondermatitic. Meets FDA 175.105.

A-103	Infinite	Insoluble	–	–	34	0.02

Chemical Type: Disodium ethoxylated nonyl phenol half ester of sulfosuccinic acid surfactant, containing nonionic and anionic groups in 34% liquid form.

Key Properties: Recommended for use in acrylate-based systems. Imparts small particle size (0.02-0.05 μ). May be used with Aerosol 22, MA-80, A-102 or other surfactants to modify particle size. Effective as post additive. Tasteless in polymers. Emulsifying, solubilizing, foaming, dispersing, lime-soap dispersing, fluidizing, cleaning, reducing surface tension. Compatible with divalent and trivalent cations. Nondermatitic. FDA petition submitted.

501	Infinite	Insoluble	–	–	28	0.10

Chemical Type: Proprietary composite composition (patent pending) surfactant, containing nonionic and anionic groups in 50% liquid form.

Key Properties: Recommended to replace nonionic surfactants in acrylate polymerizations. Particle size of polymer of latex is easily varied during polymerization when used in conjunction with other Aerosol surfactants. Emulsifying, dispersing and wetting. FDA petition submitted.

413	Infinite	Insoluble	Complete	–	38	0.06

Chemical Type: Disodium alkyl amidoethanol sulfosuccinate, an anionic surfactant in 35% liquid form.

Key Properties: Effective for emulsion polymerization of vinyl acetate and acrylates. Foaming agent for latexes. Excellent foaming agent for shampoos, wallboard, insulation, etc. Nondermatitic, nontoxic, good lubricity. Emulsifying, dispersing, wetting, foaming.

200	Infinite	Insoluble	Complete	–	35	0.02

Chemical Type: Disodium alkyl amido polyethoxy sulfosuccinate, an anionic surfactant in 30% liquid form.

Key Properties: Effective for emulsion polymerization of vinyl acetate and acrylates. Foaming agent for latexes. Excellent foaming agent for shampoos, wallboard, insulation, etc. Nondermatitic, nontoxic, good lubricity. Nonirritating, gentle on skin. Human "eye sting" comparable to amphoterics. Biodegradable.

Aerosol N-Alkyl Sulfosuccinate Surfactants

22	Infinite	Insoluble	Complete	0.5	41	0.06

Chemical Type: Tetrasodium N-(1,2-dicarboxy-ethyl)-N-octadecyl sulfosuccinate, an anionic surfactant in 35% liquid form.

Key Properties: Imparts small particle size for all systems except vinyl acetate. Used in all monomer systems including vinyl chloride. Excellent as post additive for mechanical stabilization. Imparts excellent pigment-latex rheological properties. Recommended for agricultural products, industrial and household cleaners, cosmetics and metal cleaners. For solubilizing, foaming, dispersing, fluidizing. High electrolyte compatibility. Nondermatitic. EPA approved under 180.1001(d). Meets FDA Regulations 178.3400 and 176.170 with limitations. Meets 175.105.

Product Name	..... Solubility ...... Water at 25°C (g/100 ml)	Organic Solvents	Biodegrad- ability	Non- Phytoxic Range	Surface Tension in Water (dynes/cm) (min)	CMC (approx.)
18	18@40°C	Insolbule	Complete	0.5	41	0.04

Chemical Type: Disodium N-octadecyl sulfosuccinamate, an anionic surfactant in 35% paste form.

19	18@40°C	Insoluble	Complete	0.5	41	0.04

Chemical Type: Disodium N-octadecyl sulfosuccinamate, an anionic surfactant in 35% liquid form.

Key Properties (Both of the Above): Imparts excellent pigment-latex rheological properties. Used in emulsion and suspension polymerization. Effective for vinyl chloride polymerization and others. Tasteless in polymers. Foamed latexes and plastics. Foamed insulation, cement, wallboard, resins, etc. Cleaning, washing and lubrication. Cosmetics. For foaming, suspending, lubricating and dispersing. Excellent for agricultural foams. Meets FDA Regulation 176.170 and 176.180 with limitations.

Aerosol Sodium Di-Alkyl Sulfosuccinate Surfactants

TR-70	0.1	Very	Complete	0.01-0.1	26	0.001

Chemical Type: Sodium bistridecyl sulfosuccinate, an anionic surfactant in 70% liquid form.

Key Properties: Recommended for suspension polymerization; particle size control; stabilizes surface tension of latexes. Modifies latex surface active properties. Allows polymer recovery. Dispersing and flushing pigments and colors into organic media. Preparation of printing inks and rust inhibitors. Dispersing pigments and colors into plastics. Flushing, suspending, dispersing, emulsifying, solubilizing. A highly hydrophobic surfactant. EPA exempt 180.1001(d).

OT-75	1.5	Very Soluble	Complete	0.01-0.1	26	0.07

Chemical Type: Sodium dioctyl sulfosuccinate, an anionic surfactant in 75% liquid form.

Key Properties: Recommended for emulsion and suspension polymerization. Latexes for paint, textile and adhesives applications. Widely used by virtually all industries; textile, paper, petroleum, rubber, metal, paint, plastic, cosmetic and agricultural. Wetting, rewetting, increasing absorbency and penetration, emulsifying, dewatering and reducing surface tension. Modifies many surface properties. EPA exempt 180.1001(c).

OT-80PG	1.6	Soluble	Complete	0.01-0.1	26	0.07

Chemical Type: Sodium dioctyl sulfosuccinate, an anionic surfactant in propylene glycol.

Key Properties: A high flash point form of Aerosol OT surfactant. Same uses as for Aerosol OT-75 surfactant where no water desired. High viscosity.

MA-80	34	Soluble	Slowly	0.25	28	1.0

Chemical Type: Sodium dihexyl sulfosuccinate, an anionic surfactant in 80% liquid form.

Key Properties: Recommended for emulsion polymerization. Effective in all monomer systems. Yields complete conversion, coagulum-free latexes. Imparts good adhesion on porous substrates. Allows polymer recovery. Wetting agent in batteries, strong electrolyte solutions, electroplating and leaching operations. Penetration, emulsifying, high salt tolerance, dispersing, solubilizing. Enhances bactericidal activity. EPA exempt 180.1001(d).

OT-B	1.0	Partially Soluble	Complete	0.01-0.1	26	0.06

Chemical Type: Sodium dioctyl sulfosuccinate, an anionic surfactant in 85% active powder form.

Key Properties: Recommended for dispersing pigments and dyes in polyethylene, polypropylene and other plastics. Preparation of wettable and dispersible powders. Agricultural chemical, pigment, paint, and other industries. Adjuvant for wettable products. Dispersing, wetting, solubilizing, improving color value, penetration, surface tension reduction. EPA exempt 180.1001(c).

Product Name	.....Solubility...... Water at 25°C (g/100 ml)	Organic Solvents	Biodegrad-ability	Non-Phytoxic Range	Surface Tension in Water (dynes/cm) (min)	CMC (approx.)
GPG	1.6	Soluble	Complete	0.01-0.1	26	0.07

Chemical Type: Sodium dioctyl sulfosuccinate, an anionic surfactant in 70% liquid form.

Key Properties: General purpose grade of Aerosol OT surfactant. Widely used by a variety of industries for application in wetting, spraying, fire fighting, dust laying, emulsion lubricants and coolants, degreasing, dry cleaning, etc.

OT-100	1.5	Very Soluble	Complete	0.01-0.1	26	0.06

Chemical Type: Sodium dioctyl sulfosuccinate, an anionic surfactant in 100% wax form.

Key Properties: Recommended for mold release for polymethyl methacrylate. Paint formulations, dispersing colors and dyes in plastics. Dispersing colors and pigments in polyethylene and polypropylene. Principally for solvent and nonaqueous systems. Paints, dry cleaning and spotting, rust inhibitors and lubricants. Wetting and antistat for polyethylene. Release properties, lubricity, wetting, emulsifying, water displacement, surface tension reduction. EPA exempt 180.1001(c).

OT-S	—	Very Soluble	Complete	0.01-0.1	26	0.06

Chemical Type: Sodium dioctyl sulfosuccinate, an anionic surfactant in 70% liquid form in light petroleum distillate.

Key Properties: An organic soluble form of Aerosol OT surfactant for incorporation into plastics, organosols, lacquers, varnishes and all organic media. Designed for instant solubility in all organic systems, such as dry cleaning solvents, rust inhibitors, degreasers and lubricants. Excellent wetting agent and emulsifier. EPA exempt 180.1001(c).

OT-70PG	1.6	Soluble	Complete	0.01-0.1	26	0.07

Chemical Type: Sodium dioctyl sulfosuccinate, an anionic surfactant in 70% liquid form in propylene glycol and water.

Key Properties: A high flash point form of Aerosol OT surfactant. Same uses as for Aerosol OT surfactant. Designed for shipment and use where a very high flash point is mandatory.

A-196	10	Soluble Warm	Complete	—	39	3.6

Chemical Type: Sodium dicyclohexyl sulfosuccinate, an anionic surfactant in 85% pellet form.

Key Properties: Unsurpassed for styrene-butadiene latexes. Imparts excellent water resistance and adhesion. Excellent surfactant for various systems. Dispersing, high surface tension, high CMC, displaces oil from surfaces. Promotes adhesion. EPA exempt 180.1001(d).

A-196-40	25	Soluble Warm	Complete	—	39	3.6

Chemical Type: Sodium dicyclohexyl sulfosuccinate, an anionic surfactant in 40% water form.

Key Properties: Same as A-196 (above).

AY-65	40	Insoluble	Complete	0.5	29	1.2

Chemical Type: Sodium diamyl sulfosuccinate, an anionic surfactant in 65% liquid form.

Key Properties: Emulsifier for emulsion polymerization systems, especially where polymer recovery is desirable. Soluble in strong electrolyte systems. Leaching, electroplating, wetting and dispersing. High compatibility with electrolytes. Forms large droplets which do not coalesce. Effective with systemics. EPA exempt 180.1001(d).

Product Name	. . . . . Solubility . . . . . Water at 25°C (g/100 ml)	Organic Solvents	Biodegrad-ability	Non-Phytoxic Range	Surface Tension in Water (dynes/cm) (min)	CMC (approx.)
AY-100	40	Soluble	Complete	0.5	29	1.2

Chemical Type: Sodium diamyl sulfosuccinate, an anionic waxy solid.

Key Properties: Emulsifier for emulsion polymerization systems, especially where polymer recovery is desirable. Used only where the presence of water is not desirable. High compatibility with electrolytes. EPA exempt 180.1001(d).

AY-B	40	Partially Soluble	Complete	0.5	29	1.2

Chemical Type: Sodium diamyl sulfosuccinate, an anionic 85% active powder.

Key Properties: Same as AY-100 (above).

IB-45	60	Insoluble	Complete	–	49	18.0

Chemical Type: Sodium diisobutyl sulfosuccinate, an anionic 45% liquid.

Key Properties: Emulsion polymerization of styrene, styrene-butadiene and other styrene or butadiene-based systems. Surface activity maintained in saturated electrolyte solution for leaching, wetting, electroplating and dispersing. Excellent compatibility with electrolytes. EPA exempt 180.1001(d).

F.D.A. Regulation Compliance:

TR-70:   178.3400
          176.180
          175.105

OT-75:   178.3400
          176.170

OT-B, GPG, OT-100, OT-S, OT-70PG, OT-80PG: 178.3400
                                              177.1200
                                              175.300
                                              176.210
                                              175.105

MA-80:   177.1210
          176.180
          178.3400
          176.170
          175.105

A-196, A-196-40, AY-65, AY-100, AY-B, IB-45:   178.3400
                                                176.170

BASF CORP., 491 Columbia Ave., Holland, MI 49423

Lutensit A-ES Surfactant

Chemical Type: Proprietary anionic surfactant, supplied in the form of a 40% aqueous solution.

Key Properties: Exerts a beneficial effect on the workability of textured finishes. The finishes become more supple, although the viscosity is not greatly reduced.

Leophen RBD Wetting Agent

Chemical Type: Proprietary anionic rapid wetting agent, supplied as a 65% concentration.

Key Properties: Clear, yellow viscous liquid which is readily soluble in water and has a faint aromatic odor. The solution is opaque at low temperatures and clear at high temperatures. It is practically neutral. Readily compatible with anionic and nonionic auxiliaries. Gives rise to very little foam and is largely resistant to all types of alkalies. Main feature of this product is its outstanding and rapid wetting action. It is precipitated by aluminum and salts of heavy metals. Can be stored indefinitely. If Leophen RBD, as supplied, is added in proportions of about 1%, expressed in terms of the finished paint, it allows aqueous binders to wet slightly greasy substrates more easily. Also, the adhesion of emulsion paints on galvanized steel surfaces can thus be improved.

Lutensol AT 25 Wetting and Dispersing Agent

Chemical Type: Nonionic fatty alcohol ethoxylate (nearly 100% assay).

Key Properties: White powder. Its 10% aqueous solution is clear and has a neutral reaction. Readily compatible with colorants, protective colloids and thickeners. Since it is nonionic, it is also compatible with other nonionic, anionic or cationic auxiliaries. This general-purpose product exerts a good wetting, emulsifying and dispersing action. Displays outstanding resistance to the salts causing hardness in water.

Lutensol AF 6 Surfactant

Chemical Type: Nonionic alkyl-phenolethoxylate with a degree of ethoxylation of about 6, which is almost 100% pure.

Key Properties: Clear, slightly yellowish medium-viscosity liquid. Displays pronounced surface activity, since it contains a hydrophobic and a hydrophilic group in the same molecule. Chemically indifferent and stable to precipitation. Readily compatible with anionic and cationic surfactants and with anionic and cationic compounds with a high molecular weight, such as colorants, protective colloids and thickeners. Forms opalescent 10% solutions in water at 25°C. It is advisable to dissolve the product in warm water first, and dilute with cold water, if necessary. Efficient surfactant with very little tendency to foam in aqueous solution. Used as a grinding auxiliary and dispersing agent in producing and processing aqueous pigment dispersions and colorant systems. Prevents floating and flocculation of pigments in aqueous coating systems, such as emulsion paints.

Pigment Disperser A

Chemical Type: Ammonium salt of a low-polymer polyacrylic acid which is supplied in the form of a 30% aqueous solution.

Pigment Disperser N

Chemical Type: Sodium salt of a low-polymer polyacrylic acid which is supplied in the form of a powder.

Key Properties: The addition of Pigment Dispersers A and N in proportions of 0.1-0.5% to the water used for pasting ensures rapid and thorough dispersing of pigments. (The percentages refer to the dispersing agents in the solid form and are expressed in terms of the total amounts of pigments and fillers.) The two products also reduce the risk of premature coagulation when dispersions are mixed with pigments.

Pigment Dispersers A and N do not take effect at pH values of 7 or less. Their efficiency can be increased even further by adding water-soluble polyphosphates and some alkali.

W.A. CLEARY CHEMICAL CORP., P.O.Box 10, 1049 Somerset St., Somerset, NJ 08873

Clearate Refined Soya Lecithins

Product Name	Viscosity (Gardner-Holdt) @77°F	Color (Gardner)	Acid Insoluble Lecithin (%)	Weight per Gallon (lb)	Specific Gravity @77°F
Special (Regular)	Z-6	17	63.5	8.75	1.05

Product Name	Viscosity (Gardner-Holdt) @77°F	Color (Gardner)	Acid Insoluble Lecithin (%)	Weight per Gallon (lb)	Specific Gravity @77°F
Extra (Bleached)	Z-6	14	63.5	8.75	1.05
LV (Low Viscosity)	K-M	14	50 max	8.00	0.963
WD and Super WD (Water Dispersible)	Z-4+	14	56	8.70	1.045

All of the above have the same following properties:

Acid Value	30 maximum
Moisture	1.0% maximum
Nonvolatile Content	99% minimum

Chemical Type: Refined soya lecithin.

Key Properties (All Clearate Grades):

Surface active.

Reduces surface and interfacial tension permanently.

Soluble in all vehicles, used in paints, enamels, stains, putty and caulking compounds.

Key Properties of Clearates in Paints and Enamels:

Mixing and grinding time are reduced.

Formulation can be altered with the amount of pigment used.

Will reduce tendency towards vehicle separation.

Flooding or floating will be prevented.

Sagging will be avoided.

Continuity of film formation is improved.

Penetration is increased.

Gloss or sheen is improved.

Key Properties of Clearates in Stains:

Drying is speeded.

Stain is more durable.

Bleeding and leaking are avoided.

Key Properties of Clearate WD for Use in Paints:

Keeps pigments permanently deflocculated.

Prevents hard settling of all pigments.

Will not create foam.

The washability of the dried film is not impaired.

Efflorescence is minimized.

Stable in most latex systems.

Key Properties (All Clearate Grades):

Surface active.

Reduces surface and interfacial tension permanently.

Soluble in all vehicles, used in paints, enamels, stains, putty and caulking compounds.

Key Properties of Clearates in Putty and Caulking Compounds:

A softer product results, which is easier to apply and slips clean from the putty knife.

It is possible, if desired, to reduce the over-all percentage of oil or vehicle, if the present consistency of the putty should be maintained.

Pigments, where and if present, are more thoroughly and uniformly dispersed.

The putty and caulking compound will readily fill in all cracks, holes, and crevices when applied and will tend to cling with more tenacity after application is complete.

The dispersion of solids in the vehicle is maintained, so brittleness and cracking are reduced.

DIAMOND SHAMROCK CORP., P.O. Box 2386R, Morristown, NJ 07960

Hyonic PE Series

	PE-40	PE-90	PE-100	PE-120
Appearance	Clear liquid	Clear liquid	Clear liquid	Opaque semisolid-clear liquid melt
Activity (%)	99	99	99	99
Cloud Point, °F/°C (1% Solution)	127/53	129/54	154/68	196/91
Weight per Gallon (Pounds)	8.5	8.8	8.9	8.9
Specific Gravity	1.02	1.06	1.07	1.07
Mols EtO	4	9	10	12
Surface Tension-Dynes/Cm (0.01% Soln. @ 25°C)	28	31	32	36

Chemical Type: Group of nonionic surfactants made by the ethoxylation of nonylphenol. The ratio of ethylene oxide per mol of nonylphenol is carefully adjusted.

Key Properties: Range from insoluble to dispersible or soluble in water. Water solubility increases with the length of the ethylene oxide chain. External factors can also influence the water soluble characteristics of the Hyonic PE surfactants.

Hyonic PE-40 "oil" soluble surfactant is the most hydrophobic member of the series. It is miscible in all proportions with most aliphatic hydrocarbons. It is recommended for the following applications:

Preparation of solvent emulsions cleaners.
Color development aid for tint colorants in latex paints.
Color development aids in latex paints.
Pigment wetting agents in latex paints.
Freeze-thaw stabilizers for latex paints and adhesives.
Stabilizers for PVA latex adhesive formulations. Will also improve wetting action.

Nopalcol Surfactants

Grades Available:

1-L	4-S
1-S	6-L
2-L	6-O
4-L	6-R
4-O	6-S

Chemical Type: Polyethylene glycol fatty ester nonionic surfactants. High purity materials, completely free of inorganic salts and organic sulfonates or sulfates.

Key Properties: Function as emulsifiers, plasticizers and lubricants, as well as agents for wetting, dispersing, binding and thickening. In particular, they are designed to fulfill those industrial requirements which anionic or cationic agents cannot meet. A general description follows:

They are outstanding for their surface activity and dispersibility in mediums containing highly concentrated dissolved solids.
They are stable to acids and alkalies at room temperature, but may hydrolyze at higher temperatures or in concentrated acid and alkaline solutions.
Being nonionics, they are compatible with cationic and anionic surfactants.
Their odors are typical of the fatty acid in combination.

FDA Clearances of Nopalcol Fatty Acid Esters

. . . . .FDA Section and Title.	1-L	1-S	2-L	4-L	4-O	4-S	6-L	6-O	6-R	6-S
175.105 Adhesives	X	X	X	X	X	X	X	X	X	X
175.300 Resinous and Polymeric Coatings	O	O	O	O	X	O	O	X	O	O

. . . . .FDA Section and Title.	. . . . .Grades								
		1-L	1-S	2-L	4-L	4-O	4-S	6-L	6-O	6-R	6-S
176.170	Components of Paper and Paperboard in Contact with Aqueous and Fatty Foods	O	O	O	X	X	X	X	X	O	X
176.180	Components of Paper and Paperboard in Contact with Dry Food	O	O	O	X	X	X	X	X	O	X
176.200	Defoaming Agents Used in Coatings	X	X	O	X	X	X	O	X	X	O
176.210	Defoaming Agents Used in the Manufacture of Paper and Paperboard	X	X	X	X	X	X	X	X	X	X
177.1200	Cellophane	X	O	O	X	X	X	X	X	O	X
177.1210	Closures with Sealing Gaskets for Food Containers	O	O	O	X	X	X	O	O	O	O
177.2260	Filters, Resin Bonded	O	O	O	X	X	X	X	X	X	O
177.2800	Textile & Textile Fibers	X	X	O	X	X	X	X	X	X	O

X: Approved
O: Not Approved

GAF CORP., Chemical Division, 140 West 51 St., New York, NY 10020

Igepal CO Surfactants

Product Name	Viscosity @77°F (cp)	Specific Gravity (77°F)	Solidification Point (°F)	Pour Point (°F)	Cloud Point (1% Soln. in Dis. Water)
CO-210	350	0.99	–8	–3	Insol.
Odor and Appearance: Aromatic, yellow liquid.					
CO-430	288	1.02	–20	–15	Insol.
Odor and Appearance: Aromatic, pale yellow liquid.					
CO-520	270	1.03	–29	–24	Insol.
Odor and Appearance: Aromatic, pale yellow liquid.					
CO-530	265	1.04	–31	–26	Insol.
Odor and Appearance: Aromatic, pale yellow liquid.					
CO-610	260	1.05	32	37	77
Odor and Appearance: Aromatic, pale yellow liquid.					
CO-630	263	1.06	26	31	130
Odor and Appearance: Aromatic, almost colorless liquid.					
CO-660	250	1.06	41	46	145
Odor and Appearance: Aromatic, pale yellow liquid.					
CO-710	270	1.06	45	49	162
Odor and Appearance: Aromatic, pale yellow liquid.					
CO-720	300	1.06	57	62	–
Odor and Appearance: Aromatic, dispersed opaque liquid.					
CO-730	500	1.07	68	71	208
Odor and Appearance: Aromatic, dispersed yellow liquid.					
CO-850	Solid	1.08@ 50°C	86	91	Clear at 212°F

Product Name	Viscosity @77°F (cp)	Specific Gravity (77°F)	Solidification Point (°F)	Pour Point (°F)	Cloud Point (1% Soln. in Dis. Water)
Odor and Appearance: Aromatic, pale yellow wax.					
CO-880	Solid	1.08@ 50°C	104	109	Clear at 212°F
Odor and Appearance: Aromatic, pale yellow wax.					
CO-887	1.09	–	25	34	Clear at 212°F
Odor and Appearance: Aromatic, pale yellow liquid (70% aqueous solution).					
CO-890	1.09@ 50°C	Solid	106	112	Clear at 212°F
Odor and Appearance: Aromatic, off-white wax.					
CO-897	1.10	–	41	46	Clear at 212°F
Odor and Appearance: Aromatic, pale yellow liquid (70% aqueous solution).					
CO-970	1.10@ 50°C	Solid	108	114	Clear at 212°F
Odor and Appearance: Aromatic, off-white wax.					
CO-977	–	1.10	46	52	Clear at 212°F
Odor and Appearance: Aromatic, pale yellow liquid (70% aqueous solution).					
CO-990	Solid	1.12@ 50°C	116	122	Clear at 212°F
Odor and Appearance: Aromatic, off-white wax.					
CO-997	–	1.11	63	68	Clear at 212°F
Odor and Appearance: Aromatic, pale yellow liquid (70% aqueous solution).					

Percent Ethylene Oxide in Igepal CO Surfactants:

Igepal Surfactant	Mol Ratio "n"	% Ethylene Oxide
CO-210	1½	23
CO-430	4	44
CO-520	5	50
CO-530	6	54
CO-610	7-8	60
CO-630	9	65
CO-660	10	66
CO-710	10-11	68
CO-720	12	71
CO-730	15	75
CO-850	20	80
CO-880	30	86
CO-887*	30	86
CO-890	40	89
CO-897*	40	89
CO-970	50	91
CO-977	50	91
CO-990	100	95
CO-997*	100	95

*70% Active.

Chemical Type: The Igepal CO surfactants are all derived from the same hydrophobic material, nonylphenol. By increasing the amount of hydrophilic substance, ethylene oxide, combined with the nonylphenol, a series of products with different hydrophobic-hydrophilic balances is obtained. Their chemical structure is that of a polyoxyethylated nonylphenol.

Since changes in the hydrophobic-hydrophilic balance produce important variations in wetting, detergency, emulsification, solubility, or foam, the selection of
the proper balance becomes important. The Igepal CO-series offers a wide range of
balances, but in some applications it may be advantageous to mix two or more of
the products for a specific use.

Key Properties: Igepal CO surfactants do not ionize in water. Therefore, they are nonionic and
nonelectrolytic and are not subject to hydrolysis by aqueous solutions of acid or
alkali. They cannot form salts with metal ions and are equally effective in hard and
soft water. Their nonionic nature makes them useful with either anionic or cationic
agents, and with positively charged colloids.

In appearance, the Igepal CO surfactants vary from slightly viscous oils to low
melting waxes.

Surface Activity: Igepal CO-520 through CO-730, inclusive, exhibit outstanding
suface-modifying properties as shown by their surface- and interfacial-tension
measurements. Those from CO-610 to CO-730, inclusive, are excellent wetting
agents, detergents and emulsifiers for aromatic solvents. Igepal CO-530 and lesser
ethoxylated products are useful as emulsifiers for mineral oils, kerosene and
chlorinated hydrocarbons. Products in the series above CO-710 are useful as water-
soluble emulsifiers for vegetable and essential oils. The water-soluble members of
this series exhibit excellent lime-soap dispersion.

Stability: The entire series of Igepal CO products is stable to storage. They are
stable to, and can be used safely with, acid and alkali and dilute solutions of many
oxidizing and reducing agents.

Nekal Surfactants

	Activity (%)
BA-77	75

Chemical Type: Sodium alkylnaphthenate sulfonate in powder form.

Key Properties: Dispersing agent in lake, paint and printing ink formulations; in plastic and synthetic
latices. Stabilizer in latex formulations; prevents coagulation in SBR and other synthetic rubbers. Used in the leather industry as a dyeing and leveling assistant. In
textile processing, as a wetting, dispersing and penetrating agent. In the agricultural
chemical industry, as a wetting agent in powder formulations.

BX-78   75

Chemical Type: Sodium alkylnaphthalene sulfonate in powder form.

Key Properties: Used as a wetting agent in the textile industry. Wetting and dispersing agent in the
leather industry. Wetting agent in the agricultural industry and in paper manufacturing. Used as a surfactant for rubber latex polymerization and emulsification.

WS-25-1   48

Chemical Type: Sulfonated aliphatic polyester in liquid form.

Key Properties: Wetting and rewetting agent and penetrant in the textile industry.

WT-27   70

Chemical Type: Sulphonated aliphatic polyester in aqueous solution.

Key Properties: Excellent wetting, rewetting, and penetrating agent. Used in dry cleaning detergents, emulsion polymerization, glass cleaners, wallpaper removers and battery
separators. Used as wetting and rewetting agent in paper manufacturing.

INTERSTAB CHEMICALS, INC., 500 Jersey Ave., P.O. Box 638, New Brunswick, NJ 08903

Interwet #212

Chemical Type: Proprietary composition nonionic ester type surfactant.

Product Name	Color (Gardner) (max)	Specific Gravity (75°F)	Viscosity (Gardner-Holdt) @77°F (max)	Weight per Gallon (lb)	Form
Interwet #212	4	0.988	A	8.23	Liquid

Key Properties: Performs the following:

Reduces the viscosity of plastisols and also lowers the interfacial tension between plasticizer and air bubbles. Aids in removing entrapped air from the plastisol. In deaerated plastisols, it will reduce the tendency of bubble formation during casting in the molds.

Aids in reducing initial viscosity, minimizing viscosity buildup, and aids in the removal of air from plastisols prepared with high viscosity polymeric plasticizers, and/or plasticizers possessing high solvating characteristics.

Unlike other surfactants, Interwet #212 does not exhibit deleterious effects on the heat stability of the plastisol.

Developed specifically for rigid and semi-rigid formulations to reduce the viscosity of systems having a low plasticizer to resin ratio and particularly for plastisols containing a large amount of filler or pigment.

MONA INDUSTRIES, INC., 65 East 23 St., Paterson, NJ 07524

Monawet Surfactants

Product Name	Active Content (%)	Color (APHA) (max)	Acid Number (max)	pH	Weight per Gallon (lb)	Molecular Weight
MB-45	45	75	2.5	6.0	9.3	332

Chemical Type: Colorless clear liquid sodium diisobutyl sulfosuccinate with anionic nature. Meets F.D.A. 178.3400. Contain 7 ppm heavy metals maximum.

MM-80	80	75	2.5	6.0	9.2	388

Chemical Type: Colorless clear liquid sodium dihexyl sulfosuccinate with anionic nature. Contains 10 ppm heavy metals maximum. Meets F.D.A. 175.105, 176.170 and 178.3400.

MO-70	70	100	2.5	6.0	9.0	444

Chemical Type: Colorless clear liquid sodium dioctyl sulfosuccinate with anionic nature. Contains 10 ppm heavy metals maximum.

MO-70E	70	100	2.5	6.0	9.0	444

Chemical Type: Colorless clear liquid sodium dioctyl sulfosuccinate with anionic nature. Contains 10 ppm heavy metals maximum.

MO-70R	70	100	2.5	6.0	8.8	444

Chemical Type: Colorless clear liquid sodium dioctyl sulfosuccinate with anionic nature. Contains 10 ppm heavy metals.

MO-75E	75	100	2.5	6.0	9.0	444

Chemical Type: Colorless clear liquid sodium dioctyl sulfosuccinte with anionic nature. Contains 10 ppm heavy metals maximum.

MO-84R2W	83	450	2.5	5.5@10%	9.2	444

Chemical Type: Light yellow viscous liquid sodium dioctyl sulfosuccinate with anionic nature. Contains 10 ppm heavy metals maximum.

MT-70	70	150	2.5	6.0	8.5	584

Chemical Type: Light straw clear liquid sodium ditridecyl sulfosuccinate with anionic nature. Contains 10 ppm heavy metals maximum. Meets F.D.A. 175.105, 176.180 and 178.3400.

Product Name	Active Content (%)	Color (APHA) (max)	Acid Number (max)	pH	Weight per Gallon (lb)	Molecular Weight
MT-70E	70	150	2.5	6.0	8.4	584

Chemical Type: Light straw clear liquid sodium ditridecyl sulfosuccinate with anionic nature. Contains 10 ppm heavy metals maximum. Meets F.D.A. 175.105, 176.180 and 178.3400.

MT-80H2W	80	250	2.5	5.5@10%	8.5	584

Chemical Type: Light yellow viscous sodium ditridecyl sulfosuccinate with anionic nature. Contains 10 ppm heavy metals maximum.

SNO-35	35	8	2	7.5	9.5	653

Chemical Type: Clear light amber liquid tetrasodium n-alkyl (C_{18}) sulfosuccinate with anionic nature.

TD-30	30	—	6	5.5	9.0	—

Chemical Type: Light yellow liquid half ester of sulfosuccinic acid based on an ethoxylated fatty alcohol.

Key Properties (of All Grades):

Are all liquid materials.

Are a group of powerful anionic surfactants, which have the ability to perform the following functions:

Reduce surface and interfacial tension.
Wetting.
Dispersing.
Emulsifying.
Penetrating.
Solubilizing.

Are indefinitely stable in neutral hot and cold aqueous systems and are also excellent in solutions with pH ranges from 2-10. They are subject to hydrolysis in systems outside of this range.

The aqueous solubility increases as the molecular weight decreases.

All MO grades meet the following F.D.A. Regulations:

171.105 Component in food-packaging adhesives.
175.300 Resinous and polymeric coatings.
175.320 Resinous and polymeric for polyolefin films.
176.210 Defoaming agent in paper and paperboard for food packaging.
177.1200 Additive in cellophane food packaging.
177.2800 Additional adjuvant employed in the production of textiles and textile fibers used as articles or components of articles that contact dry food only.
178.3400 Emulsifier or surface active agent in articles or compounds of articles that contact food.

Monazoline Surface Active Agents

Product Name	Imidazoline @ Mfg. (% min)	Color (Gardner) (max)	Acid Number (max)	pH (1.0% Solids Slurry)	Weight per Gallon	Molecular Weight
C	90	11	1	11.3	7.75	282

Chemical Type: 1-hydroxyethyl-2-alkylimidazoline coconut fatty acid cationic surface active agent in amber liquid form.

Key Properties: Functions as a corrosion inhibitor in latex primer paints where it prevents rust formation and bleed-through in the final paint coat.

Product Name	Imidazoline @ Mfg. (% min)	Color (Gardner) (max)	Acid Number (max)	pH (10% Solids Slurry)	Weight per Gallon	Molecular Weight
O	90	9	1	10.8	7.66	345

Chemical Type: 1-hydroxyethyl-2-alkylimidazoline oleic fatty acid cationic surface active agent in amber liquid form.

Key Properties: When used in combination with paint solvents and soft clays (such as Attagel 50), it functions as an excellent economical thickener for oil-based paints and sealants. Improves gloss and adhesion and acts as an anti-skinning agent in alkyd paints.

| T | 90 | 12 | 1 | 10.8 | 7.75 | 350 |

Chemical Type: 1-hydroxyethyl-2-alkylimidazoline tall oil fatty acid cationic surface active agent in amber liquid form.

Key Properties: When used in combination with paint solvents and soft clays (such as Attagel 50), it functions as an excellent economical thickener for oil-based paints and sealants.

RAYBO CHEMICAL CO., Huntington, WV 25722

Raybo 56-Magna Mix

Chemical Type: Proprietary composition film-forming surface active agent.

Physical Properties:

Viscosity (Gardner)	F-J
Acid Number	13-16
Lieberman-Storch	Positive
% Solids Content	78-80
Weight per Gallon	8.2 lb
Volatile	Mineral Spirits

Key Properties: Offers the formulator the broadest range of effectiveness to improve wetting, dispersion and anti-settling while contributing the least to discoloration, loss of dry, puffing, film softening and dulling. Raybo 56 shortens grinding time and produces finer dispersions and better color development. Higher pigment loadings in pastes are possible. More efficient pebble mill grinds result from the elimination of puffy structures. Pebble mills are discharged more rapidly and more thoroughly due to better flow of the pastes. Paints containing Raybo 56 have less tendency toward gelation. It is beneficial in systems which are affected by moisture. Putty, caulking compounds and wood fillers are improved and costs are lowered since additional quantities of inerts may be used. Where silicones are undesirable, Raybo 56 offers control of silking and floating. It benefits production, stability, application and economy. Conforms to Rule 66.

Raybo 82-AntiStat

Chemical Type: Proprietary composition cationic surface active agent.

Physical Properties:

Viscosity (Gardner)	A3
Weight per Gallon	7.4 lb

Key Properties: Characterized by a great affinity for metals and textiles, a high order of compatibility with film formers, and resistance to evaporation or removal. Eliminates the static electricity charge, which is a major problem in many industries.

TEXACO, INC., 2000 Westchester Ave., White Plains, NY 10650

Surfonic Surface-Active Agents

Product Name	Cloud Point	Color (Pt-Co) (max)	Viscosity (SUS) @210°F	Molecular Weight (Theory Average)	Flash Point (C.O.C.) (°F)	Weight Per Gallon (lb)
N-10	12.5 ml	400	45.5	264	355	8.10
N-31.5	25.5 ml	300	49.5	358	410	8.40
N-40	33.0 ml	200	52.5	396	435	8.50
N-60	78.5 ml	200	58.5	484	475	8.70
N-95	54°C	100	68.5	632	>500	8.80
N-100	65°C	100	72.0	660	>500	8.80
N-102	71°C	100	68.5	668	>500	8.80
N-120	52°C	100	75.5	748	>500	8.90
N-150	65°C	200	89.0	880	>500	8.95
N-200	72.5°C	200	–	1100	>500	9.00
N-300	–	200	160	1540	>500	9.10
N-400	–	300	–	1980	>500	9.10

Chemical Type: The "N" series of Surfonic Surface-Active Agents are nonionic reaction products of ethylene oxide with nonylphenol. The products are designated by a number following the letter "N". The number is a tenfold multiple of the molar ratio of ethylene oxide in the adduct. Their surface-active properties result from the combination of the hydrophilic polyoxyethylene chain and the hydrophobic nonyl phenol. These groups combine to form a molecule which "crosses" the oil-water interface and breaks down the surface tension so as to promote a dispersion. Surfonic products are supplied as essentially 100% active, clear, slightly viscous liquids up to N-120, which is slightly turbid. N-150 and higher are waxy solids at room temperature. Water solubility exists with N-95 and higher numbered products. N-60 is water dispersible while the lower numbered products are water-insoluble and oil-solubles.

Key Properties: Useful as emulsifiers, wetting agents, detergents, penetrants, solubilizing agents, and dispersants in the household detergent, textile, agricultural, metal cleaning, petroleum, cosmetic, latex paint, cutting oil and janitorial supply industries. Lower numbered, non-water-soluble products may be used as foam reducing agents. Surfonic N-95 has the best wetting time, greatest detergency and is the most versatile product. Higher numbered, water-soluble products are suited for elevated temperature operations and are relatively high foaming.

THOMPSON-HAYWARD CHEMICAL CO., P.O. Box 2383, Kansas City, KS 66110

T-DET Nonionic Surfactants

Product Name	Active Content (%)	Cloud Point (1% Solution) (°C)	Flash Point (Open Cup) (°F)	pH (1% Solution)	Specific Gravity (60°/60°F)
A-O26	100	36	250	7.0	0.98

Chemical Type: Anhydrous, liquid nonionic surface active agent produced by the reaction of ethylene oxide and a blend of straight-chain primary alcohols ranging from C_{10} to C_{12}. Clear liquid with mold odor.

Key Properties: Compatible with a wide range of anionics, cationics, acids, alkalies and other builders. Excellent wetting agent. Considered a low foamer.

| C-40 | 99.5 min | – | >200 | 6.0 | 1.05 |

Chemical Type: Liquid nonionic surface active agent produced by the reaction of castor oil with 40 mols of ethylene oxide. Yellow liquid with mild, oily odor.

Key Properties: Compatible with anionic and cationic surfactants and is moderately stable in the presence of acids, bases and salts. A valuable addition to a variety of water-based paints, using polyvinyl acetates, acrylics, alkyds and oleoresinous binders. Useful in dispersing pigment slurries, improving freeze-thaw resistance, improving gloss, leveling agent, defoaming and stabilization. These properties make it useful for universal colorants because it is compatible in most solvent-based and water-based paints.

Product Name	Active Content (%)	Cloud Point (1% Solution) (°C)	Flash Point (Open Cup) (°F)	pH (1% Solution)	Specific Gravity (60°/60°F)
DD-5	99.5 min.	Insol. in Water to 1%	>300	7.0	1.01

Chemical Type: Anhydrous, liquid nonionic surface active agent produced by the reaction of dodecyl phenol with 5 mols of ethylene oxide. Pale yellow liquid with mild, aromatic odor.

Key Properties: Compatible with anionic and cationic surfactants and is stable in the presence of acids, bases and salts.

DD-7	99.5 min	Insol. in Water to 1%	>300	7.0	1.04

Chemical Type: Anhydrous, liquid nonionic surface active agent produced by the reaction of dodecyl phenol with 7 mols of ethylene oxide. Pale yellow liquid with mild, aromatic odor.

Key Properties: Compatible with anionic and cationic surfactants and is stable in the presence of acids, bases and salts. Water dispersible surface active agent having excellent emulsifying properties for certain solvents.

DD-9	99.5 min	Water-dispersible	>300	7.0	1.05

Chemical Type: Anhydrous, liquid nonionic surface active agent produced by the reaction of dodecyl phenol with 9 mols of ethylene oxide. Pale yellow liquid with mild, aromatic odor.

Key Properties: Compatible with anionic and cationic surfactants and is stable in the presence of acids, bases and salts. An excellent surfactant in many applications where detergency, emulsification, liquid-solid contact, or wetting with low foaming are prime factors.

EPO-61	99.5 min	23	>200	—	1.01

Chemical Type: Anhydrous, liquid nonionic surface active agent produced by the reaction of ethylene oxide and propylene oxide. Essentially clear liquid.

Key Properties: Compatible with anionic and cationic surfactants and is stable in the presence of acids, bases and salts. Used in a variety of industries to reduce or eliminate foam. Can be used by itself or can be incorporated in compounded formulations.

N-4	99.5 min	Insol. in Water	>400	7	1.02

Chemical Type: Anhydrous, liquid nonionic surface active agent produced by the reaction of nonylphenol with 4 mols of ethylene oxide. Pale yellow, oily liquid with mild, aromatic odor.

Key Properties: Compatible with anionic and cationic surfactants and is stable in the presence of acids, bases and salts. Recommended for latex paint as a color development aid. Stabilizer and emulsifying agent for formulation additives such as alkyd vehicles.

N-6	99.5 min	Not Soluble to 1% in Water	>400	7	1.04

Chemical Type: Anhydrous, liquid nonionic surface active agent produced by the reaction of nonylphenol with 6 mols of ethylene oxide. Pale yellow liquid with mild, aromatic odor.

Key Properties: Compatible with anionic and cationic surfactants and is stable in the presence of acids, bases and salts. Has many applications.

Product Name	Active Content (%)	Cloud Point (1% Solution) (°C)	Flash Point (Open Cup) (°F)	pH (1% Solution)	Specific Gravity (60°/60°F)
N-8	99.5 min	26	500	6.0	1.05

Chemical Type: Anhydrous, liquid nonionic surface active agent produced by the reaction of nonylphenol with 8 mols of ethylene oxide. Essentially clear viscous liquid with mild, aromatic odor.

Key Properties: Compatible with anionic and cationic surfactants and is stable in the presence of acids, bases and salts. Used in emulsion paint systems. Recommended as a wetting agent in latex paint milling. As a dispersing agent of pigments, its use prevents pigment flocculation. This use lends itself to application and making universal color formulation. Has many other applications.

N-9.5	99.5 min	57	500	7.0	1.06

Chemical Type: Anhydrous, liquid nonionic surface active agent produced by the reaction of nonylphenol with 9.5 mols of ethylene oxide. Essentially clear viscous liquid with mild, aromatic odor.

Key Properties: Compatible with anionic and cationic surfactants and is stable in the presence of acids, bases and salts. Used in emulsion paint systems. Recommended as a wetting agent in latex paint milling. As a dispersing agent of pigments, its use prevents pigment flocculation. This use lends itself to application and making universal color formulations. Has many other applications.

N-10.5	99.5 min	68	500	7.0	1.06

Chemical Type: Anhydrous, liquid nonionic surface active agent produced by the reaction of nonylphenol with 10.5 mols of ethylene oxide. Essentially clear viscous liquid with mild, aromatic odor.

Key Properties: Compatible with anionic and cationic surfactants and is stable in the presence of acids, bases and salts. Recommended for a variety of applications.

N-12	99.5 min	82	>400	7.0	1.07

Chemical Type: Anhydrous, liquid nonionic surface active agent produced by the reaction of nonylphenol with 12 mols of ethylene oxide. Opaque liquid with mild, aromatic odor

Key Properties: Compatible with anionic and cationic surfactants and is stable in the presence of acids, bases and salts. Recommended for a variety of applications.

N-14	99.5 min	95	>400	6.0	1.07

Chemical Type: Anhydrous, nonionic surface active agent produced by the reaction of nonylphenol with 14 mols of ethylene oxide. Off-white semi-solid with mild, aromatic odor.

Key Properties: Compatible with anionic and cationic surfactants and is stable in the presence of acids, bases and salts. Recommended as a wetting agent and emulsifying agent. Has many other applications.

N-30	99.5 min	Clear at 212°F	>500	7.25	1.07

Chemical Type: Anhydrous, nonionic surface active agent produced by the reaction of nonylphenol with 30 mols of ethylene oxide. Pale yellow waxy solid with mild, aromatic odor.

Key Properties: Compatible with anionic and cationic surfactants and is stable in the presence of acids, bases and salts. Recommended for use as a detergent, wetting agent and emulsifier. Has many other applications.

N-40	99.5 min	Clear at 212°F	>500	7.25	1.07 @ 132°F

Chemical Type: Anhydrous, nonionic surface active agent produced by the reaction of nonylphenol with 40 mols of ethylene oxide. Pale yellow waxy solid with mild, aromatic odor.

Key Properties: Compatible with anionic and cationic surfactants and is stable in the presence of acids, bases and salts. Recommended for use as a detergent, emulsifier and wetting agent. Has many other applications.

Product Name	Active Content (%)	Cloud Point (1% Solution) (°C)	Flash Point (Open Cup) (°F)	pH (1% Solution)	Specific Gravity (60°/60°F)
N-50	99.5 min	Clear at 212°F	>500	7.25	1.07 @132°F

Chemical Type: Anhydrous nonionic surface active agent produced by the reaction of nonylphenol with 50 mols of ethylene oxide. Pale yellow waxy solid with mild, aromatic odor.

Key Properties: Compatible with anionic and cationic surfactants and is stable in the presence of acids, bases and salts. Recommended for use as a detergent, emulsifier and wetting agent.

Product Name	Active Content (%)	Cloud Point (1% Solution) (°C)	Flash Point (Open Cup) (°F)	pH (1% Solution)	Specific Gravity (60°/60°F)
N-70	99.5 min	Clear at 212°F	>500	6.5	1.08 @135°F

Chemical Type: Anhydrous, nonionic surface active agent produced by the reaction of nonylphenol with 70 mols of ethylene oxide. Pale yellow, waxy solid with a mild, aromatic odor.

Key Properties: Compatible with anionic and cationic surfactants and is stable in the presence of acids, bases and salts. Recommended for use as a detergent and emulsifier.

Product Name	Active Content (%)	Cloud Point (1% Solution) (°C)	Flash Point (Open Cup) (°F)	pH (1% Solution)	Specific Gravity (60°/60°F)
N-100	99.5 min	Clear at 212°F	>500	6.5	1.08 @135°F

Chemical Type: Anhydrous, nonionic surface active agent produced by the reaction of nonylphenol with 100 mols of ethylene oxide. Pale yellow, waxy solid with a mild, aromatic odor.

Key Properties: Compatible with anionic and cationic surfactants and is stable in the presence of acids, bases and salts. Recommended for use as a detergent, emulsifier and wetting agent. Recommended for use with asphalt emulsions.

Product Name	Active Content (%)	Cloud Point (1% Solution) (°C)	Flash Point (Open Cup) (°F)	pH (1% Solution)	Specific Gravity (60°/60°F)
N-307	70	>100	>212	–	1.09

Chemical Type: 70% solution of T-DET N-30, a nonionic surface active agent produced by the reaction of nonylphenol with 30 mols of ethylene oxide.

Key Properties: Compatible with anionic and cationic surfactants and is stable in the presence of acids, bases and salts. A highly water soluble nonionic used as a high temperature detergent and as an emulsifier-stabilizer in various applications.

Product Name	Active Content (%)	Cloud Point (1% Solution) (°C)	Flash Point (Open Cup) (°F)	pH (1% Solution)	Specific Gravity (60°/60°F)
N-407	70	>212°F	>212	–	1.08 @25°C

Chemical Type: 70% solution of T-DET N-40, a nonionic surface active agent produced by the reaction of nonylphenol with 40 mols of ethylene oxide.

Key Properties: A highly water soluble nonionic used as a high temperature detergent and as an emulsifier-stabilizer in various applications. Used as an emulsifier in various paint systems.

Product Name	Active Content (%)	Cloud Point (1% Solution) (°C)	Flash Point (Open Cup) (°F)	pH (1% Solution)	Specific Gravity (60°/60°F)
N-507	70	>100	>212	–	1.09 @25°C

Chemical Type: 70% active solution of T-DET N-50, a nonionic surface active agent produced by the reaction of nonylphenol with 50 mols of ethylene oxide.

Key Properties: A highly water soluble nonionic used as a high temperature detergent and as an emulsifier-stabilizer in various applications. T-DET N-507 is stable and effective in concentrated electrolyte solutions. Used as an emulsifier in various paint systems.

Product Name	Active Content (%)	Cloud Point (1% Solution) (°C)	Flash Point (Open Cup) (°F)	pH (1% Solution)	Specific Gravity (60°/60°F)
N-705	50 min	Clear at 212°F	>500	5.75	1.08 @68°F

Chemical Type: 50% active solution of T-DET N-70, a nonionic surface active agent produced by the reaction of nonylphenol with 70 mols of ethylene oxide.

Key Properties: Compatible with anionic and cationic surfactants and is stable in the presence of acids, bases and salts. Recommended for use as an emulsifier and detergent.

Product Name	Active Content (%)	Cloud Point (1% Solution) (°C)	Flash Point (Open Cup) (°F)	pH (1% Solution)	Specific Gravity (60°/60°F)
N-1007	70	Clear at 100°C	>500	6.5	1.10

Chemical Type: 70% active solution of T-DET N-100, a nonionic surface active agent produced by the reaction of nonylphenol with 100 mols of ethylene oxide.

Key Properties: Compatible with anionic and cationic surfactants and is stable in the presence of acids, bases and salts. Recommended for use as an emulsifier, detergent and wetting agent.

Toxicity (All T-DET Nonionic Surfactants): All grades should be regarded as nonhazardous.

UNIROYAL CHEMICAL, Division of Uniroyal, Inc., Naugatuck, CT 06770

Polywet Anionic Oligomeric Surfactants

Product Name	Viscosity (Brookfield) (cp)	% Solids Content	pH	Specific Gravity @25°C
AX-4	8500	40	5.5	1.12

Chemical Type: Ammonium salt of a functionalized oligomer in water. Ammonium analogue of KX-4. Yellow to amber liquid with slight thiol odor.

Key Properties: Anionic emulsifier that can be used in the preparation of acrylate and vinyl acetate latexes.

| KX-3 | 380 | 45 | 7.25 | 1.19 |

Chemical Type: Potassium salt of polyfunctional oligomer in water. Yellow to amber liquid with mild thiol odor.

| KX-4 | 2500 | 40 | 7.25 | 1.175 |

Chemical Type: Potassium salt of polyfunctional oligomer in water. Yellow to amber liquid with mild thiol odor.

Key Properties (KX-3 and KX-4): The advantages of both KX-3 and KX-4 are unique and not available in any other surfactant. Latices prepared with them will have these properties:

Very high surface tension with consequent low foaming.
Controlled particle initiation with consequent temperature and viscosity control.
Excellent reactor stability.
Excellent mechanical stability.
Ability to incorporate high levels of dry inorganic fillers such as calcium carbonate without preslurrying the filler.
Faster stripping cycles due to low foam.
Higher surface tension reducing carpet wet through by carpet latices.
Uniform particle size.
Elimination of antifoam.
Superior oxidation resistance.
Reduced level of molecular weight regulators.
Compatibility with highly efficient polymeric thickeners which operate via a bridging mechanism.
Surfactant is cocurable by adding resins such as melamine formaldehyde.

Acrylic paint latex can be formulated to a full range of gloss. It will have fine particle size and good water resistance.

Toxicity (KX-3 and KX-4): Both are classified as not a "toxic substance" by ingestion as defined in Section 1500.3 of the Federal Hazardous Substance Act. Their acute oral LD_{50} (rats) is over 10 g/kg.

Section XLII

Talcs

CYPRUS INDUSTRIAL MINERALS CO., 555 South Flower St., Los Angeles, CA 90071

Beaverwhite 200

Chemical Type: High purity, platy talc pigment, free of asbestos-type impurities.

Key Properties: Used by the paint, plastics, rubber, building products and other related industries in applications requiring whiteness and high quality, but which do not require the extreme fineness of "325 mesh" or finer talcs. The cost savings is significant.

Particle Size Distribution (% by wt):

% Minus 74 μ (200 Mesh)	99.7
% Minus 44 μ (325 Mesh)	98
% Minus 20 μ	91
% Minus 10 μ	68
% Minus 5 μ	34
% Minus 2 μ	15
% Minus 1 μ	6
% Minus 0.5 μ	1

Median Particle Size: 7½ μ

Beaverwhite 325

Chemical Type: High purity, platy talc pigment, free of asbestos-type impurities.

Key Properties: Among the low cost, 325 mesh, "workhorse" fillers and extenders which exhibit versatility in a wide variety of applications within the plastics, paint, rubber, building products and related industries.

Particle Size Distribution (% by wt):

% Minus 74 μ (200 Mesh)	99.7
% Minus 44 μ (325 Mesh)	98
% Minus 20 μ	91
% Minus 10 μ	68
% Minus 5 μ	41
% Minus 2 μ	18
% Minus 1 μ	8
% Minus 0.5 μ	3

Median Particle Size: 6½ μ

Beaverwhite 200 and 325 (properties of both):

Chemical Analysis (% by wt):

Magnesium Oxide (MgO)	32.0
Silica (SiO_2)	61.0
Calcium Oxide (CaO)	0.5
Alumina (Al_2O_3)	1.0
Potassium Oxide (K_2O)	0.0
Ferric Oxide (Fe_2O_3)	0.5
Titanium Dioxide (TiO_2)	0.0

(continued)

234

Loss on Ignition
$\left.\begin{array}{l}\text{Carbon Dioxide (CO}_2\text{)}\\ \text{Water (H}_2\text{O)}\end{array}\right\}$ Chemically Combined

Loss on Ignition	
Carbon Dioxide (CO_2) Chemically Combined	0.5
Water (H_2O)	5.0
Lead (Pb)	*
Antimony (Sb)	*
Arsenic (As)	*
Cadmium (Cd)	0.3
Mercury (Hg)	0.03
Selenium (Se)	1
Barium (Water Soluble)	1

*Not detected

Mineral Analysis (% by wt):

Talc ($3MgO\text{-}4SiO_2\text{-}H_2O$)	95
Chlorite ($5MgO\text{-}Al_2O_3\text{-}3SiO_2\text{-}4H_2O$)	4
Dolomite [$CaMg(CO_3)_2$]	1
Quartz	Trace
Asbestos-type Minerals	Not detected

Physical Properties:

Specific Gravity	2.8
Acid Solubles (% as CaO)	0.4
Surface Area, m^2/g (N_2 BET)	8
Adsorbed Moisture (%)	0.2
pH (1:5 Dilution)	9
Refractive Index	1.6

Product Name	Brightness (G.E.)	Hegman Fineness of Grind	Oil Absorption (lb/100 lb)	Density— Loose (lb/ft³)	Surface Area (m²/g)
Cyprufil 200	82+	1½–2½	24	24	7

Chemical Type: Platy magnesium alumino-silicate.

Key Properties: Especially suited to filling of polyester formulations as its very low oil absorption allows high level filling while insuring system stability. Easy to wet out in water and has low oil absorbency. These characteristics make it an ideal choice as a filler in emulsion systems where using the minimum amount of wetting agent is an advantage.

Particle Size Distribution (% by wt):

% minus 74 μ (200 Mesh)	100
% minus 44 μ (325 Mesh)	98
% minus 20 μ	79
% minus 10 μ	58
% minus 5 μ	44
% minus 2 μ	29
% minus 1 μ	18

Median Particle Size: 7 μ

Cyprufil 325	83+	3–3½	27	20	9

Chemical Type: Platy magnesium alumino-silicate.

Key Properties: Easy to wet out in water and has low oil absorbency. These characteristics make it an ideal choice as a filler in emulsion systems where using the minimum amount of wetting agent is an advantage.

Particle Size Distribution (% by wt):

% minus 44 μ (325 Mesh)	100
% minus 20 μ	92
% minus 10 μ	67
% minus 5 μ	48
% minus 2 μ	31
% minus 1 μ	19

Median Particle Size: 5.4 μ

Cyprufil 200 and 325 (properties of both):

Chemical Analysis (% by wt):

Magnesium Oxide (MgO)	32
Silica (SiO_2)	32
Alumina (Al_2O_3)	20
Calcium Oxide (CaO)	Trace
Ferric Oxide (Fe_2O_3)	3
Loss on Ignition	
(Water Chemically Combined)	12.5

Physical Properties:

pH (1:5 Dilution)	8.4
Refractive Index	1.57
Specific Gravity	2.8

Product Name	Hegman Fineness of Grind	Oil Absorption (lb/100 lb)	Density— Loose (lb/ft³)	Surface Area (m²/g)
Glacier 200	1½–2½	28	28	9

Chemical Type: Platy talc filler or extender.

Particle Size Distribution (% by wt):

% minus 74 μ (200 Mesh)	99.7
% minus 44 μ (325 Mesh)	96
% minus 20 μ	84
% minus 10 μ	57
% minus 5 μ	36
% minus 2 μ	18
% minus 1 μ	10
Median Particle Size: 8 μ	

Product Name	Hegman Fineness of Grind	Oil Absorption (lb/100 lb)	Density— Loose (lb/ft³)	Surface Area (m²/g)
Glacier 325	3–3½	30	22	11

Chemical Type: Platy talc filler or extender.

Particle Size Distribution (% by wt):

% minus 44 μ (325 Mesh)	99.6
% minus 20 μ	94
% minus 10 μ	70
% minus 5 μ	46
% minus 2 μ	23
% minus 1 μ	13
Median Particle Size: 6 μ	

Key Properties: The fillers or extenders of choice in dark-colored paints and primers, and for many other applications without high brightness requirements. Both are offered at a substantial cost savings compared to higher brightness talcs.

Chemical Analysis (% by wt):

Magnesium Oxide (MgO)	30
Silica (SiO_2)	60
Alumina (Al_2O_3)	0.5
Calcium Oxide (CaO)	2
Ferric Oxide (Fe_2O_3)	1
Loss on Ignition	
Carbon Dioxide (CO_2) } Chemically Combined	2.5
Water (H_2O) }	4.0

Mineral Analysis (% by wt):

Talc ($3MgO \cdot 4SiO_2 \cdot H_2O$)	94
Dolomite [$CaMg(CO_3)_2$]	2
Calcite ($CaCO_3$)	2
Quartz (SiO_2)	2
Asbestos-type Minerals	Not detected

Physical Properties:

Adsorbed Moisture (%)	0.2
pH (1:5 Dilution)	9
Refractive Index	1.6
Specific Gravity	2.8

Product Name	Brightness (G.E.)	Hegman Fineness of Grind	Oil Absorption lb/100 lb	Density— Loose (lb/ft³)	Surface Area (m²/g)	Specific Gravity
Mistron Cyprusperse	88+	5½–6	39	11	12	2.8

Chemical Type: Magnesium alumino-silicate of a platy nature.

Key Properties: Easily wet out in water and has low oil absorbency. These characteristics make it an ideal choice for use with emulsion systems where minimal wetting agent addition is advantageous.

Chemical Analysis (% by wt):

Magnesium Oxide (MgO)	32
Silica (SiO$_2$)	32
Calcium Oxide (CaO)	Trace
Alumina (Al$_2$O$_3$)	20
Potassium Oxide (K$_2$O)	—
Ferric Oxide (Fe$_2$O$_3$)	3
Titanium Dioxide (TiO$_2$)	Trace
Loss on Ignition (Water Chemically Combined)	2.5

Particle Size Distribution (% by wt):

% minus 74 μ (200 Mesh)	100
% minus 44 μ (325 Mesh)	100
% minus 20 μ	100
% minus 10 μ	99
% minus 5 μ	89
% minus 2 μ	55
% minus 1 μ	30
% minus 0.5 μ	12

Median Particle Size: 1.7 μm

Product Name	Brightness	Hegman Fineness	Oil Absorption	Density Loose	Surface Area	Specific Gravity
Mistron Monomix	89+	6–7	43	7	14	2.8

Chemical Type: Ultra-fine, high purity, platy talc pigment, free of asbestos-type impurities.

Key Properties: Used by the plastics industry in liquid resin systems for resistance to settling and sagging and for viscosity control. Used by the paint industry for maximum fineness of grind, flatting, film smoothness, ease of sanding and for resistance to hard settling. The product of choice in any application requiring the finest particle size talc with high whiteness and purity.

Particle Size Distribution (% by wt):

% minus 74 μ (200 Mesh)	100
% minus 44 μ (325 Mesh)	100
% minus 20 μ	100
% minus 10 μ	94
% minus 5 μ	87
% minus 2 μ	59
% minus 1 μ	23
% minus 0.5 μ	6

Median Particle Size: 2 μ

Product Name	Brightness	Hegman Fineness	Oil Absorption	Density Loose	Surface Area	Specific Gravity
Mistron RCS	88+	6	35	15	10	2.8

Chemical Type: High purity, platy talc pigment, free of asbestos-type impurities.

Key Properties: Specially processed to provide a fineness of grind similar to "ultra-fine stir-in" talcs, but with significantly lower "binder demand." Designed for application requiring smoothness and the freedom from coarse particles that only an ultra-fine talc can offer, but giving more loadability, at a surprisingly moderate price.

Particle Size Distribution (% by wt):

% minus 74 μ (200 Mesh)	100
% minus 44 μ (325 Mesh)	100
% minus 15 μ	98
% minus 10 μ	90
% minus 5 μ	60
% minus 2 μ	25
% minus 1 μ	11
% minus 0.5 μ	5

Median Particle Size: 4 μ

Chemical Analysis (% by wt):

Magnesium Oxide (MgO)	32
Silica (SiO_2)	61
Calcium Oxide (CaO)	0.5
Alumina (Al_2O_3)	1
Potassium Oxide (K_2O)	—
Ferric Oxide (Fe_2O_3)	0.5
Titanium Dioxide (TiO_2)	—
Loss on Ignition	—
Carbon Dioxide (CO_2) } Chemically Combined	0.5
Water (H_2O)	5
Lead (Pb)	None detected
Antimony (Sb)	None detected
Arsenic (As)	None detected
Cadmium (Cd)	0.3
Mercury (Hg)	0.03
Selenium (Se)	1
Barium (Ba) (Water Soluble)	1

Mineral Analysis (% by wt):

Talc ($3MgO-4SiO_2-H_2O$)	95
Chlorite ($5MgO-Al_2O_3-3SiO_2-4H_2O$)	4
Dolomite [$CaMg(CO_3)_2$]	1
Quartz	Trace
Asbestos-type Minerals	Not detected

Physical Properties:

Acid Solubles (As CaO) (%)	0.4
Adsorbed Moisture (%)	0.2
pH (1:5 Dilution)	9
Refractive Index	1.6

Product Name	Brightness (G.E.)	Hegman Fineness of Grind	Oil Absorption lb/100 lb)	Density— Loose (lb/ft^3)	Surface Area (m^2/g)	Specific Gravity
Mistron Vapor	85+	5½–6	46	6	17	2.8

Chemical Type: High purity, asbestos-free, natural magnesium silicate pigment.

Key Properties: The surfaces of its ultra-fine platy particles have a strong affinity for many types of organic molecules and, conversely, a strong hydrophobicity. However, in aqueous systems, excellent wetting may be achieved by use of proper surface active agents. Major uses are as a white reinforcing filler and processing aid in elastomers and plastics, as a rheological control agent in liquid resins, plastisols, adhesives and as a scavenger of unwanted organic matter from water slurries.

Cyprubond	85+	5½–6	46	6	17	2.8

Chemical Type: Ultra-fine particle size, platy talc which has been surface modified to improve its polymer wettability.

Key Properties: All of the desirable characteristics of Mistron Vapor have been retained, while gaining significantly in ability to provide reinforcement in rubber compounds. Polymeric compounds containing Cyprubond usually show higher modulus, higher tear strength and lower compression set when compared to other mineral fillers, due to a greater degree of polymer/filler interaction.

Particle Size Distribution (% by wt):

% minus 74 μ (200 Mesh)	100
% minus 44 μ (325 Mesh)	100
% minus 20 μ	100
% minus 10 μ	97
% minus 5 μ	92
% minus 2 μ	65
% minus 1 μ	30
% minus 0.5 μ	7

Median Particle Size: 1.5 μ

Chemical Analysis (% by wt):

Magnesium Oxide (MgO)	31
Silica (SiO_2)	62
Calcium Oxide (CaO)	1
Alumina (Al_2O_3)	Trace
Potassium Oxide (K_2O)	—
Ferric Oxide (Fe_2O_3)	1
Titanium Dioxide (TiO_2)	—
Loss on Ignition Carbon Dioxide (CO_2) } Water (H_2O) } Chemically Combined	5.5
Lead (Pb)	None detected
Antimony (Sb)	None detected
Arsenic (As)	None detected
Cadmium (Cd)	0.5
Mercury (Hg)	0.02
Selenium (Se)	4
Barium (Ba) (Water Soluble)	1.5

Mineral Analysis (% by wt):

Talc ($3MgO \cdot 4SiO_2 \cdot H_2O$)	98
Chlorite ($5MgO \cdot Al_2O_3 \cdot 3SiO_2 \cdot 4H_2O$)	—
Dolomite [$CaMg(CO_3)_2$]	2
Quartz	1
Asbestos-type Minerals	Not detected

Physical Properties:

Acid Solubles (As CaO) (%)	1
Adsorbed Moisture (%)	0.2
pH (1:5 Dilution)	1.6

Product Name	Brightness (G.E.)	Hegman Fineness of Grind	Oil Absorption (lb/100 lb)	Density— Loose (lb/ft^3)	Surface Area (m^2/g)	Specific Gravity
Mistron ZSC	85+	6–7	46	6	12	2.8

Chemical Type: Zinc stearate coated, ultra-fine platy talc. Free of asbestos contaminations.

Key Properties: The surface modification significantly alters the characteristics of the talc particle, resulting in a product with even greater hydrophobicity than unmodified talc. Particularly useful as a flatting and antiblocking agent in plastic films and organic solvent-based paints. It is a rheological control agent in liquid resins, plastisols and adhesives, for promoting more uniform cell size distribution in plastic foams, and as a zinc stearate carrier for mold release, rubber dusting and pharmaceutical applications.

Particle Size Distribution (% by wt):

% minus 74 μ (200 Mesh)	100
% minus 44 μ (325 Mesh)	100
% minus 20 μ	100
% minus 10 μ	97
% minus 5 μ	88
% minus 2 μ	60

(continued)

% minus 1 μ	29
% minus 0.5 μ	6
Median Particle Size: 1½ μ	

Chemical Analysis (% by wt):

Magnesium Oxide (MgO)	31
Silica (SiO_2)	62
Calcium Oxide (CaO)	>1
Alumina (Al_2O_3)	Trace
Potassium Oxide (K_2O)	—
Ferric Oxide (Fe_2O_3)	1
Titanium Dioxide (TiO_2)	—
Loss on Ignition	
Carbon Dioxide (CO_2)⎫ Chemically Combined	5.5
Water (H_2O) ⎭	
Lead (Pb)	*
Antimony (Sb)	*
Arsenic (As)	*
Cadmium (Cd)	0.5
Mercury (Hg)	0.02
Selenium (Se)	4
Barium (Ba) (Water Soluble)	1.5

  *None detected.

Mineral Analysis (% by wt):

Talc (3Mgo-4SiO_2-H_2O)	98
Chlorite (5MgO-Al_2O_3-3SiO_2-4H_2O)	—
Dolomite [CaMg$(CO_3)_2$]	2
Quartz	1
Asbestos-type Minerals	Not detected

Physical Properties:

Acid Solubles (As CaO) (%)	1
Adsorbed Moisture (%)	0.2
pH (1:5 Dilution)	9
Refractive Index	1.6

MINERALS AND CHEMICALS DIVISION, Menlo Park, Edison, NJ 98817

Emtal Platy Talcs

Chemical Type: Naturally occurring hydrous magnesium silicates, a class of nonmetallic minerals exhibiting high brightness and low abrasion. These talcs are platy in character with no trace of any fibrous or tremolytic talc or other asbestiform minerals. The raw mineral is improved through various processing techniques to reduce the quantity of impurities. The material is pulverized to produce Emtal Talcs of varying particle size distribution.

Chemical Analysis (% by wt):

	41	42	43	44	4190	500	549	599
Magnesium Oxide (MgO)	31.7	33.3	33.5	33.5	31.7	30.8	30.8	30.7
Silicon Dioxide (SiO_2)	47.4	37.9	35.5	37.6	47.4	56.2	56.7	58.2
Calcium Oxide (CaO)	0.6	1.0	1.1	0.7	0.6	0.4	0.3	0.3
Ferric Oxide (Fe_2O_3)	5.0	6.4	6.7	6.3	5.0	3.9	3.9	3.4
Aluminum Oxide (Al_2O_3)	1.0	0.7	0.8	0.7	1.0	0.5	0.5	0.4
Loss on Ignition (L.O.I.)	14.2	20.3	21.9	20.6	14.2	8.0	7.7	6.9
MgO + SiO_2 + CaO	79.7	72.2	70.1	71.8	79.7	87.4	87.8	89.2
Carbon Dioxide (CO_2)	9.2	17.5	19.4	17.4	9.2	3.9	3.8	2.0

Physical Properties:

	41	42	43	44	4190	500	549	599
Particle Shape	. . . . Platy							
Retained on 200 Mesh (%)	Trace	0.2	0.3	4.5	—	0.01	—	—
Retained on 325 Mesh (%)	0.15	4.0	6.0	17.0	0.01	0.2	0.01	0.01
Particle Size (μ)	9.0	14.0	16.0	19.0	3.5	9.0	3.5	2.7
Photovolt Brightness (%)	77	73	73	73	81	83	86	87
Oil Absorption								
(Gardner-Coleman)	45-50	38-42	36-40	34-38	75-83	60-66	78-86	85-93
(ASTM D281)	32-35	29-32	28-31	24-27	40-45	33-36	43-48	44-49
Hegman Fineness of Grind	5.5-6.0	—	—	—	6.5-7.0	4.5-5.0	6.0-6.5	6.5-7.0
pH	. . . . .9.3-9.8							
Specific Gravity	2.8	2.9	2.9	2.9	2.8	2.8	2.8	2.8
Weight per Gallon (lb)	23.3	24.2	24.2	24.2	23.3	23.3	23.3	23.3
Bulking Value (gal/lb)	0.043	0.041	0.041	0.041	0.043	0.043	0.043	0.043
Bulking Density (lb/ft^3)								
Loose	29	39	42	44	16	25	26	14
Packed	45	68	77	78	25	39	24	21
Free Moisture (Max %)	. . . . 0.5							

Key Properties (features and benefits):

> Soft—Low abrasion and equipment wear.
>
> Platy—Lubricity, excellent sanding and trowelling, soft settling.
>
> Low Soluble Salts—Good chemical and water resistance.
>
> Organophilic—Excellent dispersion in organic medium.

Properties in Paints: The soft, platy structure of the Emtal Talcs offers a variety of interesting and useful properties for the paint formulator: in general, superior brushability, flow, freedom from gritty particles, film smoothness, enamel holdout and slip. Emtal 41, 42 and 4190 are used extensively in metal primers, including: low-cost rust-inhibitive primers, sanding primer/surfacers and automotive refinish primers. These grades promote adhesion to metal substrates, provide "tooth" for good inter-coat adhesion, and exhibit good salt-spray resistance. Emtal 500, 549 and 599 are grades produced by a unique flotation process that improves brightness and results in low water soluble salts and freedom from gritty particles. These grades are used in industrial primers and enamels including: primer/surfacers, automotive refinish primers, semigloss enamels, low gloss and lusterless enamels, interior automotive finishes and pigmented lacquers.

> Emtal 599 meets the following specifications: ASTM D605-65T, TT-P-403a, 52-MA-523b, MIL-P-15173A.

Properties in Plastics: Automotive Body Patch Compound—Emtals are considered the standard of this industry. They offer excellent trowelability, control off-flow, sanding and feathering and shelf-life stability. Emtals are used in polyester body putty compounds, not only because they offer the required characterisitcs, but also because they offer controlled uniformity.

> Also can be used in polypropylene, vinyl floor tile and wire and cable.

Properties in Rubber: Used as a dusting agent.

Properties in Other Products: Emtal Talcs find application in a number of other products, such as caulks and sealants, dry-wall compounds, adhesives and printing inks.

PFIZER MINERALS, PIGMENTS & METALS DIVISION, 640 North 13th St., Easton, PA 18042

Platy Montana Talcs

Product Name	Particle Size (μ)	Surface Area (BET m^2/g)	Oil Absorption (lb/100 lb)	Hegman Fineness of Grind	Dry Brightness
MP10-52	0.90	17.8	52	6.5+	92.0

Product Name	Particle Size (μ)	Surface Area (BET m²/g)	Oil Absorption (lb/100 lb)	Hegman Fineness of Grind	Dry Brightness
MP12-50	1.5	14.5	50	6.0+	92.5
MP15-38	2.3	15.3	42	6.0	91.5
MP25-38	2.6	12.8	41	5.5+	91.0
MP40-27	4.6	9.5	27	4.0	91.0
MP44-26	5.0	8.7	26	3.5	91.0
MP45-26	5.0	8.7	26	3.0	87.0

Platy California Talcs

Product Name	Particle Size (μ)	Surface Area (BET m²/g)	Oil Absorption (lb/100 lb)	Hegman Fineness of Grind	Dry Brightness
CP10-40	1.8	19.2	40	6.5	96.0
CP14-35	2.5	12.9	35	5.5	95.0
CP20-30	2.3	11.7	30	5.5	95.0
CP38-33	3.5	8.1	30	4.5	94.0
CP44-31	7.5	3.2	24	3.5	93.0

Chemical Type: Talcs produced in a wide range of particle sizes from numerous California mines and from Montana ore bodies. Talc is a hydrated magnesium silicate with the theoretical formula: $3MgO \cdot 4SiO_2 \cdot H_2O$ or 31.7% MgO, 63.5% SiO_2 and 4.8% H_2O. Composition of industrial grades will vary according to the locality in which the talc is mined.

In the coding system, the first letter indicates the source: C designates California and M designates Montana. All of the talcs listed are of the platy type.

Key Properties: The major advantages that the coatings formulator will realize by selecting a certain talc or talc combinations are:

Ease of Dispersion—talcs can be easily incorporated into hydrophilic and organophilic binder systems by today's high-speed paint mixing equipment.

Stir-In Grades Available—Hegman fineness of 6.0 and over are easily attained with the finer grades.

Gloss Control—Flatting effects can be readily obtained at modest loadings. Montana talcs are especially efficient in this respect.

Dry Brightness—The high dry brightness of Pfizer talcs may permit the use of lower quantities of titanium dioxide resulting in reduced formulating costs.

Film Porosity—Platy talcs produce less porous films and permit higher loadings than the fibrous talcs. Salt spray tests show platy talc to be superior to limestone, clay and diatomaceous silicas.

Settling—Talcs not only suspend well themselves, but prevent hard settling. The platy talc is outstanding in this respect.

Sanding Properties—Platy talcs are especially useful in metal primers because of their excellent sanding characteristics.

Scrubbability—Talcs are superior to most other extenders in providing burnishing resistance and scrubbability to interior coatings.

The white, soft crudes used in making these talcs have been selected for their chemical composition and physical properties to attain high uniformity and excellent working properties. These include better leveling, flow, brushability, grindability, and suspension properties than normally found in extender pigments of this type.

Section XLIII
Thickeners and Thixotropic Agents

ALLIED COLLOIDS, INC., 161 Dwight Place, Fairfield, NJ 07006

Product Name	Viscosity (Brookfield) (as Supplied) (cp)	Percent Solids Content	pH	Viscosity (Brookfield) (1% Solution) (cp)	Specific Gravity	Odor
Viscalex HV30	5	30	4.0	125,000	1.05	Slight Acrylic

Chemical Type: Novel, acrylic copolymer containing carboxyl groups, supplied in the form of an acidic, low viscosity emulsion, of very fine particle size, thickening agent.

Key Properties: Able to carefully control the rheology of the product. In adhesives use, it is easily handleable, nonstringing and a pastelike product which is easy to spread. In paint use, it is a nondrip product with good application and flow properties. Has rheology and thickening power ideal for modern requirements. The consistent viscosity and stability specifications of conventional polyacrylates are maintained while offering advantages over natural based products.

AMERICAN CYANAMID CO., Wayne, NJ 07470

Cyanamer Polyacrylamides

Product Name	Viscosity (Brookfield) (cp)	Percent Solids Content	Screen Analysis	Screen Analysis
P-250	2.1 @ 0.1%	89	12% max on 20 mesh	30% max on 100 mesh

Chemical Type: A homopolymer of acrylamide, which is essentially nonionic and has a molecular weight of five million to six million in white powder form.

Key Properties: High viscosity can be obtained in aqueous systems with very small quantities of this material. Recommended as a thickening, flocculating, suspending and thixotropic agent. Also, provides lubrication, drag reduction and flowability.

P-26	550 @ 10%	—	100% thru 10 mesh	95% thru 30 mesh

Chemical Type: Modified polyacrylamide with a molecular weight of approximately 200 hundred thousand and a low carboxyl content.

Key Properties: The carboxyl function offers interesting possibilities for application. Provides lubrication, drag reduction and flowability.

A-370	—	93% min	—	—

Chemical Type: Modified polyacrylamide with a molecular weight of approximately 200 hundred thousand and substantial carboxylate content, in tan powder form.

Key Properties: Has excellent thickening power and stability over a wide pH range. Product uniformity and ease of handling. Suggested where thickening, stabilization and film-forming properties are desired. Provides lubrication, drag reduction and flowability.

Product Name	Viscosity (Brookfield) (cp)	Percent Solids Content	Screen Analysis	Screen Analysis
P-35	8.5 @ 15%	93% min	20 max on 200 mesh	5 max on 30 mesh

Chemical Type: Modified polyacrylamide, which is an effective dispersant and anti-precipitant in powder form, recommended for use in aqueous systems. P-35 is 50% solution in form.

Key Properties: Excellent where particles and colloids of practically all types, such as pigments. Fillers, dyes, surfactants, emulsions and polymer dispersions must be solubilized, dispersed, peptized, suspended or controlled in various ways during their use, manufacture and processing. Excellent for use as antiprecipitant, flocculating, suspending and thixotropic agent.

BASF CORP., 491 Columbia Ave., Holland, MI 49423

Collacral VL

Chemical Type: Vinyl pyrrolidone copolymer supplied as a 30% aqueous solution.

Key Properties: Nonionic thickener which improves the ability to shear and the flow out, particularly in finely divided polymer dispersions. Very small proportions suffice to exert a beneficial effect on the brushability.

Collacral P

Chemical Type: Acrylic high polymer containing amide groups, supplied in the form of a 16% aqueous solution.

Key Properties: The addition of very small amounts to alkaline or neutral dispersions is sufficient to ensure a pronounced thickening effect. Acidic dispersions must be made alkaline with ammonia before the thickener is added. However, the efficiency of Collacral P within the alkaline range is independent of pH. The viscosity curve for the thickened versions is very flat. Adding water decreases the viscosity only very gradually. Does not affect the coatings' resistance to water very much. It improves the brushability and the stability of the dispersions against coagulation.

Latekoll AS

Chemical Type: The ammonium salt of a polyacrylic acid with a high molecular weight, supplied in the form of a 10% aqueous solution.

Key Properties: The thickening effect exerted by this product is similar to that of Collacral P, but slightly less pronounced. The product also improves the brushability and compatibility with pigments.

Latekoll D

Chemical Type: Acrylic-methacrylic copolymer in 25% aqueous dispersion form.

Key Properties: The dispersion has a large number of free carboxyl groups in the copolymer, causing an acidic reaction. The emulsifying agent is anionic. When the low-viscosity dispersion is neutralized with ammonia, it is converted into a highly viscous solution that is an extremely efficient thickener for aqueous polymer dispersions and emulsion paints produced from them. Very slight amounts of this ammoniacal solution bring about such a pronounced increase in viscosity that the dispersion assumes the character of a pseudoplastic liquid. Systems thickened with this can be readily applied by the brush yet do not sag on vertical surfaces. The addition of Latekoll D slightly impairs the flow-out, but even large proportions of it do not exert very much effect on the resistance to wet abrasion of the coatings.

Key Properties (all of the above): Used to increase the viscosity of polymer dispersions, emulsion paints and texture finishes. They also improve the brushability and act as stabilizers. Collacral VL, Collacral P and Latekoll D can be particularly recommended as thickeners. They can be used either by themselves or together with cellulose ethers.

Luron Binder U

Chemical Type: Protein condensation product similar to a polyamide, a nonthermoplastic, faintly alkaline binder in the form of a colloidal aqueous solution of about 20% concentration.

Key Properties: The product is not sensitive to frost and has good freeze-thaw stability. It is an off-white to pale brownish, opaque liquid of average viscosity. It can be readily poured and forms a yellowish, slightly dull, transparent film that is ductile and flexible and has fairly good resistance to tearing. It can be cured with formaldehyde, since it is a condensation product of protein. The reaction does not take place in the diluted aqueous solution, but sets in during drying of the coating. The product can be diluted in all proportions with water. It allows homogenous films to be formed by virtue of its good compatibility with aqueous polymer binders. Increasing the proportion of Luron Binder U reduces the thermoplasticity of the dispersion binder. The product contains a preservative and can be stored for up to 12 months if the drums are kept tightly closed. The difficulties that frequently occur with the storage of casein compounds need not be feared. Luron Binder U is used as a thickener. It greatly improves the flow in conjunction with polymer dispersions containing styrene.

Luviskoll K90

Chemical Type: Polyvinylpyrrolidone supplied in the form of a powder that is readily soluble in water, although it is not electrolytic.

Key Properties: The powder has a slight acidic reaction of pH of 6.0. Aqueous solutions of Luviskoll K90 Powder are miscible with water-soluble cellulose derivatives and are compatible with conventional pigment solutions. Luviskoll K90 is used to thicken polymer dispersions and to stabilize suspensions of inorganic and organic pigments. It improves the compatibility of styrene-based emulsion paints with polyvinyl ester dispersions and their derivatives particularly if they contain very little or no wetting agent.

CABOT CORP., 125 High St., Boston, MA 02110

Cab-O-Sil Grades

Product Name	Surface Area (m^2/g)	Density (lb/ft^3)	pH (4% Solids Slurry)	Nominal Particle Size Diameter (μ)	Ignition Loss $(1000°C)$ (%)
M-5	200	2.3 max	3.8	0.014	1.0
MS-7	200	4.5	3.9	0.014	1.0
MS-75	255	4.5	3.9	0.011	1.5
HS-5	325	2.3 max	3.9	0.008	2.0
EH-5	390	2.3 max	3.8	0.007	2.5
S-17	400	4.5	3.8	0.007	2.5

Chemical Type (of all grades): Fumed silicon dioxide, a material which is generally regarded as unique in industry because of its unusual particle characteristics. Grades of Cab-O-Sil are produced with differing particle sizes, surface areas and bulk densities, through variations in the manufacturing process. These modifications do not affect the silica content, specific gravity, refractive index, color or amorphous form.

Physical Properties (of all grades)

Residue on 325 Mesh	0.02% max	Silica Content (Ignited Sample)	>99.8%
Bulking Value	5.5 gal/100 lb	Refractive Index	1.46
Specific Gravity	2.2	X-Ray Form	
Color	White		Amorphous

Key Properties (in paints): Cab-O-Sil is a remarkable ingredient used to produce or control four major properties in protective coatings and paints:

(1) Flow control, including thickening and thixotropy.
(2) Suspension, or antisettling.
(3) Gloss reduction.
(4) Free flow, anticaking and fluidization of powders.

Cab-O-Sil is a fumed silica having particles of submicroscopic size with a total surface area of 200 to 400 m^2/g, depending on grade. The combination of its unusually small particle size, tremendous surface area, purity and chain-forming tendencies makes Cab-O-Sil unique among coating additives as well. Cab-O-Sil is effective in solvent, aqueous, pure-resin and powder polymer systems.

Regulatory Status and Toxicity: The use of Cab-O-Sil (99.80% SiO$_2$) has been approved by the Food and Drug Administration for many food applications as both a direct food additive and as a substance allowed in the manufacture of materials that come into direct contact with food in various producing, manufacturing, packing, processing, preparing, transporting and holding operations.

EPA Status: Registered in the EPA inventory as Silicon Dioxide, C.A.S. #7631-86-9; EPA Code A-343-7356.

Hazardous Materials Act (P.L. 93-633): Classified as a nonhazardous, nonrestricted substance under the provisions of the Act of 1977.

Federal Labelling Act: Cab-O-Sil meets all of the requirements of this Act with respect to exemptions from hazardous labelling.

Key Properties in Adhesives: Cab-O-Sil is a versatile, efficient raw material used by the adhesives chemist to create desirable performance characteristics in his products. The precision which Cab-O-Sil brings to the control of viscosity and other properties matches the sophistication of today's engineered adhesives, which are far more efficient and specialized than yesterday's products. Wide use of Cab-O-Sil increases control over such critical properties as extrusion, spreadability, penetration, bond strength, cohesion, temperature stability and flexibility and corrosion resistance. Cab-O-Sil is used to impart six major properties in adhesives:

(1) Flow control, including thickening and thixotropy.
(2) Cohesion (reinforcement).
(3) Suspension.
(4) Shear resistance.
(5) Antiflocking.
(6) Free flow and anticaking of powders.

Cab-O-Sil is fumed silica having particles of submicroscopic size with a total surface area of 200 to 400 m^2/g, depending on grade. The combination of its unusually small particle size, tremendous surface area, purity and chain-forming tendencies makes Cab-O-Sil unique among silicas and unique among adhesive additives.

DEGUSSA CORP., Route 46 at Hollister Road, Teterboro, NJ 07608

Aerosil Fumed Silica (Special Grades)

Product Name	BET Surface Area (m^2/g)	Particle Size (μ)	Density (Standard) (g/ℓ)	Density (Densed) (g/ℓ)	Moisture (2 hr @ 105°C) (%)	Ignition Loss (12 hr @ 1000°C) (%)
OX50	50	40	130	—	<1.5	<1

Key Properties: Especially used for low thickening purposes, having relatively small surface area and larger primary particles with less tendency to agglomerate.

Product Name	BET Surface Area (m^2/g)	Particle Size (μ)	Density (Standard) (g/ℓ)	Density (Densed) (g/ℓ)	Moisture (2 hr @ 105°C) (%)	Ignition Loss (12 hr @ 1000°C) (%)
TT600	200	40	40	—	<2.5	<2.5

Key Properties: Preferably used for flatting purposes because of pronounced secondary structure.

| MOX80 | 80 | 30 | 60 | 160 | <1.5 | <1 |

Key Properties: A mixed oxide consisting of silica with approximately 1% alumina, designed for aqueous dispersions and special purposes.

| MOX170 | 170 | 15 | 50 | 130 | <1.5 | <1 |

Key Properties: A mixed oxide (the same as MOX80), but with a smaller particle size.

| COK84 | 170 | — | 50 | — | <1.5 | <1 |

Key Properties: A mixture of Aerosil and highly dispersed aluminum oxide in 5:1 ratio. It is particularly suited for thickening aqueous and other polar systems.

| R972 | 110 | 16 | 50 | — | <0.5 | <2 |

Key Properties: Hydrophobic behavior makes this type especially suitable for improving the flowability of powders and for many other applications where it is necessary to increase the hydrophobicity.

Appearance (of all): Fluffy white powder.

Chemical Analysis (special grades)

	OX50	TT600	MOX80	MOX170	COK84	R972
pH Value (4% Aqueous Suspension)	4.2	4.0	4.0	4.0	4.0	3.8
Silica (SiO_2), %	>99.8	>99.8	98.3	98.3	84	>99.8
Alumina (Al_2O_3), %	<0.08	<0.05	0.8	0.8	16	<0.05
Ferric Oxide (Fe_2O_3), %	<0.01	<0.003	<0.01	<0.01	<0.1	<0.01
Titanium Dioxide (TiO_2), %	<0.03	<0.03	<0.03	<0.03	<0.03	<0.03
Hydrogen Chloride (HCl), %	<0.01	<0.025	<0.025	<0.025	<0.1	<0.05
Sieve Residue (45 μm)	<0.1	<0.05	<0.1	<0.1	<0.1	—

Aerosil Fumed Silica (Standard Grades)

Product Name	BET Surface Area (m^2/g)	Particle Size (μ)	Density (Standard) (g/ℓ)	Density (Densed) (g/ℓ)	Moisture (2 hr @ 105°C) (%)	Ignition Loss (12 hr @ 1000°C) (%)
130	130	16	50	120	<1.5	<1

Key Properties: Low thickening action. Particularly efficient reinforcing filler in cold-curing silicone rubbers.

| 150 | 150 | 14 | 50 | 120 | <0.5 | <1 |

Key Properties: Standard grade for RTV-silicone rubber sealants. Special packing.

| 200 | 200 | 12 | 50 | 120 | <1.5 | <1 |

Key Properties: Standard grade. Preferred for thickening, suspension, reinforcing and thixotropic purposes generally.

| 300 | 300 | 7 | 50 | 120 | <1.5 | <2 |

Key Properties: Special type with a smaller particle size and high surface area, which results in greater thixotropy.

| 380 | 380 | 7 | 50 | 120 | <1.5 | <2.5 |

Key Properties: Special type with small particle size and very high surface area, designed for highly thixotropic systems.

Appearance (of all): Fluffy white powder.

Chemical Analysis (standard grades)

pH (4% Aqueous Suspension)	3.6–4.3
Silica (SiO$_2$)	>99.8%
Alumina (Al$_2$O$_3$)	0.05%
Ferric Oxide (Fe$_2$O$_3$)	<0.003%
Titanium Dioxide (TiO$_2$)	0.03%
Hydrogen Chloride (HCl)	<0.025%
Sieve Residue (45 μm)	<0.05%

FERRO CHEMICAL DIVISION, 7050 Krick Road, Bedford, OH 44146

Product Name	Melting Point (0°C)	Specific Gravity @ 100°C/25°C
Viscatrol-A	86	0.980

Chemical Type: Proprietary composition fine, white organic powder.

Key Properties: Thixotropic thickening agent for use in paints, caulks, plastisols and allied products.

GAF CORP., Chemical Division, 140 West 51 St., New York, NY 10020

Vinyl Ether Polymers

Gantrez AN-119 Powder (Low-viscosity type)
Gantrez AN-139 Powder (Medium-viscosity type)
Gantrez AN-149 Powder (Medium-viscosity type)
Gantrez AN-169 Powder (High-viscosity type)
Gantrez AN-179 Powder (High-viscosity type)

Key Properties: Water-soluble copolymers with thickening, dispersing and stabilizing properties. Produce clear films of high tensile and cohesive strength. Compatible with plasticizers. Soluble in acid, caustic and certain organic solvents; can be insolubilized by crosslinking with polyfunctional compounds such as polyols and polyamines. Protective colloids, particle size regulators and mechanical stabilizers in emulsion polymerization of vinyl acetate and in suspension polymerization of vinyl chloride. Gelling agents, thickeners, emulsion stabilizers, explosive stabilizers, anticorrosion coatings and suspending aids. When modified with nonionics, they gel or thicken concentrated acid and caustic solutions.

Adhesives: Improved wet grab, shear strength and specific adhesion. Form highly polar films on difficult surfaces. Also recommended for us in detergents and textiles.

Gantrez AN-8194 Polymer

Chemical Type: Poly(octadecyl vinyl ether/maleic anhydride) in 40% toluene solution.

Key Properties: Hydrocarbon-soluble copolymer that forms waxy films with good water resistance and thermal properties and is of special interest in adhesives and coatings. Compatible with silicone release agents and modifies release level for pressure sensitive labels and tapes. Also used in water-repellent coatings for porous and nonporous substrates.

Product Name	Active Content (%)	pH
Thickener L Polymer	15	9.0

Chemical Type: Modified form of poly(methyl vinyl ether/maleic anhydride) in liquid form.

Key Properties: Compatible with most latices without adversely affecting dyeing and aging properties for films. Forms smooth, free-flowing latices to facilitate uniform application. Versatile thickener primarily designed for natural or synthetic latices used in upholstery and rug backings. Suitable also for other latex and emulsion systems such as adhesives, printing pastes and paints.

Product Name	Active Content (%)	pH
Thickener LN Polymer	15	9.0

Chemical Type: Modified form of poly(methyl vinyl ether/maleic anhydride) in liquid form.

Key Properties: Developed specifically as a thickening agent for water-based PVC, acrylic and polyvinyl acetate emulsion paints. Also, for thickening other emulsion and latex systems as adhesives and water-phase print pastes. Also, functions as additive for flowout and leveling improvements of emulsion flat, semigloss, gloss, interior and exterior paints.

Gaftex PT Thickener

Chemical Type: Modified copolymer of methyl vinyl ether and maleic anhydride in powder form.

Key Properties: Thickener in aqueous media. Particularly effective in textile print paste systems.

Gantrez B-773 Polymer

Chemical Type: Poly(vinyl isobutyl ether) (70% solids solution in hexane).

Key Properties: Tacky polymer, soluble in aromatic, aliphatic and chlorinated hydrocarbons. Widespread utility in a variety of adhesive systems. Excellent adhesion to plastic, metal coated surfaces. Plasticizer and leveling agent for surface coatings.

Gantrez S-95

Chemical Type: Poly(methyl vinyl ether/maleic acid) hydrolyzed low molecular weight polymer.

Gantrez S-97

Chemical Type: Poly(methyl vinyl ether/maleic acid) hydrolyzed high molecular weight polymer.

Key Properties: Water-soluble polyelectrolyte similar to Gantrez AN polymers in chemical and physical properties and in areas of application. Major advantage is rapid cold-water solubility over the entire pH range. A 5% aqueous solution has a pH of approximately 2.

Gantrez ES-225 Polymer

Chemical Type: Monoethyl ester of poly(methyl vinyl ether/maleic acid) in 50% solids ethanol solution.

Gantrez ES-335 Polymer

Chemical Type: Monoisopropyl ester of poly(methyl vinyl ether/maleic acid) in 50% solids isopropanol solution.

Gantrez ES-425 Polymer

Chemical Type: Monobutyl ester of poly(methyl vinyl ether/maleic acid) in 50% solids ethanol solution.

Gantrez ES-435 Polymer

Chemical Type: Monobutyl ester of poly(methyl vinyl ether/maleic acid) in 50% isopropanol solution.

Key Properties: Form tough, clear, glossy films which dry tack-free and exhibit good adhesion and moisture resistance. Soluble in esters, ketones, glycol ethers, alcohol; insoluble in acidic water, soluble in alkali. Films have enteric and controlled-release properties. Compatible with various plasticizers and modifiers.

Coatings: Film-formers in shoe polishes, decorative and protective sprays, paper and leather coatings; temporary protective metal and plastic coatings. Enteric and sustained- (controlled-) release coating agents for tablets. Film-formers in aerosol and spray bandages.

Cosmetics: Film-formers in hair sprays and grooming aids.

Gantrez M-154 Polymer

Chemical Type: Poly(methyl vinyl ether) (50% solids solution in water).

Gantrez M-555 Polymer

Chemical Type: Poly(methyl vinyl ether) (50% solids solution in toluene).

Gantrez M-556 Polymer

Chemical Type: Poly(methyl vinyl ether) (50% solids solution in toluene).

Gantrez M-574 Polymer

Chemical Type: Poly(methyl vinyl ether) (70% solids solution in toluene).

Key Properties: Soluble in water and diverse organic solvents. Thermally reversible solubility in aqueous systems. Function as tackifiers, binders and plasticizers. Used in printing ink, textile sizes and finishes; latex modification.

Vinylpyrrolidone Polymers

Chemical Type: High-pressure acetylene derivatives; water-soluble and crosslinked homopolymers, of the polyvinylpyrrolidone chemical type.

PVP K-15

Chemical Type: Average molecular weight of 10,000 (powder form).

Key Properties: Soluble in water and most organic solvents. Binder, stabilizer, protective colloid. Forms hard, transparent, lustrous films. Detoxifies many poisons and irritants. Viscosity modifier for aqueous systems. Precipitates tannins and polycarboxylic acids.

Adhesives: Used in adhesives to bond glass, metal and plastics; as film-former in pressure-sensitive, water-rewettable type; as viscosity modifier in polymer-based adhesives.

Coatings: Pigment and dyestuff dispersant and film-leveler in paints and varnishes.

PVP K-30

Chemical Type: Average molecular weight of 40,000 (powder form).

PVP K-60

Chemical Type: Average molecular weight of 160,000 (45% aqueous solution).

PVP K-90

Chemical Type: Average molecular weight of 360,000 (powder or 20% aqueous solution).

HENKEL CORP., 425 Broad Hollow Road, Melville, Long Island, NY 11746

Product Name	Melting Point (°C)	Hydroxyl Value	Acid Value (max)	Saponification Value	Iodine Number (Wijs) (max)	Specific Gravity @ 68°F
Rilanit Special	83	160	4	178	6	0.9

Chemical Type: Hydrogenated castor oil thickening agent for nonaqueous surface coatings and printing inks, in powder form.

Key Properties: Produces a thixotropic viscosity increase. This thixotropy offers a number of advantages, the following of which are the most important:

(1) Reduction of sagging and dripping.
(2) Application of thick coatings.
(3) Prevention of pigment settling and flooding and floating.
(4) Improvement of flow.

The following binders are among the most important used in conjunction with Rilanit Special:

(1) All kinds of alkyd resins including modified types, such as styrenated alkyds.
(2) Chlorinated rubber.
(3) Binders containing chlorine such as polyvinyl chloride, its copolymers and post-chlorinated products, as well as chlorinated polyethylene.
(4) Curable epoxy coatings including solvent-free systems.
(5) Epoxy ester.
(6) Tar or bitumen and their combinations.
(7) Two-component polyurethane systems.

HERCULES, INC., 910 Market St., Wilmington, DE 19899

Ethyl Cellulose

Chemical Type: Ethyl cellulose is a cellulose ether made by the reaction of ethyl chloride, with alkali cellulose, as expressed by the type reaction:

$$RONa + C_2H_5Cl \rightarrow ROC_2H_5 + NaCl$$

where R represents the cellulose radical.

There are three ethoxyl types of Hercules Ethyl Cellulose. They are listed below, along with the degree of substitution.

Type	Ethoxyl Content	Degree Substitution Ethoxyl Groups per Anhydroglucose Unit
K-Type	45.5–46.8%	2.28–2.38
N-Type	47.5–49.0%	2.42–2.53
T-Type	49.0+%	2.53+

Each of these ethoxyl types is subdivided into viscosity types, as indicated below. Thus, ethoxyl and viscosity types may be readily specified, as K-100, N-7, T-100, etc. Ethoxyl and viscosity types of ethyl cellulose:

K	N	T	Designation (cp)	Limits (cp)
—	X	—	4	3–5.5
—	X	—	7	6–8
—	X	X	10	8–11
X	X	—	14	12–16
—	X	—	22	18–24
X	X	X	50	40–52
X	X	X	100	80–105
X	X	X	200	150–250
—	X	—	300	250–350
X	—	—	5,000+	5,000

....Ethoxyl Types*.... / *.....Viscosity Types** ....*

*X refers to types now produced. Dashes in the table indicate no demand at present for the particular type. However, this does not mean that these types cannot be produced.

**Viscosity on all types is run at 5% concentration by weight and 25°C. Viscosity is determined in 80:20 toluene:ethanol by weight on sample dried 30 minutes at 100°C.

Note: Ethanol refers to ethyl alcohol S.D., 2B, 190-proof.

Physical Properties of N-Type Ethyl Cellulose

Bulking Value in Granular Form (Pounds/Gallon)	2.6–2.8
Bulking Value in Solution (Gallons/Pound)	0.099–0.104
Color (Hazen) in Solution	2–5
Discoloration (By Sunlight)	Very slight
Elongation at Rupture (% 3-Mil Film, Conditioned @ 77°F and 50% R.H.)	7–30

Flexibility (Folding Endurance, M.I.T. Double Folds, 3-Mil Film)	160–2000
Hardness Index (Sward, 3-Mil Film)	52–61
Light Transmission (Practically Complete)	3100–4000 Å
Light Transmission (Better than 50% Complete)	2800–3100 Å
Moisture Absorption (By Film in 24 Hours @ 80% R.H.)	2
Odor	None
Refractive Index	1.47
Softening Point	152°–162°C
Specific Gravity	1.14
Specific Volume (in³/lb in Solution)	23.9
Taste	None
Tensile Strength (lb/in², 3-Mil Film, Dry)	6800–10500
Tensile Strength Wet (% of Dry Strength)	80–85
Water Vapor Transmission (g/m²/24 Hours, 3-Mil Film, ASTM E96-66E)	890

Physical Properties: Effects of Ethoxyl Content and Viscosity—As in the case of other cellulose derivatives, there are certain properties of ethyl cellulose which depend somewhat on the degree of substitution (ethoxyl content). For example, it affects softening point, hardness, water absorption, solubility in ethanol and solubility in 80:20 toluene:ethanol. There are certain other properties, such as tensile strength, elongation and flexibility which are not greatly affected by the degree of substitution, but which depend largely upon the degree of polymerization, which can be measured by viscosity.

Key Properties: The outstanding physical and chemical properties, together with some of the indicated uses, are described, as follows:

  (1) Color: Ethyl cellulose is practically colorless and retains this condition, under a wide range of uses. Neither sunlight nor ultraviolet light affects the color.
  (2) Compatibility: Ethyl cellulose is compatible with an unusually wide range of resins and plasticizers, including oils and waxes.
  (3) Low Density: The low density makes it possible to get greater coverage and greater volume per unit weight than with the other cellulose derivatives.
  (4) Electrical Properties: The excellent electrical properties of ethyl cellulose combined with its good thermal stability and outstanding flexibility and toughness led to its early and continued use in cable lacquers where conditions requiring these properties are encountered.
  (5) Flexibility: The great flexibility of ethyl cellulose over a wide range of temperatures is one of its most marked characteristics.
  (6) Flammability: Offers no fire hazards.
  (7) Softening Point: The softening point of ethyl cellulose is relatively low and can be made lower by proper adjustment of plasticizers.
  (8) Solubility: Soluble in a wide variety of solvents, making it easy to formulate this versatile material for any purpose where solvent application is desirable.
  (9) Stability to Chemicals: Of all cellulose derivatives, none is more stable to chemicals than ethyl cellulose.
  (10) Stability to Water: The types of ethyl cellulose discussed here are not affected by water.
  (11) Stability to Light: Light, visible or ultraviolet, has no discoloring action on ethyl cellulose.
  (12) Stability to Heat: Application of heat up to its softening point has little effect on ethyl cellulose.
  (13) Taste: Has no taste.
  (14) Thermoplasticity: Possesses excellent plastic flow characteristics. It is possible to process plastics completely in heated Banbury mixers or on heated two-roll mills without the aid of volatile solvents.
  (15) Toughness: The high tensile strength is worthy of note.

Applications: Experience has shown that ethyl cellulose may be used in any number of ways. The following notes are intended to aid potential users in more definitely determining where and how ethyl cellulose may be used:

  (A) Lacquers:
    (1) Hard lacquers for rigid surfaces.
    (2) Tough lacquer.

(3) Bronzing lacquer.
(4) Lacquer for polystyrene plastic.
(5) Lacquer for rubber.
(6) Gel lacquers.
(7) Specialty wood finishes.
(8) Water-white wood finish.
(9) Alkali-resistant lacquer.
(10) Paper lacquer.
(11) Flow-back high-gloss lacquer.
(12) Solvent-based strip coatings.

(B) Inks:
(1) Silk-screen inks.
(2) Gravure and flexographic inks.

(C) Varnishes

(D) Adhesives: Ethyl cellulose contributes—
(1) Low-temperature flexibility.
(2) A broadening of the critical melting range and a resultant decrease in plastic flow.
(3) Strength.
(4) Increase in melting point of the mixture.
(5) Decrease in sweating of plasticizers.
(6) Beter control of tackiness in adhesive film.

(E) Coated fabrics.

(F) Additive to calendered vinyl resin films.

(G) Hot melt applications:
The use of ethyl cellulose has received considerable attention. This is because ethyl cellulose is a product with an unusual combination of properties, which makes it easily adaptable to this mode of application. It can be made stable to heat, it has excellent thermoplasticity, it dissolves readily in many hot resins, plasticizers, oils and wax mixtures. Of particular interest is the fact that it imparts to such mixtures a remarkable toughening action.

(H) Casting plastics.

(I) Pigment-grinding base.

(J) Film and foil.

(K) Plastics.

(L) Binder for pigments and fillers.

FDA Status: Hercules Ethyl Cellulose has many clearances for use in food and articles that contact foods under the Federal Food, Drug and Cosmetic Act, as amended. A summary of the more important current clearances with references is as follows:

Regulation	Description
21 CFR 121.101h ⎫ 21 CFR 121.2526 ⎬ 21 CFR 121.2571 ⎭	Coatings and inks for paper and paperboard products used in food packaging
21 CRF 121.2514	Resinous and polymeric coatings
21 CFR 121.2550	Closures with sealing gaskets for food containers

Klucel

Chemical Type: Hydroxypropyl cellulose is a nonionic water-soluble cellulose ether.

Grades and Viscosity Types: Klucel is produced in two grades, determined by intended use:

Grade	Designation	Intended Use
Standard	—	Industrial
Food	F	Food, Pharmaceutical, Cosmetic

.. Viscosity Types..		.... Concentration in Water by Weight			
Standard	Food	1%	2%	5%	10%
H	HF	1,500–2,500	—	—	—
M	MF	—	4,000–6,500	—	—
G	GF	—	150–400	—	—
J	JF	—	—	150–400	—
L	LF	—	—	75–150	—
E	EF	—	—	—	300–700

Physical Properties

Viscosity	The ranges shown above comprise the specifications
Physical Form	White to off-white, granular solid
Particle Size	95% through 30 mesh
	99% through 20 mesh
Ash Content (Calculated as Na$_2$SO$_4$)	0.5% max*
Moisture Content (As Packed)	5.0% max
pH (Water Solution)	5.0–8.5

*Lower ash content obtainable if required

Key Properties: Has a remarkable combination of properties. Combines organic solvent solubility, thermoplasticity and surface activity with the thickening and stabilizing property characteristics of other water-soluble cellulose polymers available from Hercules. Recommended for:

Applications	*Properties Utilized*
Adhesives	
Solvent-based	Thickener
Hot melt	Thermoplastic
Coatings	
Edible food coatings	Glaze, oil- and oxygen-barrier
Textile and paper coatings	Solvent-soluble film former, oil- and fat-barrier,
Film coatings	heat sealable
Paint Removers	
Acid-based	Thickener, acid resistant
Scrape-off and flush-off	

Natrosol

Chemical Type (all grades): Hydroxyethyl ether of cellulose. Cellulose is treated with sodium hydroxide and reacted with ethylene oxide to introduce hydroxyethyl groups which yield hydroxyethyl ether. The reaction product is purified and ground to a fine white powder.

Types and Grades: Natrosol is produced in four levels of hydroxyethyl molar substituion: 1.5, 1.8, 2, 5 and 3.0. These are designated as **Natrosol 150, 180, 250 and 300** repectively. Natrosol 250 is produced in the ten viscosity types listed in the table. Natrosol 150, 180 and 300 are made in several, but not all viscosity types.

R-Types: All grades and types of Natrosol can be treated to provide a powder that displays fast dispersion without lumping when added to water. The grades so treated are designated by the letter R. This treatment does not alter the solution viscosity. Some grades and viscosity types are available without this treatment, if desired.

B-Types: Certain medium- and high-viscosity types of Natrosol 250 are available in a grade that had superior biostability in solution. These grades are designated by the letter B.

Particle Size: Natrosol can generally be supplied in any one of the three particle sizes listed under Physical Properties. All types and grades are available in the regular grind and most are available in the X and W grinds also.

Viscosity Types	Brookfield Viscosity @ 25°C, cp @ Varying Concentrations		
	1%	2%	5%
HRR	3,400–5,000	—	—
H4R	2,600–3,300	—	—
HR	1,500–2,500	—	—
MHR	800–1,500	—	—
MR	—	4,500–6,500	—
KR	—	1,500–2,500	—
GR	—	150–400	—
ER	—	25–105	—
JR	—	—	150–400
LR	—	—	75–150

Note: The ranges given are not necessarily the viscosity specifications for this type.

Physical Properties

Polymer:	
Powder	White to light tan
Odor	None
Moisture Content (As Packed)	5.0% max
Ash Content (Calculated as Na_2SO_4)	4.0% max
Effect of Heat:	
Softening Range	135°–140°C
Browning Range	205°–210°C
Bulk Density	0.6 g/ml
Particle Size:	
Regular Grind on U.S. 40 Mesh	10% max
X Grind on U.S. 60 Mesh	0.5% max
W Grind on U.S. 80 Mesh	0.5% max
Biological Oxygen Demand (B.O.D.):	
Type H (250 or 180)	7,000
Type L (250 or 180)	18,000
Solutions:	
Specific Gravity (2% Solution)	1.0033
Refractive Index (2% Solution)	1.336
pH	7
Surface Tension (Dynes/cm):	
Natrosol 250L at 0.1%	66.9
Natrosol 250L at 0.001%	67.3
Natrosol 180L at 0.1%	66.7
Natrosol 180L at 0.001%	69.8
Bulking Value in Solution (gal/lb)	0.1
Films:	
Refractive Index	1.51
Specific Gravity at 50% Relative Humidity	1.34

Key Properties: Dissolves readily in water, either cold or hot. Its solutions have somewhat different flow properties from those obtained with other water-soluble polymers. It is used to produce solutions having a wide range of viscosity. Such solutions are pseudo-plastic—that is, they vary in viscosity depending upon the amount of stress applied. Uses of Natrosol are as a thickener, protective colloid, binder, stabilizer, and suspending agent in a variety of industrial applications, including pharmaceuticals, textiles, paper, adhesives, decorative and protective coatings, emulsion polymerization, ceramics and many miscellaneous uses. For a guide to these uses, see below:

Applications	*Properties Utilized*
Adhesives	
Wallpaper adhesives	Thickening and lubricity
Latex adhesives	Thickening and water-binding
Plywood adhesives	Thickening and solids holdout
Coatings	
Latex paint	Thickening and protective colloid
Texture paint	Water-binding
Miscellaneous	
Joint cements	Thickening
Hydraulic cements	Water-binding and set retarder
Plaster	Water-binding
Caulking compound and putty	Thickening
Printing inks	Thickening and rheology control
Asphalt emulsions	Thickening and stabilizing
Paper	
Coating colors	Water-binding and rheology control
Textiles	
Latex-back sizes	Thickening

FDA Status: The U.S. Food and Drug Administration has included Natrosol on the list of materials that have been cleared for use in adhesives and in resinous and polymeric coatings employed on the food-contact surfaces of metal, paper, or paperboard articles, and other suitable substrates intended for use in food packaging under the following Code of Federal Regulations. Natrosol R-grades are cleared only under Sections 175.105 and 176.180.

EPA Regulation: Meets EPA Regulation 180.1001(c).

Toxicology and Dermatology: Acute oral toxicity tests, using Natrosol 250, showed no adverse effects on rats. Natrosol 250 was found to produce no more skin irritation in tests than commercial wheat starch.

INTERSTAB CHEMICALS, INC., 500 Jersey Ave., New Brunswick, NJ 08903

Product Name	Color (Gardner) (max)	Specific Gravity (75°F)	Viscosity (Gardner-Holdt) @ 77°F (max)	Weight per Gallon (lb)	Acid Value (min)	Form
Interlite VCA	5	1.35	I	11.25	300	Liquid

Chemical Type: Proprietary viscosity control additive.

Key Properties: Designed for use in vinyl solution coatings and organosols containing an active resin solvent. (Not recommended for plastisol dispersed systems.) Its advantages are:

(1) Controls viscosity of vinyl solutions and resin-solvent containing organosol coatings formulations and prevents progressive bodying during storage.
(2) Permits the use of a wider range of stabilizers, including barium-cadmiums.
(3) Will not adversely affect the heat or light stability of the coating during processing or after application.

RAYBO CHEMICAL CO., Huntington, WV 25722

Raybo 77-AlGel

Chemical Type: Aluminum octoate that gels without heat.

Physical Properties

Viscosity (Gardner)	*Weight per Gallon*
A	7.50 lb

Key Properties: Raybo 77 produces aluminum octoate in most petroleum products without the use of heat. The inconvenience, added costs and hazards of heating are avoided. In some formulas, where the percentage of aluminum octoate is high and heat is used, the developed viscosity interfers with uniform production of the gel. Raybo 77 prevents this difficulty. Raybo 77 facilitates aluminum octoate gels in petroleum products, such as aliphatic solvents, paraffin oil, spindle oil, lubricating oil, mineral oil and petroelum resinous oil (Piccocizer R). The use of Raybo 77 allows a wide latitude. The following are a few of the many uses for aluminum octoate gels:

(1) Suspension agent in paints.
(2) Viscosity producer for paint and varnish removers, oils and paints.
(3) Flow retarder for caulking compounds and stipple finishes.
(4) Penetration reducer for coatings on porous surfaces.
It conforms to Rule 66.

ROHM AND HAAS CO., Independence Mall West, Philadelphia, PA 19105

Acrysol Thickeners

Product Name	Percent Solids Content	pH	Specific Gravity
ASE-60	28	2.9	1.05
ASE-75	40	2.5	1.08
ASE-95	20	2.9	1.05
ASE-108	20	3.0	1.05

Chemical Type (all of the above): Acrylic emulsion copolymers.

Key Properties (all of the above): Similar in properties to natural gums of high viscosity water-soluble cellulosic derivatives, but with greatly simplified handling techniques. Possible to incorporate directly into binder system. Thickening action begins with addition of base to the proper pH. Should be diluted 1:1 with water when added thus, "in-situ." Their properties are the following:

Acrysol ASE-60: Imparts a short buttery rheology.
Acrysol ASE-95: Gives a long, leggy rheology.
Acrysol ASE-75: Is somewhere in-between the two above.
Acrysol ASE-108: Imparts a long, stringy rheology.

These thickeners effectively suspend pigment particles and other finely divided solids.

Product Name	Percent Solids Content	pH	Specific Gravity
G-110	22	9.0	1.06

Chemical Type: Ammonium polyacrylate solution.

Key Properties: Superior thickening and stabilizing effect on synthetic and natural latices. May be used as supplied to produce thickened latices of a smooth, consistent texture. Insensitive to sodium, calcium and magnesium ions. Not susceptible to microbial attack often encountered with cellulosics.

| GS | 12.5 | 9.1 | 1.06 |

Chemical Type: Sodium salt of an acrylic polymer.

Key Properties: Excellent for natural and synthetic latices. Produces smooth formulations that maintain constant viscosity at elevated temperatures. Not recommended for self-reactive acrylic emulsions.

| HV-1 | 10.0 | 9.0 | 1.10 |

Chemical Type: High molecular weight sodium polyacrylate solution.

Key Properties: Excellent all purpose thickening agent where thixotropic flow properties are desired.

Section XLIV

Waxes

BARECO DIVISION, 6910 East 14th St., Tulsa, OK 74112

Bareco Chemically Reacted Microcrystalline Waxes

Product Name	Melting Point (ASTM D127) (°F)	Needle Penetration @ 77°F	Color (ASTM D1500)	Acid No.	Saponification No.
Cardis One	198	4.0	2.5	19.0	45
Cardis 314	195	8.0	2.5	20.0	45
Cardis 320	193	10.0	3.0	35.0	75
Cerethane 63	200	2.0	2.5	30.0	58
Ceramer 67	208	3.0	1.0	47.0	78
Petronauba C	195	6.5	2.5	25.0	55
Petrolite C-23	195	6.5	3.5	23.0	55
Petrolite C-36	195	7.5	4.0	33.0	80
Petrolite C-400	220	2.5	2.5	12.5	25
Petrolite C-7500	208	2.0	0.5	13.5	30
Petrolite C-8500	205	5.5	2.0	7.5	20
Petrolite C-9500	203	5.5	0.5	30.0	53

Bareco Hard Microcrystalline Waxes (Hard)

Product Name	Melting Point (ASTM D127) (°F)	Needle Penetration @ 77°F	Brookfield Viscosity @ 210°F (cp)	Specific Gravity @ 75°F (ASTM D792)	Color (ASTM D1500)
Mekon White	199	5	10.7	0.93	17*
Fortex	202	5	30.0	0.93	1.0
Petrolite C-1035	199	5	10.5	0.93	0.5
Petrolite C-700	196	6	10.9	0.93	1.0
Be Square 195	196	7	10.7	0.93	0.5**
Be Square 185	190	11	11.1	0.92	1.0

Bareco Hard Microcrystalline Waxes (Plastic)

Be Square 175	182	19.0	–	–	1.0***
Victory	175	29.0	–	–	1.0**
Starwax 100	187	16.0	11.5	0.92	1.0

*ASTM D156.
**White and brown colors also available.
***Black color also available.

Chemical Type: A series of high melting point waxes. Produced by the solvent recrystallization of selected petroleum fractions and consist of n-paraffinic, branched paraffinic and naphthenic hydrocarbons in the C_{36} to C_{60} range. Proprietary solvent refining techniques permit the close control of these fractions into a series of high melting, hard petroleum waxes with unique physical and functional properties.

Key Properties: Hot Melt Adhesives – Compatible in adhesive systems containing the more widely used copolymer resins and amorphous hydrocarbon resins. The proper grade selection permits adjusting the high temperature adhesive functionality. For high temperature functioning adhesives, the highest melting grades are most useful, while the lower melting grades are most useful for general purpose hot melts. All of the Bareco Hard Waxes stabilize adhesive strength upon aging and their lower surface tension aids in wetting many different surfaces.

A partial list of the other extensive uses in a wide variety of products follows:

Hot Melt Coatings – 10 to 20% stabilizes gloss and improves the sealing strength of copolymer-based coatings.

Cup and Paper Coatings – 1 to 10% in paraffin wax reduces friction, improves gloss and blocking point.

Rubber and Elastomers – Plasticizer, anti-sunchecking and anti-ozonant.

Emulsion and Latex Coatings – A major ingredient of nonionic emulsions for sizes and as a coating modifier in polymer latex systems.

Toxicity: May be refined to meet the purity requirements of the FDA. Refer to the manufacturer for further information.

Bareco Unmodified 2000 Polywaxes

Product Name	Melting Point (ASTM D127) (°F)	Softening Point (ASTM D36) (°F)	Molecular Weight	Specific Gravity @ 77°F	Brookfield Viscosity @ 300°F (cp)	Needle Penetration @ 77°F
500	187	187	500	0.93	3	7
655	215	215	700	0.96	6	3
1000	235	235	1,000	0.96	11	1
2000	257	257	2,000	0.96	50	0.5

Standard Physical Constants:

Melt index (ASTM D1238/FR–A)	>5000
Color (ASTM D156)	5

Chemical Type: A 2,000 molecular weight homopolymer of ethylene, which is a crystalline aliphatic hydrocarbon.

Key Properties: Exhibits a low order of solubility in organic solvents, particularly at room temperature. Outstanding in their ability to harden and increase the melting point of formulated hot melt adhesives without increasing the melt viscosity of the blend. Exhibits outstanding heat stability. Useful modifier for flow and lubrication in plastics. Impart slip and mar resistance in coatings and inks.

Recommended for use as a component in ethylene-vinyl-acetate-based hot melt adhesives, as a plastics processing aid, mold release, sanding aid, slip and antimar aid in coatings and inks and as an antiblocking agent.

FDA Status: Meets the quality requirements set by the FDA under the olefin polymers regulation 21CFR 177.1520.

Modified Hydrocarbon Waxes

Product Name	Melting Point (ASTM D127) (°F)	Needle Penetration @ 77°F	Color (ASTM D1500)	Acid No.	Saponification No.
Polymekon	200	4.0	3.0	–	–
Petrolite WB-5	195	4.0	brown	18.0	70
Petrolite WB-7	195	4.0	brown	9.0	48
Petrolite WB-10	168	6.0	brown	18.0	75
Petrolite WB-11	170	4.5	brown	9.0	45
Petrolite WB-14	172	3.0	brown	5.0	48
Petrolite WB-16	185	4.0	brown	4.0	25

INTERNATIONAL WAX REFINING CO., INC., 181 East Jamaica Ave., Valley Stream, NY 11582

Product Name	Melting Point (°F)	Acid Number	Saponification Value	Ester Value
Beesyn White Wax	149+	20	93	73

Key Properties: USP White Beeswax can now be completely replaced by Beesyn White Wax. This complete beeswax substitute is made from domestic USP and NF materials. Readily available. Uniform quality. Low cost. Makes a glossy, smooth, stable, soft cream. Much lower cost than White Beeswax.

Microcrystalline Waxes

Product Name	Melting Point (°F)	Acid No.	Saponification Value	Needle Penetration @ 77°F	Ester Value	Color
White Micro Wax 163/169	166	0	0	9	0	white

Key Properties: Excellent, superior oil retention.

Toxicity: FDA approved. Odorless and tasteless.

#15 Micro Wax	187	0	0	–	0	white

Key Properties: Excellent for polishes, carbon papers, hot melt compounds, adhesives, surface waxing and textile compounds.

Chemical Analysis and Physical Properties (both of the above):

Flash point	500°-580°F
Specific gravity @ 25°C	0.915-0.935
Refractive index @ 80°C	1.437-1.440
Unsaponifiable matter	100%
Source	petroleum

Toxicity: FDA approved. Odorless and tasteless.

Synthetic Beeswax

Product Name	Melting Point (°F)	Acid No.	Saponification Value
"B" Wax White (extra white bleached)	150	20	93
"B" Wax Yellow (light refined yellow)	150	20	93

Chemical Analysis and Physical Properties (both of the above):

Ester No.	73
Ratio No.	3.6
Specific gravity @ 25°C	0.950

Key Properties (both of the above): Uniform quality. Faint and pleasant odor. Dull and plastic fracture. Excellent emulsifying properties. Excellent stability. Has the following advantages:

All the characteristics of genuine beeswax.
Excellent emulsifying qualities.
Gives a fine stable emulsion in creams and pastes.
Does not contain harmful ingredients.
Produces creams and pastes of excellent texture.
Will not become rancid.
Does not contain any foreign matter or sediment.
Priced far below genuine beeswax.

Used in many industries including cosmetics, salves, ointments, protective coatings, polishes, emulsions, soaps, greases, rubber, inks, textile and leather treating, etc.

Toxicity: Nontoxic. Will not become rancid.

MICRO POWDERS, INC., 1730 Central Park Ave., Yonkers, NY 10710

Product Name	Melting Point (°F)	Congealing Point (°F)	Needle Penetration @ 77°F	Acid Value	Saponification Value	Specific Gravity @ 77°F
MP-12 Wax	219	201	2	0	0	0.94

Chemical Type: Special grade of micronized synthetic wax with an average particle size of 5 μ and maximum particle size of 25 μ.

Key Properties: Has been processed for use in printing inks, paints and coatings where grinding or milling are part of the production process. Economical, easy to use, and will impart a high degree of lubricity and abrasion resistance when added at levels of 1 to 3% by weight to all types of inks, paints and coatings.

MP-12 Wax has been micronized to a particle size distribution that corresponds to a Hegman fineness of 6.5 (NPIRI of 5 to 6) and is easily incorporated into formulations that use sand mills, ball mills, or roller mills for processing of dry color pigments. This grade of wax can be added directly to the grind phase with the pigment or it can be added during the let-down phase and dispersed.

It is recommended for use in industrial lacquers, paints, coil coatings, wood coatings, as well as many types of inks. Due to its excellent lubricity and hardness it will exhibit exceptional antiblocking properties when added to the above coatings.

MP-22 Wax	219	201	2	0	0	0.94

Chemical Type: Micronized synthetic wax with an average particle size of 4 μ.

Key Properties: Has been formulated to meet the needs of the coatings industry. Will impart a high degree of abrasion resistance and slip into almost any type of surface coating. Prior compounding and grinding are eliminated by the use of this pigment-like powder which can be readily dispersed without the need for any special processing equipment.

MP-22 is supplied in a particle size distribution that corresponds to a Hegman fineness of 7 (NPIRI of 4 to 5).

Recommended for use in paints, coil coatings, paper coatings and other types of surface coatings. Works exceptionally well in water- and alcohol-based flexographic and gravure inks.

MP-22VF Wax

Key Properties: Finer grind of MP-22. This grade corresponds to a Hegman fineness of 7.5 (NPIRI of 2 to 3). It is designed for applications in which particle size and ease of dispersion are critical. It is especially recommended for use with flushed colors in the manufacturing of offset and heat-set printing inks.

MP-22C Wax	219	201	2	0	0	0.94

Chemical Type: Special grade of micronized synthetic wax that has been processed to an extremely fine and narrow particle size distribution.

Key Properties: Easily dispersed into paints, coatings and printing inks. Prior compounding or grinding are eliminated with the use of MP-22C which can be readily dispersed into most coatings with high speed mixing. It has been processed to a Hegman fineness of 7.5 (NPIRI of 1 to 2). Recommended for overprint varnishes and industrial finishes, as well as many types of inks. The MP-22C will impart excellent abrasion resistance, slip and antiblocking properties when added at levels of 1 to 3%.

MP-22XF	219	201	2	0	0	0.94

Chemical Type: Extra fine grade of micronized synthetic wax that has been specially processed for optimum ease of dispersion.

Key Properties: Recommended for all types of coatings and liquid and paste inks. Will impart excellent slip and rub resistance without the need for prior compounding or grinding. Due to its extremely fine particle size it is ideal for moderate or high speed mixing systems. Supplied in a particle size distribution that corresponds to a NPIRI grind gauge reading of 1.5 to 2. Has excellent properties in overprint varnishes and is recommended for many types of inks.

Product Name	Melting Point (°F)	Congealing Point (°F)	Needle Penetration @ 77°F	Acid Value	Saponification Value	Specific Gravity @ 77°F
MP-26 Wax	219	209	1	0	0	0.95

Chemical Type: Premium grade of micronized synthetic wax with respect to hardness and particle size. Average particle size of 3 μ.

Key Properties: This wax is designed for use in surface coatings where an extremely fine particle size along with optimum slip is required. Supplied as a micronized powder corresponding to a Hegman fineness of 7.5 (NPIRI of 3). Recommended for use in oil-based and solvent-based paints and printing inks.

MP-26VF

Key Properties: MP-26 is also available in a finer particle size called MP-26VF. This grade has been processed to a 2 NPIRI grind gauge reading and is recommended for applications where an extremely fine grind is required. It works especially well with flushed colors in letterpress and offset inks.

Polyethylene Waxes

Product Name	Melting Point (°F)	Congealing Point (°F)	Needle Penetration @ 77°F	Acid Value	Specific Gravity @ 77°F	Particle Size (μ)
MPP-123 Polyethylene Wax	233	218	1.5	0	0.92	4

Chemical Type: Micronized polyethylene wax.

Key Properties: Will impart excellent abrasion resistance when incorporated at low levels into all types of printing inks, paints and other types of surface coatings. Can be directly added to most systems by simply dispersing it with a sand mill or similar equipment. It has been micronized to a particle size of Hegman fineness 7 (NPIRI of 5) grind gauge reading.

Product Name	Melting Point (°F)	Congealing Point (°F)	Needle Penetration @ 77°F	Specific Gravity @ 77°F	Maximum Particle Size (μ)	Particle Size (μ)
MPP-620F	241	237	1	0.95	13	3

Chemical Type: Finely micronized polyethylene.

Key Properties: Has been processed to a Hegman fineness of 7.0 (4.5 to 5.0 NPIRI). Additions of 1 to 3% by wt will impart excellent rub and mar resistance and surface slip to printing inks, paints and coatings. Due to the fine particle size of MPP-620F, prior compounding or grinding are not necessary. Incorporation into most coatings can be readily achieved with high speed mixing equipment.

Has been formulated to provide excellent heat resistance and will impart greater resistance to solvent absorption and swelling than micronized synthetic waxes. Recommended for use in the following: many types of inks, coil coatings, powder coatings and high gloss overprint varnishes. Provides excellent antiblocking properties when added to water- and solvent-based wood finishes, as well as flexographic and gravure inks.

| MPP-620VF | 241 | 237 | 1 | 0.95 | 10 | 2.5 |

Chemical Type: Finely micronized polyethylene.

Key Properties: Has been processed to a Hegman fineness of 7.5 (2.0 to 2.5 NPIRI) grind gauge reading. Additions of 1 to 3% by wt will impart excellent rub and mar resistance and surface slip to paints, printing inks and coatings. Due to fine particle size of MPP-620VF, prior compounding or grinding are not necessary. Incorporation into most coatings can be readily achieved with high speed mixing equipment. Has been formulated to provide excellent heat resistance and will impart greater resistance to solvent absorption and swelling than micronized synthetic waxes. Recommended for use in coil coatings, powder coatings, high gloss overprint varnishes and many types of inks. Has been shown to provide excellent antiblocking properties when added to water- and solvent-based finishes and flexographic and gravure inks.

Product Name	Melting Point (°F)	Congealing Point (°F)	Needle Penetration @ 77°F	Specific Gravity @ 77°F	Maximum Particle Size (µ)	Particle Size (µ)
MPP-620XF	241	237	1	0.95	8	2

Chemical Type: Finely micronized polyethylene.

Key Properties: Has been processed to a Hegman fineness of 7.5 (1.0 to 1.5 NPIRI). Additions of 1 to 3% by wt will impart excellent rub and mar resistance and surface slip to paints, coatings and printing inks.

Due to the extra fine particle size of MPP-620XF, prior compounding or grinding is not necessary. Incorporation into most coatings can be readily achieved with high speed mixing equipment. Has been formulated to provide excellent heat resistance and will impart greater resistance to solvent absorption and swelling than micronized synthetic waxes. Recommended for use in the following: coil coatings, powder coatings, high gloss overprint varnishes, metal decorating paints and many types of inks. Has been shown to provide excellent antiblocking properties, as well as gloss retention when added to water- and solvent-based wood finishes and flexographic and gravure inks.

The MPP polyethylene waxes are also available in the following grades.

Product Name	Melting Point (°F)	Congealing Point (°F)	Needle Penetration @ 77°F	Acid Value	Saponification Value	Specific Gravity @ 77°F
MPP-620	241	237	1.0	0	0.95	−
MPP-622	230	223	1.5	0	0.95	−
MPP-622F	−	−	−	−	−	3
MPP-622VF	−	−	−	−	−	2.5

Chemical Type (all of the above): Micronized polyethylene wax.

Polyfluo Waxes

Chemical Type (of all grades): The Polyfluo grades are unique combinations of low density polyethylene waxes and polytetrafluoroethylene (PTFE). Due to their high degree of toughness as well as lubricity, the Polyfluo grades will provide an excellent combination of properties and increased flexibility to the ink or coatings manufacturer. Polyfluo products may be used as the sole additive, but more common is their combination with micronized or polyethylene waxes.

Product Name	Congealing Point (°F)	Specific Gravity @ 77°F
Polyfluo 190	260	0.99

Chemical Type: Specially modified micronized fluorocarbon, with an average particle size of 3 µ.

Key Properties: Will impart excellent abrasion resistance and slip to most printing inks. Has been processed to a 3 to 4 NPIRI grind gauge reading to eliminate the need for compounding or excessive grinding. Has been designed to provide the toughness associated with polyethylene along with the slip provided by fluorocarbon when incorporated into printing inks, paints or other types of surface coatings. Optimum properties are usually achieved by adding 1 to 3% by wt.

Polyfluo 302	260	1.05

Chemical Type: Special grade of micronized modified fluorocarbon with an average particle size of 2 µ.

Key Properties: Developed to provide improved slip and lubricity when added to printing inks, paints and surface coatings. Has been micronized to a Hegman fineness of 7.5 (1 to 2 NPIRI) grind gauge reading for easier dispersion and better gloss properties. Incorporation into most coatings can be easily achieved with conventional high speed mixing equipment. There is no need for prior compounding or grinding due to the fine particle size of the Polyfluo 302. At low levels it will impart a high degree of lubricity along with excellent rub resistance when added to paints, coatings and inks. Recommended areas of usage are: overprint varnishes, industrial paints such as coil and can coatings and many types of inks. Optimum properties are usually achieved by adding 1 to 2% by wt when the Polyfluo 302 is used as the sole slip additive. In some cases more desirable slip properties are achieved by adding the Polyfluo 302 in combination with micronized waxes or polyethylenes.

Product Name	Congealing Point (°F)	Specific Gravity @ 77°F
Polyfluo 400	260	1.35

Chemical Type: Special grade of micronized modified fluorocarbon with an average particle size of 2 μ.

Key Properties: Developed to provide maximum slip and lubricity when added to paints, surface coatings and printing inks. Polyfluo 400 has been micronized to a Hegman fineness of 7.5 (1 to 2 NPIRI) grind gauge reading for easier dispersion and better gloss properties. Incorporation into most coatings can be easily achieved with high speed mixing equipment. There is no need for prior compounding or grinding due to the fine particle size of Polyfluo 400. At low levels it will impart a high degree of lubricity along with excellent rub resistance when added to paints, coatings and inks. Recommended areas of usage are: industrial paints such as coil and can coatings, overprint varnishes, heat-set offset, quick set, ultraviolet and metal decorating inks.

Polyfluo 540	260	0.99

Chemical Type: Very fine grade of micronized modified fluorocarbon which has been specially processed to an average particle size of 2 μ.

Key Properties: Provides easier dispersion, better gloss and faster particle migration. Has been micronized to a Hegman fineness of 7.5 (1 to 2 NPIRI) grind gauge reading which eliminates the need for prior compounding, grinding or melting. Will impart an excellent combination of film toughness (rub resistance) and lubricity (slip) when added into paints, surface coatings and printing inks.

Recommended for use in solvent-based paints and coatings and many types of inks. Rub and slip properties of overprint varnishes are greatly enhanced when added at ½ to 1% loadings. UV coatings and inks have shown improvement in rub resistance and slip due to the fast particle migration of Polyfluo 540 to the ink surface before curing.

Toxicity (of all grades): At temperatures below 600°F the Polyfluos are completely inert. However, at temperatures above 600°F some polymer fumes may be emitted due to oxidation of the fluorocarbon polymer. Therefore, smoking should not be permitted when working with the material.

MOORE-MUNGER MARKETING, INC., 140 Sherman St., Fairfield, CT 06430

Paraflint RG

Chemical Analysis and Physical Properties:

Congealing point, °F	205	Color	white
Needle penetration @ 77°F	1.5	Viscosity @ 250°F, cp	9.5
Molecular weight, ~ aver	750	Specific gravity @77°F	0.945
Acid number	nil	Dielectric strength @ 60 cp, V/mil	760
Saponification number	nil	Dielectric constant	2.33
Bromine number	nil	Ash content, max %	0.1
Odor:	essentially none		

Chemical Type: Exceptionally hard, high melting point hydrocarbon wax of high purity. Mixture of saturated, straight-chain paraffin hydrocarbons having an average molecular weight of approximately 750. With the exception of the longer chain length, Paraflint is similar in molecular structure to conventional paraffin waxes derived from petroleum. Also, it resembles them in its white color and low melt viscosity.

Paraflint differs from conventional paraffin waxes in its high temperature properties and crystal structure. The congealing point (205°F) and the hardness is much greater, particularly at elevated temperatures where it is comparable to carnauba. Paraflint's coefficient of friction is substantially lower than conventional paraffins and its resistance to scuffing and blocking is superior. The crystal size is also finer and more like microcrystalline waxes. It is synthesized by the Fischer-Tropsch process.

Key Properties: Hot Melt Adhesives — In hot melt adhesive formulations, Paraflint can be used to control viscosity, set-up time and hardness. It exhibits good compatibility with most hot melt resins and copolymers.

Hot Melt Coatings — In hot melt coatings, Paraflint provides improved blocking resistance, gloss stability, and grease resistance while maintaining barrier properties. The lower friction of Paraflint is beneficial in high speed carton overwrap applications.

Paper Coatings — Improves scuff, blocking, grease and dust resistance, hardness, tensile strength, gloss and gloss stability. It lowers friction and decreases tack.

It is also used in polish formulations, as a polyethylene modifier and chemical raw material, and many other end uses.

Toxicity: Highly refined, pure and white. Meets FDA regulations for use in food coatings.

NATIONAL WAX CO., 3650 Touhy Ave., Skokie, IL 60076

Microcrystalline Waxes

Product Name	Melt Point (°F)	Color	Oil Content (%)	FDA Status
Crude Petrolatum	115–160	dark brown	20–60	not approved
Refined Petrolatum	90–130	white to amber	10–25	approved for some applications
Hardening Waxes	180–195	white to amber	1–3	approved

Note: National Wax has the facilities to manufacture custom blends of waxes, along with custom blends of waxes with polymers to meet a customer's individual requirements.

Paraffin-Type Petroleum Waxes

Slack Waxes	115–150	amber to off-white	5–25	usually not approved
Crude Scale Wax	110–130	off-white to white	2–5	approved
Semi-Refined Wax	125–135	off-white to white	1.0–1.5	approved
Fully-Refined Wax*	125–160	white	0.5–1.0	approved

*According to U.S. Pharmacopoeia Standards, a fully-refined wax must have no odor nor foreign impurities, must have at least a 28 Saybolt color and no more than 0.5% oil.

FRANK B. ROSS CO., INC., 6–10 Ash St., Jersey City, NJ 07304

Candelilla Wax

Grades: Crude, in lump
Refined, in lump, flakes and powder
Technical, modified with other ingredients for better performance
and lower cost

Chemical Type: A vegetable wax. Light brown to light yellow, hard, brittle, slightly tacky and lustrous wax with a distinctive odor. Not as hard as carnauba and does not reach its maximum hardness for several days after cooling.

Chemical Analysis and Physical Properties:

Melting point, °F (°C)	155-162 (68.5-72.5)
Flash point, °F min	465
Specific gravity @ 25°C	0.982-0.993
Acid number	12-22
Saponification number	43-65
Iodine number	19-44
Unsaponifiable matter, %	65-75
Refractive index	1.4600

Key Properties: Recommended for use in paper coatings, adhesives and in many other compounds.

Regulatory: FDA approved under Regulations 121.1059, 121.2520, 121.2569, 121.2571.

Ceresine Wax

Product Name	Color	Melting Point (Capillary Tube) (°F)	Needle Penetration @ 77°F
252	white	128-135	12
1670 Flakes	white	145-155	12
375	white	139-149	11
1154/4	white	140-150	12
101	white	154-164	7
608	white	157-164	10
1502	white	158-168	11
1530	white	160-165	11
1556	white	177-187	8
1154/2	yellow	128-138	12
20	yellow	132-142	9
101	yellow	154-164	9
1135/7R	yellow	153-163	14
609	yellow	160-165	9
1529	yellow	170-175	14
901B	tan	128-135	13
1741	tan	145-155	13
1108/5	tan	160-170	7
15	orange	132-142	9

Chemical Type: When used commercially the name is rather loosely applied. The ceresins of domestic origin are basically an upgrading of the various types of paraffins through the controlled admixing of materials that are dictated by their physical characteristics and not by their chemical specifications. Through experimentation the user finally determines the specific type necessary for his end use.

Owing to the variety of paraffins available and the choice of several materials which may be admixed, there is naturally a considerable variation in the character and physical properties of ceresins. Depending upon the additives, domestic ceresins may be hard or soft and those with identical melting points can have entirely different penetrations, softening points, plasticity ranges and properties of solvent absorption and gel.

Chemical Analysis and Physical Properties (all grades):

Melting point, °F (°C)	128-185 (53.3-87.8)
Specific gravity @ 25°C	0.880-0.935
Acid number	0
Saponification number, max	2
Unsaponifiables, %	100
Refractive index	1.425-1.435

Key Properties: Recommended for adhesives, paints and many other end uses.

Regulatory: FDA approved under Regulation 121.2520.

Japan Wax

Grades: There are several grades of genuine Japan waxes available which are distinguished by brand names.

Chemical Type: Vegetable wax derived from the protective coating on the berry kernels from several varieties of the Japanese sumac (haze) trees. The wax is prepared by crushing the aged berries and then extracting by means of pressure and/or solvent. To refine the wax is again melted, filtered and bleached by sunlight and/or chemical bleach. Pale cream colored wax with a gummy feel. Actually, it resembles a fat more than a wax.

Chemical Analysis and Physical Properties:

Melting point, °F (°C)	122-133 (50-56)
Flash point, °F	385-400
Specific gravity @ 25°C	0.975-0.984
Acid number	6-30
Ester number	210-225
Saponification number	216-255
Unsaponifiables, %	2-4
Iodine number	4-15
Refractive index	1.4550

Key Properties: Recommended for package coatings for paper, metal and fabric and many other uses.

Regulatory: FDA approved under Regulations 121.101, 121.2514, 121.2520, 121.2526 and 121.2540.

Product Name	. . . . .Melting Point.		Acid	Saponification
	(°F)	(°C)	Number	Number
Japan Wax Substitute				
525	127-133	53-56	4	185-195
966	123-131	51-55	5	185-195

Chemical Type: These substitutes have been very carefully formulated and maintained at an exact degree of purity during the various steps of manufacture.

Key Properties: Both are well balanced formulations with the chemical analysis closely approximating that of the various imported genuine grades. In the majority of cases, these grades may be used as a direct substitute for genuine Japan wax.

Montan Wax

Grades: **German** – Crude – pellets, powdered.
Domestic – Crude-Refined – flakes, powdered.

Chemical Type: Montan Wax is derived from lignite which is vegetable matter partly mineralized to a product related to bituminous coal. The lignites from which the Montan Waxes are extracted are found chiefly in Central Europe and California. The method of extraction and refining is basically the same both in this country and in Europe. The waxy part is extracted from the lignites (crushed to a powder form) by selective solvents. The wax is then purified.

Chemical Analysis and Physical Properties (German – Crude):

Softening point (R&B) °F (°C)	181-192 (83-89)
Acid number	31-38
Saponification number	87-104
Ash content, %	0.3-0.6
Rosin test	negative

Chemical Analysis and Physical Properties (Crude-Refined):

Softening point (R&B) °F (°C)	183-190 (85-88)
Acid number	40-55
Saponification number	95-125
Ash content, %	0.2-0.3
Rosin test	negative
Color	dark brown, brown, tan

Key Properties: Recommended for use in adhesives, rubber products, sizings and many other products.

Regulatory: FDA approved under Regulations 121.2519, 121.2520 and 121.2562.

Ozokerite Wax—White

Product Name	Melting Point (°F)	Needle Penetration @ 77°F
64W Flakes	145–155	12
71W Flakes	152–162	10
1543W Flakes	155–165	12
1981	155–165	15
56-1124	162–172	12
77W Flakes	164–174	8
56-1122	164–174	10
2525	165–175	10
1823	165–175	12
1544W Flakes	168–178	12
1477	170–180	25
1545W Flakes	175–185	10
1556	177–187	8
871	188–195	5

Ozokerite Wax—Yellow

2318	140–150	12
71Y	152–162	10
77Y	164–174	8
869	165–175	20
861	188–198	4
863	188–198	4

Chemical Type: True ozokerite wax is a bituminous product occurring in Miocene formations near petroleum deposits in Poland, Austria, Russia, Ukraine and also in Utah and Texas. In recent years, however, these true ozokerites have not been freely available in the United States. Today the ozokerites of commerce are blends of true waxes which produce a dry type wax of high solvent absorption and gel values similar to the formerly imported ozokerites. They have long fibers, unlike the structure of the paraffins and microcrystalline waxes.

Chemical and Physical Properties: Ozokerites are saturated hydrocarbons having neither acid nor saponification values.

Key Properties: Recommended for use in adhesives, paints, varnishes and many other products.

Regulatory: FDA approved under Regulation 121.2520.

Ross Petroleum Waxes

Chemical Type and Grade: Petroleum waxes are by-products of the petroleum industry. They are removed from the crude oil by distillation and several methods of extraction. The waxes recovered at the various stages of production have different characteristics and constants. The waxes are all hydrocarbons and are mainly classified as paraffins and microcrystallines. Roughly speaking, the paraffins come from the dewaxing of paraffin distillates and the microcrystallines from residues. In each of these classifications the waxes vary in hardness, depending on the character of the crystals and their melting points and freedom of oil.

Ross Microcrystalline Wax—White

1329/1	140–150	25–30
1365	140–150	25–30

(continued)

Product Name	Melting Point (°F)	Needle Penetration @ 77°F
1251/7	150-160	10-15
214	160-170	10-15
1275WH	165-175	60-80
1275W	170-180	25-35
1203/1	170-180	25-35
863	185-195	3-5
190	185-195	8-12
1160/14	190-195	2-7
1135/15W	195-200	2-5

Ross Microcrystalline Wax—Yellow

669	140-150	25-30
1385	140-150	25-30
1149/10	165-175	18-22
1140/10	165-175	18-22
1275ML	170-180	25-35
170 Amber	170-180	15-20
2305	170-180	25-30
1460	172-178	20-30
916	172-178	20-30
1149/14	190-195	2-7

Ross Microcrystalline Wax—Black

170 Black	165-175	18-22

Ross Microcrystalline Wax—Brown

175 Brown	170-180	13-16

Ross Microcrystalline Wax—Powdered (White)

1135/15W	195-200	2-5

Chemical Analysis and Physical Properties (Fully Refined Paraffin):

Color	white
Melting point, °F	125-165
Oil and moisture content, % max	0.5
Consistency	hard, dry solid
Structure	coarse, fibrous crystals
Purity	odorless and tasteless
Acid number	nil
Saponification number	nil

Chemical Analysis and Physical Properties (Oxidized Microcrystalline Wax):

Color	orange brown, brown yellow
Melting point, °F	180-200
Oil and moisture content	nil
Consistency	hard solid
Structure	microcrystalline
Purity	characteristic odor
Acid number	15-50
Saponification number	25-80

Chemical Analysis and Physical Properties (Scale Wax):

Color	white
Melting point, °F	125-130
Oil and moisture content, %	1-6
Consistency	soft solid, oily

Ross Wax 160

Chemical Type: Synthetic, high melting point wax. A tan, medium hard wax which remains stable under prolonged molten conditions, except the color may darken. Lump, flakes, powdered and atomized form.

Chemical Analysis and Physical Properties:

Melting point, °F (°C)	314-318 (157-159)
Flash point, min, °F (°C)	590 (310)
Acid number, max	10
Iodine value	7.5
Specific gravity	1.0232
Needle penetration (ASTM D5-61 @ 100 g)	3.5
Dielectric constant (1,000 cycles)	3.0
Power factor	0.00977

Key Properties: Recommended for use in adhesives, paints, varnishes, hot melt coatings, paper components, paper coatings, and many other uses.

Regulatory: FDA approved under Regulation 121.2520.

Spermaceti Wax

Grades: Spermaceti U.S.P.
Spermaceti Substitute No. 573

Chemical Type: Spermaceti Wax consists mainly of cetyl palmitate spermaceti wax which has been refined and forms lustrous white masses of long, glistening white crystals. The crystals are quite brittle so that the wax can easily be powdered. The wax has a very faint odor and mild taste.

Chemical Analysis and Physical Properties:

Melting point, °F (°C)	107.6-122 (42-50)
Flash point, °F	470-480
Specific gravity @ 25°C	0.940-0.946
Acid number	0.0-0.5
Ester number	116-125
Saponification number	116-125
Unsaponifiables, %	45-50
Iodine number, max	3.0

Key Properties: Recommended for use in adhesives, coatings, and many other compounds.

Regulatory: FDA approved under Regulations 121.2507, 121.2514, 121.2520, 121.2569.

Section XLV
Multifunctional and Miscellaneous Compounds

ARGUS CHEMICAL, 633 Court St., Brooklyn, NY 11231

Mark C Phosphite Chelator

Chemical Type: Phosphite chelator for vinyl compounds.

Physical Properties

Appearance	Clear colorless liquid
Specific Gravity @ 25°C	1.043
Refractive Index N_D^{25} @ 25°C	1.520

Key Properties: Most efficient phosphite chelator for vinyl compounds developed to date. An appreciable improvement in heat stability will result when Mark C is used in conjunction with just about all barium-cadmium systems, particularly those based on metallic soaps. In addition to its beneficial effect on heat stability, Mark C provides excellent resistance to the effects of outdoor weathering. Mark C, when used in conjunction with Mark M, Mark LL or Mark KCB, will impart outstanding crispness of initial color along with superior clarity. In compounds containing high amounts of phosphate plasticizer, Mark C is especially effective for controlling color development.

Mark 1413 UV Absorber

Chemical Type: 2-Hydroxy-4-n-octoxybenzophenone.

Physical Properties

Appearance	Light straw colored powder
Melting Range	46°–48°C
Specific Gravity	1.16

Key Properties: Recommended for use in vinyls, polyolefins and other polymers which require optimum light stability. Extremely effective in protecting PVC, polyethylene and polypropylene against the degradative effects of UV radiation as experienced in outdoor-weathering or fluorescent light exposure.

BARECO DIVISION, 6910 East 14th St., Tulsa, OK 74112

Bareco Emulsions and Dispersions

Product Name	Viscosity (Brookfield) @ 75°F (#1/30 rpm) (cp)	Viscosity (Brookfield) @75°F (#1/60 rpm) (cp)	Percent Solids Content	pH
Polymekon SPP-W	61	44	40	10.4

Chemical Type: Dispersion of Polymekon in water in white/opaque form.

Key Properties: Stable over a wide pH and temperature range. Used as a slip and antimar additive in the formulation of water-based inks and coatings.

Product Name	Viscosity (Brookfield) @ 75°F (#1/30 rpm) (cp)	Viscosity (Brookfield) @ 75°F (#1/60 rpm) (cp)	Percent Solids Content	pH
Westpet EM-69	63	48	40	9.4

Chemical Type: Anionic emulsion of Polymekon in white/opaque form.

Key Properties: Useful as a slip and antimar additive in the formulation of water-based inks and coatings.

Cardipol LP O-25/E	13	13	24	9.4

Chemical Type: Anionic emulsion of Cardipol LP O-25 in white/opaque form.

Key Properties: Used in the formulation of a wide variety of products, such as lubricants, inks and coatings where lubricating characteristics such as slip and antimar properties are desirable. Forms a tough film with good gloss. Recommended for use in polishes and protective coatings.

Bareco Polymers

Product Name	Melting Point (ASTM D127) (°F)	Needle Penetration @ 77°F	Color ASTM D1500)	Acid No.	Sapon. No.
Cardipol LP	240	2.0	3.0	—	—
Cardipol LP O-25	230	4.0	2.0	31.0	65

Vybar Polymers

Product Name	Melting Point (ASTM D-36 Mod) (°F)	Needle Penetration @ 77°F	Viscosity (ASTM D 2669) @ 300°F	Color (ASTM D1500)
103	162	5.0	107	0.5
260	126	13.0	104	0.5
825	<-30	—	18	0.0

BASF CORP., 491 Columbia Ave., Holland, MI 49423

Product Name	Acid Value	Solidification Point (°C)	Clarity of 10% Solution (25°C)
Emulan OC	~0	37.5	Clear

Chemical Type: Highly ethoxylated, nonionic 100% fatty alcohol, with white to yellowish color.

Key Properties: It has the nature of soft wax at room temperature. Its properties are similar to Emulan EL and it is also used as an emulsifying agent for oil-in-water emulsions. It may also be used as a stabilizer against salts that cause hardness in water to facilitate the cleanup of tools used to apply highly pigmented emulsion paints, particularly if it is used with Collacral VL.

Emulan EL	0.5	20	Clear

Chemical Type: Almost 100% of highly ethoxylated nonionic castor oil. Pale yellow with a characteristic odor.

Key Properties: Very readily compatible with other nonionic, anionic and cationic emulsifying agents and auxiliaries. Emulan EL is suitable for the production of stable oil-in-water emulsions since it is nonionic. The product also exerts dispersing and wetting actions. The polarity of the hydrophopic and hydrophilic parts of the molecule are responsible for its characteristic properties. Emulan EL can be used for all types of oil-in-water emulsions, such as to prepare emulsions of plasticizers for incorporating EMU Powder 120FD formulations. It is also used as a stabilizer for aqueous dispersions of solids.

CAREY CANADA INC., East Broughton Station, P.Q., Canada GON 1HO

Carey Asbestos Fibre Grades 7M

	7MS-1	7M-3	7M-5	7M-90
Quebec Standard Test (CATM B-1: in ounces)	. . . . 0. . . . . . . .0 . . . . 1 . . . . . . .15.			
Ro-Tap Dry Screen (CATM B-2: 100 g - 30 min - 2.4 mm; in %)				
Plus 6 Mesh	—	—	—	—
Plus 14 Mesh	2.0	3.0	2.0	3.0
Plus 20 Mesh	6.0	10.0	10.0	8.0
Plus 28 Mesh	26.0	21.0	21.0	13.0
Plus 35 Mesh	30.0	23.0	22.0	13.0
Plus 65 Mesh	20.0	18.0	20.0	14.0
Minus 65 Mesh	16.0	25.0	25.0	49.0
Bauer-McNett Wet Classification (CATM C-1: 2 g-20 min; in %)				
Plus 14 Mesh	0.5	2.0	2.0	5.0
Plus 35 Mesh	1.5	7.0	7.0	10.0
Plus 100 Mesh	10.0	18.0	16.0	18.0
Plus 200 Mesh	11.0	15.0	16.0	18.0
Minus 200 Mesh	77.0	58.0	59.0	49.0
Rapid Surface Area (CATM D-3: in dm^2/g)	185.0	97.5	75.0	65.0
Wet Volume (CATM D-1: 20 g-2,000 cm^3- 4 hr; in ml)	525	335	310	210
Grit & Spicules (CATM E-2: in %)	0.3	9.0	9.0	15.0
Color (CATM G-1)				
Amber	68.5	67.5	67.5	64.5
Blue	67.5	66.5	66.5	63.5
Green	69.0	68.0	68.0	65.0

Carey Asbestos Fibre Grades 7R

	7RS-1	7R-3	7R-33
Quebec Standard Test (CATM B-1: in ounces)	. . . . . .0 . . . . . .0 . . . . . . 0 . . . . . 16		
Ro-Tap Dry Screen (CATM B-2: 100 g-30 min-2.4 mm; in %)			
Plus 6 Mesh	—	—	—
Plus 14 Mesh	0.8	0.3	0.5
Plus 20 Mesh	4.0	4.0	5.0
Plus 28 Mesh	25.0	16.7	18.0
Plus 35 Mesh	30.0	26.0	26.0
Plus 65 Mesh	21.0	22.0	22.5
Minus 65 Mesh	19.2	31.0	28.0

	7RS-1	7R-3	7R-33
Bauer-McNett Wet Classification (CATM C-1: 20 g-20 min; in %)			
Plus 14 Mesh	0.3	1.0	1.0
Plus 35 Mesh	1.0	4.0	4.0
Plus 100 Mesh	9.7	18.0	16.0
Plus 200 Mesh	11.0	16.0	16.0
Minus 200 Mesh	78.0	61.0	63.0
Rapid Surface Area (CATM D-3: in dm^2/g)	185.0	87.0	87.0
Wet Volume (CATM D-1: 20 g-2,000 cm^3-4 hr; in ml)	500	270	290
Grit & Spicules (CATM E-2: in %)	0.3	6.0	6.0
Color (CATM G-1)			
Amber	68.5	67.0	67.5
Blue	67.5	66.0	66.5
Green	69.0	67.5	68.0

Carey Asbestos Fibre Grades 7RF

	7RF-66	7RF-10	7RF-9
Quebec Standard Test (CATM B-1: in ounces)	. . . . .No test		
Ro-Tap Screen (CATM B-2: 100 g-30 min-2.4 mm; in %)			
Plus 6 Mesh	—	—	—
Plus 14 Mesh	Trace	Trace	Trace
Plus 20 Mesh	0.1	0.3	0.2
Plus 28 Mesh	6.0	3.0	1.5
Plus 35 Mesh	49.0	22.0	6.5
Plus 65 Mesh	35.0	58.0	65.0
Minus 65 Mesh	9.9	16.7	26.8
Bauer-McNett Wet Classification (CATM C-1: 20 g-30 min; in %)			
Plus 80 Mesh	1.0	0.5	0.5
Plus 325 Mesh	12.5	12.0	11.5
Minus 325 Mesh	86.5	87.5	88.0
Rapid Surface Area (CATM D-3: in dm^2/g)	230.0	175.0	170.0
Wet Volume (CATM D-1: 20 g-2,000 cm^3-4 hr; in ml)	550	300	205
Grit & Spicules (CATM E-2: in %)	Trace	Trace	Trace
Color (CATM G-1)			
Amber	67.0	66.5	65.5
Blue	66.0	65.5	64.5
Green	67.5	67.0	66.0

Carey Asbestos Fibre Grades 7T

	7TS-1	7TS-3	7TS-7	7T-5
Quebec Standard Test (CATM B-1: in ounces)	. . . . . . 0 . . . . . . 0. . . . . . .0 . . . . . . 16			

	7TS-1	7TS-3	7TS-7	7T-5
Ro-Tap Dry Screen (CATM B-2: 100 g-30 min-2.4 mm; in %)				
Plus 6 Mesh	—	—	—	—
Plus 14 Mesh	0.1	0.3	1.0	0.1
Plus 20 Mesh	1.5	3.0	3.0	0.9
Plus 28 Mesh	17.4	13.5	15.0	5.0
Plus 35 Mesh	36.0	25.5	24.0	18.5
Plus 65 Mesh	29.0	23.5	24.0	26.5
Minus 65 Mesh	19.0	34.2	33.0	49.0
Bauer-McNett Wet Classification (CATM C-1: 20 g-20 min; in %)				
Plus 14 Mesh	0.1	1.0	0.7	1.0
Plus 35 Mesh	0.8	3.5	3.3	1.5
Plus 100 Mesh	8.1	16.0	16.0	13.5
Plus 200 Mesh	11.1	16.0	16.0	18.0
Minus 200 Mesh	80.0	63.5	64.0	66.0
Rapid Surface Area (CATM D-3: in dm^2/cg)	165.0	87.0	86.0	76.0
Wet Volume (CATM D-1: 20 g-2,000 cm^3-4 hr; in ml)	400	275	270	200
Grit & Spicules (CATM E-2: in %)	0.2	5.0	6.2	3.4
Color (CATM G-1)				
Amber	67.0	67.0	63.5	65.0
Blue	66.0	66.5	62.5	64.0
Green	67.5	67.5	64.0	65.5

Key Properties (all grades of asbestos fibers): Recommended for use in asphalt fibrated mastics and coatings and in block filler paints.

DYNAMIT NOBEL CHEMICALS, Kay-Fries, Inc., 200 Summit Ave., Montvale, NJ 07645

Dynasylan Bonding Agents

Product Name	Percent Purity	Boiling Point (°C)	Flash Point (°C)	Molecular Weight	Specific Gravity	Refractive Index
VTC	99.2	92.5	21.1	161.5	1.268	—
Chemical Type: Vinyltrichlorosilane.						
VTMO	97	123	23.5	148	0.975	—
Chemical Type: Vinyltrimethoxysilane.						
VTEO	97 min	158	42	190	0.915	—
Chemical Type: Vinyltriethoxysilane.						
VTMOEO	97 min	108	66	280	1.040	—
Chemical Type: Vinyltris(β-methoxyethoxy)silane.						
MEMO	98 min	85	125	248	1.047	1.429
Chemical Type: γ-Methacryloxypropyltrimethoxysilane.						
GLYMO	97 min	90	160	236.3	1.075	1.428
Chemical Type: γ-Glycidyloxypropyltrimethoxysilane.						
AMEO	95 min	68.5	93	221	0.95	—

Chemical Type: γ-Aminopropyltriethoxysilane.

Color and Description (all types): Clear to slightly yellowish highly mobile fluids.

Hydrolyzable Chloride Content: 0.01% max (0.05% max for AMEO).

Note: The above silanes may also, if required, be supplied with other alkoxy groups. Additional silanes, such as alkyltrialkoxysilanes, are manufactured on request.

Chemical Type (all types): Monomeric organo-functional silicon compounds. In addition to three hydrolyzable groups (alkoxy or halogen), their molecules contain a functional group that is bonded to the silicon atom by a firm Si-C bridge, which may have a number of links. Generally it may be represented by

$$R'-\underset{\underset{OR}{|}}{\overset{\overset{OR}{|}}{Si}}-OR \quad \text{or} \quad R'-SiX_3$$

Where R' is a functional group, possibly on a C bridge, R is an alkyl group and X is halogen atom.

The general formula shows the two different reaction possibilities: (1) the alkoxy or halogen groups that form the reactive silanol for the inorganic base after hydrolysis or (2) a functional group that will react with suitable resins.

Key Properties: (1) Sealing Compounds: A small addition of a reactive Dynasylan bonding agent can be added to polyurethane, polysulfide, polyester and butyl and acrylic resin-based sealing compounds. This will give them improved adhesion to building materials, earthenware, aluminum and glass, particularly where there is stress as a result of the presence of moisture. Dynasylan types that may be used here are AMEO, GLYMO and VTMOEO.

(2) Adhesives: The use of Dynasylan bonding agents in nitrilophenol resin, epoxy resin and nitrilo-rubber metal adhesives gives increases of up to 100% in shearing strengths in the case of aluminum adhesion. These improvements are particularly noticeable at high temperatures. Dynasylan types that may be used here are AMEO and GLYMO.

(3) Mineral-Filled Polyester and Epoxy Resins: In filled polyester and epoxy resins, the addition of small quantities of a Dynaslan produces improved mechanical and electrical properties (dielectric constants, loss factors, etc.) that are maintained under stress from moisture. Dynasylan types that may be used here are MEMO, VTMOEO, GLYMO and AMEO.

(4) Pretreatment of Fillers and Pigments: It is recommended that the filler (if the content is very low) be pretreated, in order to achieve an improvement in the bonding properties, at low cost. The Dynasylan produces lower moisture absorption in the filler and better adhesion to the base. Dynasylan types that may be used here are GLYMO, MEMO, VTMOEO and alkyltrialkoxysilanes.

(5) Bonding Agents for Coatings: Adhesion problems often occur where ceramic, silicate and metal surfaces have to be coated with polymers, particularly after stress through moisture. Dynasylan bonding agents, which may either be added directly to the polymer or applied beforehand in the form of a special primer, will substantially improve adhesion.

ESSENTIAL CHEMICALS CORP., Merton, WI 53056

Ecconol B: Coupled fatty acid alkanolamide, nonionic, 90% active liquid detergent base for use in formulating all purpose and heavy duty cleaners.

Ecconol 61: Foam stabilized anionic, nonionic complex of amide sulfonate, 67% active, liquid wetting agent base for high foaming detergents, crystal clear dishwash and car wash formulations and synthetic hand cleaners.

Ecconol 66: Coco fatty acid alkanolamide, nonionic, viscous, 100% active liquid wetting agent and emulsifier for medium to heavy duty cleaner formulations.

Ecconol 489: Special amine detergent emulsifier, 100% active booster with pH of 11 to 12, used in heavy duty cleaner and wax stripper formulas.

Ecconol 606: Organic, salt-free, liquid, linear alkyl sulfonate, 60% active base detergent for shampoo, dishwash, car wash, and household cleaner formulations.

Ecconol 628: Coconut oil alkanolamide, nonionic liquid, 100% active detergent base used to formulate viscous cleaners. Excellent foam stabilizer.

Ecconol 2066: Coupled nonionic complex liquid detergent base, 75% active for formulating general purpose degreasers, wax and finish strippers and heavy duty cleaners.

Wettex E1114: Polyoxyethylated nonylphenol (nonylphenoxypolyethyleneoxyethanol) organic surfactant, 100% active, clear viscous liquid, all purpose wetting agent for formulating household and industrial cleaners, etc.

The supplier states that the above raw materials are basically used in the maintenance chemical field. They are unable to advise specifically which ones are suitable for the paints and adhesives fields.

ETHYL CORP., Ethyl Tower-451 Florida, Baton Rouge, LA 70801

Product Name	Melting Point (°F)	Viscosity (cs) @ 20°C	Boiling Point (°F)	Vapor Pressure @ 20°C	Flash Point (TCC) (°F)	Weight per Gallon (lb)
Ethanox 701	97	7.3	487	<0.01 mm	>210	7.61

Chemical Type: 2,6-Di-tert-butylphenol with a molecular weight of 2,063, in crystalline light straw colored solid form.

Key Properties: Oxidation inhibitor and stabilizer for fuels, oils, plastics, rubber and other products.

Toxicity: Animal tests indicate that Ethanox 701 is slightly toxic by oral administration to rats (LD_{50} is 9,200 mg/kg) and dermal application to rabbits (LD_{50} >10,000 mg/kg). It is a mild eye irritant and a mild, but not primary, skin irritant. At temperatures up to 100°C, insufficient Ethanox 701 antioxidant is vaporized to produce any signs of illness in rats exposed for six hours.

Ethanox 702	309	–	482	<0.001 mm	>400	33 lb/ft³

Chemical Type: White to light straw crystalline powder, 4,4'-methylenebis(2,6-di-tert-butylphenol), a sterically hindered, temperature stable, phenolic antioxidant.

Key Properties: Oxidation inhibitor in natural and synthetic rubbers, polyolefin plastics, resins, adhesives, petroleum oils and waxes. Use is sanctioned by the FDA, as listed in Title 21, Code of Federal Regulations, in the following applications: Adhesives (175.005), Petroleum Hydrocarbon Resins (174, 175, 176, 177, 178 and 179.45), Polyethylene (177.1520) and Polybutadiene (177.2600).

Toxicity: Animal tests indicate Ethanox 702 is practically nontoxic by oral administration to rats (LD_{50} >24 g/mg). It is not an eye irritant or a skin irritant. Exercise good personal hygiene when handling this material.

Ethanox 703	201	354	>200	–	–	–

Chemical Type: Light yellow crystalline powder. A sterically hindered, temperature stable, phenolic antioxidant, 2,6-di-tert-butyl-α-dimethylamino-p-cresol. Molecular weight of 263.4.

Key Properties: Oxidation inhibitor in natural and synthetic rubbers, polyolefin plastics, resins, adhesives, petroleum oils and waxes.

Toxicity: The acute oral LD_{50} for Ethanox 703 is 1,030 mg/kg of body weight.

Ethanox 736	255	594	>210 (TCC)	<0.001 mm	–	39

Chemical Type: White to light straw crystalline powder. A sterically hindered, temperature stable, phenolic antioxidant, 4,4'-thiobis(6-tert-butyl-o-cresol). Molecular weight of 358.5.

Key Properties: Oxidation inhibitor in natural and synthetic rubbers, polyolefin plastics, resins, adhesives, petroleum oils and waxes.

Toxicity: The acute oral LD_{50} for Ethanox 736 Antioxidant is 6,340 mg/kg of body weight when administered in a 20% solution in peanut oil. The usual precautions such as avoidance of unnecessary personal contact and exercise of good industrial hygienic practices are recommended.

Product Name	Melting Point (°F)	Vapor Pressure @ 300°C	Bulk Density (lb/ft³)
Ethanox 330	471	<0.5	36.3

Chemical Type: Crystalline white powder, [1,3,5-trimethyl-2,4,6-tris(3,5-di-tert-butyl-4-hydroxybenzyl) benzene], a sterically hindered, temperature stable, phenolic antioxidant. Molecular weight of 775.2.

Key Properties: A highly effective, noncoloring, odorless stabilizer for plastics, resins, rubber and waxes. It has exceptional low volatility and is outstanding in applications requiring high processing temperatures.

Toxicity: Ethanox 330 is relatively harmless orally, practically nontoxic by skin absorption and no more than slightly irritating to intact or abraded skin. Acute oral LD_{50} is >15 g/kg of body weight. The use of up to 0.5% by weight of Ethanox 330 is allowed by FDA in polymers for food contact applications (Subpart F:178.2010).

Development Products:
(1) Antioxidant 724: 2,6-di-tert-butyl-4-ethylphenol.
(2) Antioxidant 744: 2,6-tert-butyl-4-n-butylphenol.
(3) Antioxidant 754: 4-hydroxymethyl-2,6-di-tert-butylphenol.

Note: Ethyl also has investigated hundreds of other potential antioxidants, which can be produced if interest warrants.

GAF CORP., Chemical Division, 140 West 51 St., New York, NY 10020

Uvinul Ultraviolet Absorbers

Uvinul Absorber 400

Chemical Type: 2,4-Dihydroxybenzophenone (100% active) in powder form.

Key Properties: Most effective for ultraviolet radiation in the 290 to 330 nm spectral region. Soluble in alcohols, ether-alcohols, cyclic ethers, ketones and esters; insoluble in water, very compatible with polystyrene, polyester, acrylic, and epoxy resins and rubbers. Used in outdoor paints and coatings, varnishes, colored liquid toiletries and cleaning agents, filters for photographic color films and prints and rubber-based adhesives.

Uvinul Absorber M-40

Chemical Type: 2-Hydroxy-4-methoxybenzophenone (100% active) in powder form.

Key Properties: Similar to 400 except for higher solubility in aromatic solvents. Has more heat stability than 400 or 490. Especially compatible with polystyrene and rubber; provides good weather resistance in resins and plastics. Stabilizes polyvinyl chloride and polyesters against ultraviolet-light degradation. Suggested for use also in nitrocellulose lacquers, varnishes and oil-based paints.

Uvinul Absorber MS-40

Chemical Type: 2-Hydroxy-4-methoxybenzophenone-5-sulfonic acid (100% active) in powder form.

Key Properties: Sulfonated, water-soluble derivative of M-40. Employed in sunscreen products and in hair sprays and shampoos for dyed and tinted hair. Also useful for leather and textile fibers.

Uvinul Absorber D-50

Chemical Type: 2,2',4,4'-Tetrahydroxybenzophenone (100% active) in powder form.

Key Properties: Commercial UV absorber with the broadest UV absorption spectrum. Most effective in the 290 to 350 nm spectral range. Stronger absorption properties than D-49 or 490 in the 360-400 nm range. Retards fading of pigments and dyestuffs; prolongs the life of polymeric materials; photostabilizes cosmetic formulations and minimizes discoloration of synthetic rubber or plastic latices.

Uvinul Absorber 490

Chemical Type: Mixture of D-49 and other tetra-substituted benzophenones (100% active) in powder form.

Key Properties: Most effective for UV radiations in the 290 to 355 nm spectral range. Used in nitrocellulose lacquer, fluorescent paint, inks, and for protecting furniture woods, colored liquid toiletries and cleaning agents, isocyanate systems and butyrate metal lacquers.

Uvinul Absorber DS-49

Chemical Type: Sodium 2,2'-dihydroxy-4,4'-dimethoxy-5-sulfobenzophenone (67% active) in powder form.

Key Properties: Sulfonated, water-soluble derivative of D-49. Employed in cosmetic formulations to prevent fading of colors and viscosity changes caused by ultraviolet light. Also used in textiles and water-based paints.

Uvinul Absorber N-35

Chemical Type: Ethyl 2-cyano-3,3-diphenylacrylate (100% active) in powder form.

Key Properties: Noncolor contributing UV absorber; does not contain aromatic hydroxyl groups; effective under varying pH conditions. Developed especially for nitrocellulose lacquers and polyvinyl chloride. Suitable for use in alkaline systerns such as urea-formaldehyde and epoxy amine formulations and in cosmetics.

Uvinul Absorber N-539

Chemical Type: 2-Ethylhexyl-2-cyano-3,3-diphenylacrylate (100% active), in powder form.

Key Properties: Light-colored, completely miscible with mineral spirits and nonreactive with metallic driers. Soluble in nonpolar plastics. Especially effective in flexible and rigid polyvinyl chloride. Suggested for use in nitrocellulose lacquers, varnishes, vinyl flooring and oil-based paints. Of interest in aerosol and oil-based suntan lotions.

HENKEL, INC., 425 Broad Hollow Road, Melville, Long Island, NY 11746

Deplastol

Chemical Type: Highly effective proprietary composition viscosity depressant and viscosity stabilizer for plastisols and organosols.

Chemical Analysis and Physical Properties

Appearance	Slightly yellowish, clear liquid
Acid Value	0.1–1.0
Saponification Value	140–150 approx.
Iodine Color Value	5 max
Specific Gravity @ 20°C (68°F)	0.985–0.988
Refractive Index @ 20°C (68°F)	1.454–1.456
Brookfield Viscosity @ 20°C (68°F)	47 cp
Solidification Point	–3°C (27°F) approx.
Flash Point	190°C (374°F) min
Fire Point	220°C (428°F) min

Key Properties: The addition of small amounts of Deplastol lowers the viscosity of PVC pastes and reduces the tendency for viscosity increases due to aging. In addition, Deplastol improves the flow, facilitates the release of entrapped air and aids the dispersion of pigments and fillers in these systems. Deplastol can be used with nearly all types of PVC resins. Has no adverse effects on the physical properties of the finished product when used at the recommended levels.

Edenol 190

Chemical Type: Highly effective proprietary composition viscosity depressant and viscosity sta- bilizer for plastisols and organosol.

Chemical Analysis and Physical Properties

Appearance	Practically colorless liquid
Acid Value	0.2 max
Saponification Value	150 approx.
Iodine Color Value	2 max
Pour Point	–10°C (14°F)
Flash Point	210°C (410°F) min
Fire Point	245°C (473°F) min
Specific Gravity @ 20°C (68°F)	0.863
Refractive Index @ 20°C (68°F)	1.452
Brookfield Viscosity @ 20°C (68°F)	13.5 cp
Volatility in the Brabender @ 90°C (194°F)	
After 24 hours	0.40% max
After 48 hours	0.60% max

Key Properties: The addition of small amounts lowers the viscosity of PVC pastes. Pastes using Edenol 190 exhibit very good viscosity stability and resistance to viscosity increases due to aging. Can be used with nearly all types of PVC resins. Edenol 190 enhances low temperature resistance, gloss, water resistance and resistance to aging and light.

HERCULES, INC., 910 Market St., Wilmington, DE 19899

Ethyl Hydroxyethyl Cellulose

Chemical Type: Ethyl hydroxyethyl cellulose (EHEC) is a mixed cellulose ether prepared by a process involving three reactions.

(1) Alkali cellulose in which cellulose is swollen by aqueous NaOH.
(2) Hydroxyethylation, an alkali-catalyzed reaction.
(3) Ethylation of hydroxyethyl cellulose.

Odorless, tasteless, white, granular powder and flake forms.

Viscosity Types: The mechanical properties of EHEC film, such as tensile strength, elongation, and flexibility, are directly proportional to its degree of polymerization (molecular weight). This is measured by the solution viscosity of the polymer. EHEC is manu- factured by Hercules in flake form and is available in three viscosity types:

EHEC-High	125–250 cp
EHEC-Low	20–35 cp
EHEC-Extra Low	10–19 cp

The term "viscosity of EHEC" means the viscosity in centipoises at 25°C of a 5% solution by weight of EHEC in a solvent mixture of 80 parts by weight toluene and 20 parts by weight of ethanol. The ethanol used is 95% ethyl alcohol, 2B grade.

Physical Properties

Bulk Density (lb/ft³)	19–22	Film Properties	
Solution Properties		Specific Gravity	1.12
Specific Volume (in³/lb)	24.5	Tensile Strength (lb/in²)	3,000–6,000
Bulking Value (gal/lb)	0.105	Elongation (%)	5–10
Color (ASTM D1209)	0.9	Flexibility (MIT Double Folds)	500–900
Refractive Index	1.47	Flow Temperature (ASTM D569)	175°C
		Volume Resistivity (ohm-cm)	6.2×10^{13}

Key Properties: Has outstanding tolerance for aliphatic hydrocarbons, broad compatibility, high flexibility and good toughness. These properties make it useful in various protec- tive coatings and decorative packaging applications. EHEC is the only commercially available cellulosic polymer substantially soluble in low cost, low odor aliphatic hydrocarbon solvents. As a result, it helps reduce formulation cost and improves application conditions. Recommended for the following uses:

(1) Printing Inks
    (A) Screen process inks.
    (B) Rotogravure inks.
(2) Clear Lacquers
    (A) For metallized plastics.
    (B) For wood floors.
    (C) For porous paper.
(3) Other Coatings
    (A) Additive to control pigment floating.
    (B) Additive to control flatting of varnishes.

FDA Status: EHEC is relatively nontoxic and has been cleared under the Federal Food, Drug and Cosmetic Act for the following uses that involve food contact.

Regulation	*Description*
21 CFR 121.2514	Resinous and polymeric coatings for metal substrates. Also, suitable substrates for repeated use.
21 CFR 121.2520	Adhesives.
21 CFR 121.2526	Components of paper and paperboard in contact with aqueous and fatty foods.
21 CFR 121.2548	Zinc-silicon dioxide matrix coatings.
21 CFR 121.2550	Closures with sealing gaskets for food containers.
21 CFR 121.2571	Components of paper and paperboard in contact with dry foods.

Nitrocellulose

Chemical Type: Hercules produces three standard types of nitrocellulose, designated RS, AS and SS.

    RS Type: Has an average nitrogen content of 12% (11.8-12.2%) and is available in a large number of viscosity grades, from 18 cp to 2,000 seconds. One special grade with a viscosity equal to about 200,000 seconds is also available. Viscosity ranges for each viscosity grade is shown in the table.
    AS Type: Has an average nitrogen content of 11.5% (11.3-11.7%). It is currently available in only two grades.
    SS Type: Has an average nitrogen content of 11% (10.9-11.2%). Available in five standard viscosity grades.

Soluble Nitrocellulose Types

Designation	... *Viscosity (Standard Method)* .... *In Seconds for a Solution of*		
	12.2%	*20%*	*25%*
RS (11.8-12.2 % N)			
RS 18-25 cp	18-25*	—	—
RS 30-35 cp	30-35*	—	—
RS ¼-second	—	—	4-5
RS ⅜-second	—	—	6-8
RS ½-second	—	3-4	—
RS ¾-second	—	6-8	—
RS 5 to 6-second	5-6.5	—	—
RS 15 to 20-second	15-20	—	—
RS 30 to 40-second	30-40	—	—
RS 60 to 80-second	60-80	—	—
RS 125 to 175-second	125-175	—	—
RS 600 to 1,000-second	600-1,000	—	—
RS 1,000 to 1,500-second	1,000-1,500	—	—
RS 1,500 to 2,000-second	1,500-2,000	—	—
Extra-high-viscosity	500+ sec	—	—
(Unbleached) (12.2-12.4 % N)	(In 4% Sol.)		
AS (11.3-11.7 % N)			
AS ½-second	—	3-4	—
AS 5 to 6-second	5-6.5	—	—
SS (10.9-11.2 % N)			
SS 30-35 cp	30-35*	—	—
SS ¼-second	—	—	4-6

Designation	... Viscosity (Standard Method)....		
	....In Seconds for a Solution of....		
	12.2%	20%	25%
SS ½-second		3-4	—
SS 5 to 6-second	5-6.5	—	—
SS 40 to 60-second	40-60	—	—

*In cp for 12.2% solution

Physical Properties of RS Nitrocellulose

Odor of Material	None	Specific Volume in Solution (in³/lb)	16.26
Taste of Material	None	Specific Gravity of Cast Film	1.58–1.65
Color of Film	Water-White	Refractive Index (Principal)	1.51
Clarity of Film	Excellent	Light Transmission (Å)	3,130
Bulking Value in Solution (gal/lb)	0.0704		

Chemical and Physical Properties of Unplasticized Clear Film of RS Nitrocellulose

Moisture Absorption @ 21°C in 24 hr @ 80% RH	1.0%
Water Vapor Permeability @ 21°C, g/cm²/cm/hr (x 10⁻⁶)	2.8
Sunlight (Effect on Discoloration)	Moderate
Sunlight (Effect on Embrittlement)	Moderate
Effect on Aging	Slight
Effect of Cold Water	Nil
Effect of Hot Water	Nil
General Resistance to:	
Weak Acids	Fair
Strong Acids	Poor
Weak Alkalies	Poor
Strong Alkalies	Poor
Alcohols	Partly soluble
Ketones	Soluble
Esters	Soluble
Hydrocarbons:	
Aromatic	Good
Aliphatic	Excellent
Oils:	
Mineral	Excellent
Animal	Good
Vegetable	Fair to Good

Key Properties: RS: Completely soluble in esters, ketones, ethyl-alcohol mixtures and glycol ethers. It has excellent tolerance for aromatic hydrocarbon diluents, such as toluene and less tolerance for aliphatic petroleum diluents. It is compatible with many resins. Recommended for lacquers for wood, metal, paper, textiles and foil; lacquer emulsions for wood, metal and architectural finishes. Also, adhesives, cements and inks. RS nitrocellulose meets Federal Specification TT-N-350B.

AS Type: Soluble in the same type of solvents as the RS type. Can also be dissolved in solvent systems containing a high proportion of low molecular weight anhydrous alcohols. Its widest use lies in lacquers for coating cellophane, paper and textiles.

SS Type: Recommended for use in lacquers for paper and foil, low-odor lacquers, sealers, fillers, printing inks and plastics. Considerably more soluble in alcohols than is the RS type.

FDA Status: All grades of Hercules RS, AS and SS nitrocellulose have the following FDA clearances for use in food-packaging:

Regulation	Description
21 CFR 175.105	Adhesives.
21 CFR 175.300	Resinous and polymeric coatings (coating on metal substrate or on any suitable substrate when intended for repeated use).
21 CFR 176.170	Components of paper and paperboard in contact with aqueous and fatty foods.
21 CFR 176.180	Components of paper and paperboard in contact with dry food.
21 CFR 177.1200	Cellophane.
21 CFR 177.1210	Closures with sealing gaskets for food containers.

Regulation	Description
21 CFR 177.1630	Polyethylene phthalate polymers.
21 CFR 181.30	Manufacture of paper and paperboard products used in food packaging and is not a food additive.

Parlon

Chemical Type: Parlon is produced in only one chemical type, but is supplied in five viscosity types. Designations for the currently available viscosity types are shown below.

Standard Viscosity Types

Type	Range, cp	Use
SS	4–7	In printing inks and in spray-applied product finishes and machinery enamels.
S10	9–14	In spray-applied product finishes, machinery enamels, chemical-resistant maintenance paints, masonry paints, fire-retardant paints and traffic paints.
S20	17–25	In spray- or brush-applied product finishes, machinery enamels, chemical-resistant maintenance paints, masonry paints, fire-retardant paints, marine finishes and traffic paints.
S125	125–200	In adhesives, paper and textile coatings and brush-applied paints for many of the above applications.
S300+	300+	In adhesives and in textile and paper coatings.

Physical Properties

General
  Form (As Shipped) — White, granular powder
  Color of Film — Water-white
  Clarity of Film — Good
  Odor and Taste — None
  Moisture (As Shipped) — 0.5% max
Physical
  Specific Gravity (Of Cast Film) — 1.63
  Specific Volume (in^3/lb) — 70
  Bulking Value (gal/lb) — 0.071
  Refractive Index — 1.554
Thermal
  Burning Rate — Nonflammable
  Effect of Dry Heat on Film
    (Continuous Exposure) — Stable up to and at 125°C
  Softening Point of Film — 140°C
Physical-Chemical (Clear, Unplasticized Film)
  Effect of Sunlight — Discolors and embrittles
  Effect of Cold Water — None
  Effect of Hot Water — Slight blush
  Moisture Absorption (80% RH in 24 hr) — 0.14%
  Water-Vapor Transfer of Free Film (g/100 in^2/mil
    in 24 hr at 95°F and 100% RH) — 1.0
Mechanical* (ASTM Methods Used for All Tests)
  Tensile Strength (lb/in^2) — 5,200
  Elongation — 1.6%
  Flexibility (MIT Double Folds) — 139
  Hardness (Sward Index, % of Glass) — 70–80

*Unmodified Parlon S125

Key Properties: All Parlon viscosity types yield, in a given formulation, films of equivalent hardness. Flexibility of the films will vary slightly, with the film from the higher viscosity having the greater flexibility. Viscosities are determined on solutions of 20% concentration in toluol at 25°C.

ICI AMERICAS, INC., Concord Pike & New Murphy Rd., Wilmington, DE 19897

H-7 Adhesive Additive

Chemical Type: 20% complex phenolic based solids solution in three normal ammonium hydroxide, with specific gravity of 1.05, in liquid form, completely soluble in water.

Key Properties: Widely used with resorcinol-formaldehyde-latex (RFL) systems to promote adhesion of polyester cord to rubber in the tire industry.

INTERSTAB CHEMICALS, INC., 500 Jersey Ave., New Brunswick, NJ 08903

Product Name	Color (Gardner) (max)	Specific Gravity (75°C)	Viscosity (Gardner-Holdt) @ 77°F (max)	Weight per Gallon (lb)	Percent Solids Content	Form
Calcium Pulp	Off-white	0.925	Smooth Paste	7.71	30	Smooth Creamy Pulp

Chemical Type: A specially processed calcium soap of linoleic acid (30% solids in 70% water with 1.1% metal content).

Key Properties: Developed for use in solvent-based systems. Performs equally well in:

(1) Putty: To prevent oil pigment separation.
(2) Paint: To aid suspension, ease of brushing and controlled penetration.

In addition, it also acts as a stabilizing agent, improves package stability and prevents sagging. Calcium Pulp can be dispersed uniformly in additional water while maintaining its creamy appearance in contrast to some pulps which repel additional water and become grainy. This characteristic permits the paint manufacturer to incorporate small doses of additional water in a formulation to improve thixotropic properties.

Polyester Accelerators

Product Name	Specific Gravity @ 75°F	Viscosity (Gardner-Holdt) @ 77°F (max)	Percent Metal Content	Weight per Gallon @ 75°F	Diluent
NL-6	0.936	A_5	6.0	7.80	Hydrocarbon
NL-12	1.040	K	12.0	8.66	Hydrocarbon
NL-49-P	0.980	A	1.0	8.16	Phthalate Plasticizer
NL-51-P	0.998	B	6.0	8.31	Phthalate Plasticizer

Chemical Type: All contain pure cobalt salt of 2-ethyl hexoic acid in various diluents as listed. They are all clear blue liquids. These are selectively produced as accelerators for peroxide initiated curing of unsaturated polyester resins.

Key Properties: These accelerators are manufactured under rigid standards of quality control in order to ensure a high level of performance consistent with commercial requirements for peroxide initiated cures of polyester resins. The quantity of accelerator that is required to promote the cure to completion depends upon several factors: principally, the gel time desired, the activity response of the polyester resin, the temperature at which curing takes place, pot life time that is necessary and the type or activity of the peroxide catalyst.

ISOCHEM, Cook St., Lincoln, RI 02865

Product Name	Percent Solids Content	Weight per Gallon (lb)	Specific Gravity
Airouts (All Three Grades)	45	7.0	0.86

Chemical Type (Airout): Silicone reacted with a tallate to produce a proprietary composition air release agent.

Key Properties: Aids epoxy resins, silicones, polyurethanes, vinyls, butyrates and most resins or rubbers in releasing air more uniformly and more rapidly. Completely stable and compatible with epoxy resins. No orange peel or cracking in coatings. No separation on standing.

Chemical Type (Airout S): Modified form of Airout that is miscible and compatible with xylol, alcohols and aromatic esters.

Key Properties: Excellent properties in organisols and plastisols.

Chemical Type (Super Airout): Silane modified Airout.

Key Properties: Has the beneficial properties of Airout, plus improved luster, surfaces, improved adhesion of resins to other surfaces, upgrading of water resistance of resins and is a fast, efficient air releaser. It speeds the wet out of fiber glass and fillers, when added to laminating resins, to give maximum uniformity and adhesion in polyester systems.

Key Properties (all grades of Airout): Eliminates air voids in resin formulations and eliminates craters in films and improves coatings films, castings, impregnations and potting applications. Airout will not sweat out or give greasy surfaces and improves finish. Will not downgrade resin formation or adhesion and will not interfere with electrical or physical properties, yielding a superior finish. The effect of Airout in the resin lasts indefinitely and, in many applications, vacuuming and deairing are eliminated. Suggested concentration is ¼ to 1½%, based on resin, but excess Airout will not interfere with cure, adhesion or finish. In epoxy work, Airout may be added to resin, hardener or mixed components. All Airouts are effective in water or solvent based systems and 100% resin formulations. The addition of 1 to 10% Airout in polyurethane or other conformal coatings will yield a transparent air-free film that will reach maximum cure in a shorter time than untreated resin. Airouts also allow a longer stability of viscosity and control the size of air cells. Has excellent properties as a foam stabilizer, as a urethane viscosity depressant and air release agent in urethane castings without loss of strength.

Note: Airout is registered in the U.S. Patent Office: No. 879334.

Product Name	Active Content (%)	pH	Percent Water (max)	Specific Gravity
Plast D	100	7.00	1	0.92

Chemical Type: Nonionic tall oil ester in tan liquid form.

Key Properties: Effective for the following uses:

(1) Exceptional leveling and fisheye prevention in aqueous or oil paints and paper coatings.
(2) Hydrophilic emulsifier for organics.
(3) Wetting agent for pigments.
(4) Efficient in emulsifying hydrophobics and stabilizing mineral oil, fats, wax, paint and solvent emulsions.
(5) Stabilizer for resin formulations.
(6) Inhibits viscosity build-up on shelf aging.
(7) Low foaming in emulsification.
(8) Base for defoamer formula.
(9) Grinding aid for pigments.
(10) Eliminates flooding and flowing of mixed pigments.

Has the following characteristics: Improves package stability, improves adhesion, reduces cracking on weathering, improves brushing and leveling, defoams and disperses and improves tint retention.

Finds usage in the following compounds: Paper deinking, stabilization, wetting agent for pigments and fillers, emulsifier and emulsion stabilizer in coatings based on oils, long chain resins, chlorinated rubber, metallic pigments, plastisols, organosols, alkyds, air and baking enamels, urethane paints and zinc powder paints.

Product Name	Viscosity (cp)	Moisture Content (%)	Hydroxyl Number	Acid Number	Dielectric Constant	Weight per Gallon (lb)
X-Air	5	0.25	20.0	0.03	10	7.50

Chemical Type: Alkyl ether reactive diluent and release agent for urethanes, in clear liquid form.

Key Properties: May be added to urethane or polyol portions of urethane systems to eliminate air voids and reduce viscosity. X-Air reacts completely with the system and does not downgrade tensile strength, chemical resistance or other properties. Pot life and cure properties remain the same as the original. X-Air is recommended for all types of urethanes (viz: polyether and polyester). May be used in polyester, epoxy or silicone formulations with good results. Recommended for potting, encapsulations, coatings, adhesives, forms and dips. X-Air is not a solvent and will not flash off in curing. It aids in dispersing of fillers and colors. X-Air readily mixes at room temperature and may be used in concentration from 1 to 10% by weight, on urethane formulas. A viscosity reduction of up to 200% is obtainable without detriment to system. It is stable in the urethane system and its presence slows crusting potential on most formulations. Enhances viscosity stability on long term storage. X-Air has no detrimental effect on films and will not cause haze. Will not lower heat distortion, scuff resistance, modulus, elasticity, water absorption or downgrade heat aging of systems.

Also Available As: X-Air Plus which is X-Air fortified with Isochem's Airout (Modified Silicone Air Releasant).

Note: X-Air is registered in the U.S. Patent Office: No. 879333.

ITT RAYONIER, INC., 1177 Summer St., Stamford, CT 06904

Raylig Lignins

Chemical Type and Analysis (Percent by Weight)

	Raylig D	Raylig DL*	Raylig 260-L*
Moisture	7.0	45.0	50.0
Sodium	4.7	4.7	4.7
Sulfur (As S)	9.0	9.0	4.9
Sulfite (As S)	0.1	0.1	0.1
Sulfate (As S)	1.7	1.7	1.1
Reducing Substances by Somogyi-Nelson Method	9.4	9.4	—
Free Sugars (By Chromatography)	1.4	1.4	—
Total Sugars (By Chromatography)	1.8	1.8	—
Total Sugars (After Hydrolysis)	—	—	25.0
Lactonizable Hydroxy Acids	4.2	4.2	2.3
Sodium Lignosulfonate (By UV Absorption)	—	—	57.0
Ash	—	—	11.9
Estimated Sulfonate Sulfur (As S)	7.3	7.3	—
pH (1% Water Solution)	6.5	6.5	6.2

*Both expressed on a dry basis

Raylig D
Raylig DL

Chemical Type: Unique, desugared and resulfonated lignosulfonate. Both are available as bulk 45% solutions and brown-colored, free-flowing powders.

Key Properties: Has a sulfonate sulfur content of 7.3% (as S) which is considerably higher than values for commercial crude and desugared lignosulfonates. Accordingly, it is expected that there will be greater compatibility with metal salts over other commercial lignosulfonates and over some synthetic polymer dispersants. Effective dispersants for hydrated silica, talc, kaolin, gypsum and carbon black. For effective use in these applications, pH control will usually be required to give slightly alkaline systems.

Raylig 260L

Chemical Type: Liquid by-product of a highly developed process for chemical cellulose made from the softwoods fo the Pacific Northwest. It is the lignin constituent when combined with the cellulose fiber which gives wood its structural strength.

Key Properties: The unique properties of this material are applicable to a vareity of uses. The binding dispersant and sequestering characteristics find utility in: Adhesive extension, wax emulsions, carbon black slurries and many other uses.

JOHNS-MANVILLE, Ken Caryl Ranch, Denver, CO 80217

Super Fine Super Floss

Chemical Type: Fine particle size, white grade diatomite.

Chemical Analysis

	Percent by Weight		Percent by Weight
Silica (SiO$_2$)	91.2	Potassium Oxide (K$_2$O)	0.8
Alumina (Al$_2$O$_3$)	3.1	Sodium Oxide (Na$_2$O)	2.1
Calcium Oxide (CaO)	0.5	Other	0.4
Ferric Oxide (Fe$_2$O$_3$)	1.0	Loss on Ignition	0.4
Magnesium Oxide (MgO)	0.5		

Particle Size Distribution (Coulter Counter)

Particle Size Finer Than	Percent by Weight	Particle Size Finer Than	Percent by Weight
20 μ	100	4 μ	67
10 μ	100	3 μ	37
8 μ	98	2 μ	12
6 μ	94	1 μ	2

Physical Properties

Color	White	Bulking Value (Weight per Gallon)	19.2 lb
Reflectance, Blue	92	Bulking Value (Gallons per Pound)	0.052
Reflectance, Green	91	Refractive Index	1.43
Reflectance, Amber	92	Moisture (As Shipped)	0.3%
Plus 325 Mesh	Trace	pH	9.4
Oil Absorption (lb/100 lb)	100	Average Particle Size	3.3 μ
Hegman Fineness of Grind	5½		

Key Properties: Designed for the coatings industry for its flatting properties. Can easily be dispersed to Hegman Fineness of over 5½. It is used in high quality paints for low luster, enamel-like smoothness and economy. It can also be used as an antiblocking agent for polyolefin film applications.

MACKENZIE CHEMICAL WORKS, INC., Rt. 2, Box 219-M, Bush, LA 70431

Product Name	Molecular Weight	Melting Point (°C)	Boiling Point (°C)
Ferric Acetylacetonate	353.17	179	—

Chemical Type: Fe(C$_5$H$_7$O$_2$)$_3$, with 15.80% iron, 51.01% carbon, 5.99% hydrogen. Bright, orange-red crystalline powder. 100% through 325 mesh.

Key Properties: Bonding agent for resin-to-metal use. Curing agent for polyurethane resins. Light-fast pigment. Resin stabilizer. Catalyst for wide range of applications. Filler in polymers. Paint drier. Pigment. Many other uses.

Product Name	Molecular Weight	Melting Point (°C)	Boiling Point (°C)
Magnesium Acetylacetonate	222.52	262	—

Chemical Type: Mg(C$_5$H$_7$O$_2$)$_2$, with 10.93% magnesium, 53.93% carbon and 6.29% hydrogen. Creamy white crystals.

Key Properties: Recommended for use as a bactericide and deodorizer. Also, color stabilizer in photography.

Manganic Acetylacetonate	352.24	172	—

Chemical Type: Mn(C$_5$H$_7$O$_2$)$_3$, with 15.6% manganese, 51.2% carbon and 5.98% hydrogen. Shiny black crystals.

Key Properties: Recommended for use as a paint drier. Many other uses.

Aluminum Acetylacetonate	324.31	189	315

Chemical Type: Al(C$_5$H$_7$O$_2$)$_3$, with 8.3% aluminum, 55.5% carbon, 6.5% hydrogen. Sulfate and iron free. White crystalline powder.

Key Properties: Recommended for use as a catalyst for a wide range of applications, bactericide, deodorizer and other uses.

Product Name	Molecular Weight	Melting Point (°C)	Boiling Point (°C)	Specific Gravity @ 25°C
Chromium Acetylacetonate	349.33	214	340	1.34

Chemical Type: Cr(C$_5$H$_7$O$_2$)$_3$, with 14.9% chromium, 51.6% carbon, 6.0% hydrogen. Remarkably stable reddish, violet crystals.

Key Properties: Recommended for use as an oxidation catalyst, stabilizer for nitroparaffins, pigment and many other uses.

Cupric Acetylacetonate	261.74	230	—	—

Chemical Type: Cu(C$_5$H$_7$O$_2$)$_2$, with 24.3% copper, 45.8% carbon, 5.4% hydrogen. Blue crystals.

Key Properties: Bonding agent for resin-to-metal use. Curing accelerator for polyurethane resins. Light-fast pigment. Resin stabilizer. Catalyst for wide range of applications. Filler in polymers. Paint drier. Pigment. Fungicide. Bactericide. Many other uses.

Nickel Acetylacetonate	256.9	228	—	—

Chemical Type: Ni(C$_5$H$_7$O$_2$)$_2$, with 22.84% nickel, 46.75% carbon, 5.49% hydrogen. Bright green crystals.

Key Properties: Recommended for use as: paint drier, pigment, filler in polymers, bactericide and deodorizer, and many other uses.

Product Name	Molecular Weight	Melting Point (°C)
Titanyl Acetylacetonate	262.1	Decomposes at 184

Chemical Type: TiO(C$_5$H$_7$O$_2$)$_2$, with 18.1% titanium, 45.8% carbon, 5.39% hydrogen. Orange crystals.

Key Properties: Recommended for use as: catalyst, filler in polymers and synthesis intermediate.

Vanadium Acetylacetonate	348.25	Distills at 182

Chemical Type: V(C$_5$H$_7$O$_2$)$_3$, with 14.7% vanadium, 51.7% carbon, 6.0% hydrogen. Greenish brown crystals.

Key Properties: Recommended for use as: paint drier and pigment. Other uses.

Product Name	Molecular Weight	Melting Point (°C)
Vanadyl Acetylacetonate	265.15	250

Chemical Type: $VO(C_5H_7O_2)_2$, with 19.3% vanadium, 45.3% carbon, 5.28% hydrogen. Prussian Blue crystals.

Key Properties: Recommended for use as: paint drier and pigment. Other uses.

Zirconium Acetylacetonate	487.62	172

Chemical Type: $Zr(C_5H_7O_2)_4$, with 18.71% zirconium, 49.26% carbon, 6.74% hydrogen. White crystals.

Key Properties: Crosslinking agent for oxygen-containing polymers.

Note: There are other types of metal acetylacetonates available for many other uses.

M. MICHEL AND CO., INC., 90 Broad St., New York, NY 10004

Cachalot Fatty Alcohols

Product Name	Types	Grade	Acid Value (max)	Boiling Range (95%) (°C)	Cloud Point (°C)	Melting Point (°C)	Solidification Point (°C)
O-3	Oleyl	NF	0.2	310–350	5–8	—	—
O-8	Oleyl	CTFA	0.2	300–350	12–18	—	—
O-15	Oleyl	Cosm	0.2	300–350	18–25	—	—
O-27	Oleyl	Tech	0.2	290–350	30–40	—	—
S-56	Stearyl	USP/CTFA	0.3	330–350	—	56–58	—
S-54	Stearyl	USP	0.5	330–350	—	56–60	—
S-53	Stearyl	Tech	0.3	315–350	—	51–54	—
C-52	Cetyl	NF/CTFA	0.3	210–330	—	48–49	—
C-50	Cetyl	NF/CTFA	0.1	310–330	—	47–49	—
C-51	Cetyl	NF	0.5	310–340	—	46–50	—
M-43	Myristyl	Cosm	0.3	280–295	—	—	36–38
L-90	Lauryl	CP	0.1	250–265	—	—	22–24
L-50	Lauryl	Tech	0.1	260–340	—	—	20–24

Product Name	Color Hazen- APHA (max)	Flash Point (°C)	Hydroxyl Value	Iodine Value	Molecular Weight	Refractive Index
O-3	80	180	205	93	268	1.4701
O-8	150	170	210	82.5	268	1.4581
O-15	150	170	213	75	268	1.4568
O-27	150	170	215	50	265	1.4487
S-56	20	190	205	1.0 max	274	1.4348
S-54	25	185	210	2.0 max	274	1.4346
S-53	80	170	213	2.0 max	265	1.4340
C-52	20	165	230	0.5 max	245	1.4320
C-50	20	165	230	1.0 max	248	1.4321
C-51	25	165	230	1.5 max	248	1.4323
M-43	20	150	260	0.3 max	216	1.4335
L-90	10	120	300	0.3 max	187	1.4316
L-50	10	130	273	0.5 max	206	1.4338

Product Name	Saponification Value	Specific Gravity	Viscosity (cp)	Specific Gravity and Viscosity Temperature (°C)
O-3	1.0	0.825	11.7	50
O-8	1.0	0.825	11.5	50
O-15	1.0	0.825	11.2	50
O-27	1.0	0.825	10.7	50
S-56	1.0	0.815	9	70
S-54	1.0	0.815	9	70
S-53	1.0	0.815	8	60/70
C-52	1.0	0.820	7	50/70
C-50	0.5	0.820	7	50/70
C-51	1.0	0.820	7	50/70
M-43	0.5	0.825	9	40/50
L-90	0.5	0.830	8	25/50
L-50	1.0	0.830	8	25/50

Key Properties: Cachalot fatty alcohols have a diversity of applications as adhesive tackifiers, antifoams, antistatic agents, bactericides, coatings, antistatic agents, corrosion inhibitors, cosmetic ingredients, detergents, emulsifiers, evaporation control agents, flotation and extraction agents, fungicides, ink solvents, lubricants, plasticizers, paint flattening agents, pharmaceutical vehicles, polymerization modifiers, quaternary ammonium compounds, soap emollients, textile processing auxiliaries, viscosity index improvers, water and soil repellents, waxes, and many more. As additives, fatty alcohols find use as defoamers, dispersants, emollients, emulsifiers, lubricants, penetrants, plasticizers, solubilizers, stabilizers, super-fatting agents, and thickeners.

Toxicity: (1) Natural, straight-chain, primary alcohols in the C_{12-24} range are considered nontoxic, nonirritating and essentially innocuous.

(2) The C_{16-18} alcohols are normal intermediates in human body fat metabolism and are constituents of human sebum.

(3) The C_{8-10} chain lengths may irritate eyes and skin in some individuals and protective glasses and gloves are recommended. Breathing of hot vapors should be avoided. Normal industrial hygiene should be practiced in handling these chemicals, using water to wash parts of the body that come in contact.

MICRO POWDERS, INC., 1730 Central Park Ave., Yonkers, NY 10710

Product Name	Melting Point (°F)	Maximum Particle Size (µ)	Particle Size (µ)	Specific Gravity @ 77°F	NIPRI Grind Gauge	Color
Fluo 300	600	9	2	2.2	1.5-2.0	White
Fluo HT	620	9	2	2.2	1.0-1.5	White
Fluo HTG	620	9	2	2.2	1.0-1.5	Gray
Fluo HBT	—	9	2	2.2	—	—

Chemical Type: Micronized polytetrafluoroethylene (PTFE), which has been processed to a Hegman fineness of 7.5.

Key Properties: Have been developed for use in areas where excellent heat resistance, slip and rub resistance are required. Easily dispersed into coatings and printing inks. Due to their very fine particle sizes, the Fluos can either be predispersed into a paste or added directly into most coatings and inks with conventional mixing equipment. The Fluos are most effective when used at low levels (½-1%) as an additive to either the MP grades of micronized synthetic waxes or the MPP grades of micronized polyethylene. A low percentage of Fluos gives the most dramatic improvement. The addition of a small amount of Fluos along with the Micro Powders grades of waxes or polyethylenes allows each material to perform with maximum efficiency. The Fluos are especially recommended for paints, coatings, overprint varnishes and many types of inks.

Toxicity: At temperatures above 600°F, some polymer fumes may be emitted due to oxidation of the fluorocarbon polymer. Therefore, smoking should not be permitted when working with this material.

Product Name	Congealing Point (°F)	Specific Gravity @ 77°F
Synfluo 150	255	0.99

Chemical Type: Micronized modified fluorocarbon with average particle size of 3 μ.

Key Properties: Has been processed to impart optimum slip and hardness to most coatings and inks. Supplied as a fine white powder that corresponds to a 3 NPIRI grind gauge reading. This fine particle size will eliminate the need for melt compounding or grinding. Can be readily dispersed into most ink vehicles by means of normal high speed mixing. Synfluo is recommended for use in lacquers, overprint varnishes and many inks. Generally, 1-3% by weight will provide optimum properties.

Toxicity: At temperatures below 600°F, Synfluo 150 will not emit any fluorocarbon polymer fumes. However, at temperatures above 600°F, some polymer fumes may be emitted due to oxidation. Therefore, smoking should not be permitted when working with Synfluo 150.

MOBAY CHEMICAL PRODUCTS, Plastics and Coatings Division, Pittsburgh, PA 15205

Mondur CB-Type Products

Chemical Type: Urethane products.

Physical Properties

	Mondur CB-60	Mondur CB-75
% NCO	10.0-10.8	12.5-13.5
Average Equivalent Weight	404	325
% Solids	60	75
Solvents:		

CB-60: Ethyl glycol acetate/xylol (25:15 parts by weight)
CB-75: Ethyl acetate

Key Properties: These products have been used as a component in the formulation of urethane coatings, adhesives and binders in many diverse and demanding applications which require long-term protection of products and equipment in corrosive exposures.

MOBIL CHEMICAL CO., P.O. Box 26683, Richmond, VA 23261

Dialkyl Alkylphosphonates

Product Name	Boiling Point (°C)	Specific Gravity 20°/4°C	Flash Point (°F)	Flash Point Method*	Solubility in Water
Dimethyl methyl	92	1.160	144	TCC	Soluble
Diethyl ethyl	82	1.025	195	TCC	Soluble
Dibutyl butyl	127	0.948	280	COC	Insoluble
Bis(2-ethylhexyl)	340	0.908	340	PM	Insoluble

*Flash Point Methods: COC, Cleveland Open Cup; TCC, Tag Closed Cup; and PM, Pensky Martens Closed Cup

Chemical Type (of all): Stable organic phosphorus compounds which are colorless liquids with mild odors.

Key Properties (of all): Miscible with alcohol, ether and most organic solvents. Has the following end uses:

(1) Heavy metal extraction and solvent separation.
(2) Preignition additives to gasoline.
(3) Antifoam agents.
(4) Plasticizers and stabilizers.
(5) Additives in solvents and low temperature hydraulic fluids.
(6) To replace organic phosphates for improved stability.

Dialkyl Phosphites

Product Name	Boiling Point (°C)	Specific Gravity 20°/4°C	Flash Point (°F)	Flash Point Method	Solubility in Water
Dimethyl Phosphite	72	1.200	140	TCC	Soluble
Diethyl Phosphite	65	1.079	165	TCC	Soluble
Diisopropyl Phosphite	70	0.985	164	TCC	Soluble
Dibutyl Phosphite	118	0.995	259	COC	Sl. Soluble
Bis(2-ethylhexyl)Phosphite	163	0.937	334	COC	Insoluble
Dilauryl Phosphite	>200	0.910	320	PM	Insoluble
Bis(tridecyl)Phosphite	>200	0.910	356	COC	Insoluble
Dioleyl Phosphite	>200	0.900	240	PM	Insoluble

Chemical Type (all types): Mobile colorless liquids with mold odors and good thermal stability.

Key Properties (all types): Miscible with alcohol, ether, and most common organic solvents. Used as additives for extreme pressure lubricants, organic phosphorus intermediates, adhesives and antioxidants.

Organic Acid Phosphates

Product Name	Acid No. (min)	Specific Gravity 20°/4°C	Flash Point (°F)	Flash Point Method	Solubility in Water
Butyl	430	1.13	265	TCC	Sl. Soluble

Form: Liquid.

| Isooctyl | 305 | 1.02 | 243 | PM | Insoluble |

Form: Liquid.

| Phenyl | 360 | 1.28 | 338 | TCC | Sl. Soluble |

Form: Light tan semisolid.

| Octyl phenyl | 200 | 1.08 | 316 | COC | Insoluble |

Form: Light tan semisolid.

| PA-75 | 270 | 1.19 | 86 | PM | Sl. Soluble |

Form: 75% solution of phenyl acid phosphate in butyl alcohol.

| 2-Ethylhexyl | 305 | 1.07 | 260 | PM | Insoluble |

Chemical Type (of all): Mixed mono- and dihydrogen phosphate esters which are strong acids and form salts with alkalies and amines.

Key Properties (of all): They have higher purity and lighter color and offer a wide range of solubilities and thermal stabilities. Recommended for the following end uses:

(1) Acid catalyst in resin curing.
(2) Chemical intermediates in the formation of rust preventatives, antistatic agents, textile lubricants, oil additives and heavy metal extraction.

Trialkyl Phosphates

Product Name	Boiling Point (°C)	Specific Gravity 20°/4°C	Flash Point (COC)	Solubility in Water
Tributyl Phosphate	139	0.98	330	Insoluble

Key Properties: Used commercially in fire resistant aircraft hydraulic fluids, metal extractions, antifoam applications for paper and protective coatings, and as a low temperature plasticizer.

Tributoxyethyl Phosphate	222	1.018	435	Insoluble

Key Properties: Widely used as a leveling agent for synthetic floor polishes.

Trioctyl Phosphate	220	0.923	220	Insoluble

Chemical Type (of all): Clear, mobile liquids with little color and odor.

Key Properties (of all): Miscible in many organic solvents. These water-insoluble products show good thermal and hydrolytic stability. Major uses include plasticizers, dispersants, flame retardants and antifoaming agents.

PACIFIC SMELTING CO., 22219 South Western Ave., Torrance, CA 90510

Pasco Zinc Oxide

Chemical Type: Six standard grades of zinc oxide to ASTM Spec. D79, which satisfy most industrial applications, using the production technique, commonly referred to as the French Process. Intermediate processing is readily available when surface-treated zinc oxide requires coating with other reagents.

Chemical Analysis (Percent by Weight)

	20	22	31	35	42	45
Zinc Oxide	99.9	99.7	99.6	99.6	99.2	99.4
Lead as PbO	0.003	0.02	0.05	0.04	0.10	0.10
Cadmium as CdO	0.001	0.005	0.005	0.005	0.008	0.005
Copper	<0.0005	<0.0005	<0.0005	<0.0005	<0.0006	<0.0005
Manganese	Nil	Nil	Nil	Nil	Nil	Nil
Iron	0.0002	0.0004	0.0004	0.0004	0.0007	0.0004
Total Sulfur as SO_3	Nil	Nil	Nil	Nil	Nil	Nil
Water-Soluble Salts	0.01	0.03	0.03	0.03	0.06	0.05
Diluted Acetic Acid Insolubles	0.01	0.07	0.09	0.09	0.12	0.31
Loss at 110°C	0.10	0.10	0.10	0.10	0.12	0.10
Ignition Loss	0.10	0.12	0.13	0.13	0.17	0.17

Physical Properties

	20	22	31	35	42	45
Mean Particle Size, μ	0.33	0.25	0.25	0.11	0.20	0.11
Specific Surface Area, m^2/g	3.2	4.3	4.3	9.5	5.4	9.5
Specific Gravity	5.6	5.6	5.6	5.6	5.6	5.6
Weight per Gallon, lb	46.7	46.7	46.7	46.7	46.7	46.7
One Pound Bulks, gal	0.0214	0.0214	0.0214	0.0214	0.0214	0.0214
Rub-Out Oil Absorption	14	18	18	20	18	20
Apparent Density, lb/ft^3	30.0	25.0	25.0	14.0	25.0	14.0
Fineness Through 325 Mesh, %	99.98	99.95	99.95	99.95	99.85	99.85
Appearance	Cream White	Cream White	Cream White	Cream White	Cream White	White Off-White

Key Properties: Consistency of quality and uniformity. Grade No. 20 Zinc Oxide is a medium particle size oxide for products where high purity is required. Besides the standard grades, a No. 200 grade which is suitable for animal feed and fertilizer industry applications is also produced.

Specifications: All grades conform to ASTM D79 and Federal TT-P-463a specifications.

RAYBO CHEMICAL CO., Huntington, WV 25722

Raybo 60-NoRust

Chemical Type: Proprietary composition to eliminate can corrosion.

Physical Properties

Appearance	Clear, straw colored liquid	pH	9.35
Viscosity (Gardner)	A-3	Weight per Gallon	9.3 lb
% Solids Content	21–23	Specific Gravity	1.14

Key Properties: Raybo 60 is designed to prevent can corrosion in all types of latices. It was developed with both contact and volatile water-soluble corrosion inhibitors in a water solution which prevents rusting both above and below the liquid level. The addition of Raybo 60 actually will eliminate the need for painted or lacquer lined containers along with the elimination of the unsightly appearance of the paint when the lid is removed. Raybo 60 solves the problem of the unsightly brownish appearance on top of a closed container. Conforms to Rule 66.

SHERWIN-WILLIAMS CO., 1500 Higgins Road, Park Ridge, IL 60068

2,6-Dimethylol-para-Cresol (DMPC)

Chemical Type (synonyms): 2-Hydroxy-5-methyl-1,3-benzenedimethanol, 2-hydroxy-α^1, α^3-mesitylenediol, 2-hydroxy-5-methyl-m-xylene-α, α'-diol and 2,6-dimethylol-4-methylphenol. Formula: $C_9H_{12}O_3$. Molecular weight: 168.19.

Physical Properties

Appearance	Beige to pink powder
Assay	90.0% minimum
Melting Range*	118°–127°C
Moisture	1.0% max

*DMPC undergoes a reaction upon melting

Key Properties: Can be used as the following:

(1) Antioxidant or stabilizer—Has been shown to work in combination with a wetting agent to stabilize rubber latexes and other stabilizer uses. Selective derivatives of DMPC, in combination with other chemicals, have also been used as antioxidants.

(2) Crosslinking or Curing Agent—Useful as a crosslinking or curing agent in many polymer systems.

(3) Polymer Component—Readily reacts with other appropriate monomers forming copolymers.

(4) Surface Treating Agent—Has been used to treat surfaces to improve bonding or adhesion between minerals and plastics or between cellulosics and rubber. Can be used to treat fibers also.

(5) Miscellaneous Uses—Can function as a chemical intermediate to react with other compounds to produce materials for many uses.

Toxicity: The following data shows that DMPC is probably not a serious industrial hazard:

LD_{50} Oral (Rats)—5,800 mg/kg.
LD_{50} Dermal (Rabbits)—>2 g/kg.
LC_{50} Inhalation (One Hour) (Rats)—>2.9 mg/ℓ (Aerosol)*
Ocular Irritation (Rabbits)—100 mg exposure up to 72 hours; no significant irritation.

*Actual concentration measured in the breathing zone.

Nacopa

Chemical Type: Monosodium-4-chlorophthalate ($C_8H_4ClNaO_4$), with a molecular weight of 222.57.

Chemical Analysis

Assay	65% minimum
Moisture	2% maximum
Sodium Chloride	7% maximum

Key Properties: Recommended for use as a modifying agent for phthalocyanine pigments to retard crystallization. Resin modifier.

Toxicity

Oral LD_{50} (Rats)—5,300 mg/kg.
Dermal LD_{50} (Rabbits)—2 g/kg.
Inhalation LC_{50} (Actual concentration measured in breathing zone) (1 Hour) (Rat)—2.63 mg/ℓ.
Eye Irritation (Rabbit)—Caused irritation.
Safe Handling—As with any chemical, good hygienic practices in handling should be maintained.

TEXACO, INC., 2000 Westchester Ave., White Plains, NY 10650

Jeffersol Glycol Ethers

Methyl Ethers:
Jefferson EM: Ethylene glycol monomethyl ether
$(CH_3OCH_2CH_2OH)$;
Molecular weight: 76.09
Jeffersol DM: Diethylene glycol monomethyl ether
$(CH_3OCH_2CH_2OCH_2CH_2OH)$
Molecular weight: 120.15

Ethyl Ethers:
Jeffersol EE: Ethylene glycol monoethyl ether
$(CH_3CH_2OCH_2CH_2OH)$
Molecular weight: 90.12
Jeffersol DE: Diethylene glycol monoethyl ether
$(CH_3CH_2OCH_2CH_2OCH_2CH_2OH)$
Molecular weight: 134.17

Butyl Ethers:
Jeffersol EB: Ethylene glycol monobutyl ether
$(CH_3CH_2CH_2CH_2OCH_2CH_2OH)$
Molecular weight: 118.17
Jeffersol DB: Diethylene glycol monobutyl ether
$(CH_3CH_2CH_2CH_2OCH_2CH_2OCH_2CH_2OH)$
Molecular weight: 162.22

All are clear liquids with mild, pleasant odors.

	EM	DM	EE	DE	EB	DB
Acidity as Acetic Acid (max weight %)	0.01	0.01	0.01	0.01	0.01	0.01
Appearance	. . . . . Clear, substantially free of suspended matter					
Boiling Range (ASTM, °C):						
Initial Boiling Point (min)	123.5	188	134	198	169	220
Dry Point (max)	125.5	198	136	205	173	235
Color (Pt-Co) (max)	10	10	10	10	10	15
Odor at Room Temperature	. . . . .Mild and not objectionable					
Specific Gravity @ 20°/20°C (min)	0.964	1.020	0.929	0.990	0.901	0.953
Specific Gravity @ 20°/20°C (max)	0.967	1.025	0.933	0.994	0.904	0.958
Water (wt %) (max)	0.2	0.2	0.1	0.1	0.2	0.2
Flash Point (°F)	115*	200**	118*	210**	165*	225*
Flash Point (TCC) (°F)	106	194	110	194	141	230
Freezing Point (°C)	-85	—	—	—	—	-68.1

	EM	DM	EE	DE	EB	DB
Viscosity, cp @ 20°C	1.7	3.9	2.0	3.85	6.4	6.49
Weight per Gallon (lb) @ 20°C	8.03	8.56	7.75	8.24	7.50	7.93

*Tag Open Cup (TOC)
**Cleveland Open Cup (COC)

Key Properties: The glycol ethers combine the solubility characteristics of alcohols, ethers and hydrocarbons, since hydroxyl, ether and alkyl groups are present in each molecule. This accounts for the unusual and desirable solvent properties on which most of the commercial applications of glycol ethers depend. They are miscible with water and most organic solvents. Widely used for their excellent solvent properties as solvents or coupling agents for surface coatings, hydraulic fluids, plasticizers, cleaning compounds, textile dyes, cutting and soluble oils and in many other aqueous organic systems.

Thanol Polyols

Product Name	Acid No. (mg KOH/g) (max)	Flash Point (Closed Cup) (°F)	Hydroxyl No. (mg KOH/g)	pH (10:6 Isopropanol: Water Solvent)	Specific Gravity (20°/20°C)	Viscosity (Brookfield) @ 77°F (cp)
Diols:						
Thanol C-150	—	330	743	—	1.261	700
Thanol E-2101	0.1	405	56.0	7.0	1.062	400
Thanol PPG-2000	0.1	370	56.0*	6.0	1.005	370
Triols:						
Thanol F-3000	0.1	385	56.2	6.5	1.009	560
Thanol FM-3511	0.1	455	48.0	6.5	1.017	600
Thanol F-3016	0.06	450	56.0	6.5	1.015	500
Thanol F-3520	0.05	415	47.0	6.5	1.019	540
Thanol SF-5505	0.03	380*	33.5	8.75	1.022	900
Thanol SF-265**	—	360	635	—	1.078	430
Thanol R-350-X**	—	315	530	—	1.116	14,500
Multifunctional Polyols:						
Thanol SF-2750**	—	345	220	10.5	1.0468	2,600
Thanol TR-380**	—	405	295	7.3	1.1409	450
Thanol R-650-X**	—	305	450	—	1.059	29,500
Thanol R-480**	—	300	530	—	1.136	15,000
Thanol R-600**	—	260	470	—	1.100	13,500

*In 10:1 methanol:water solvent
**Contains tertiary nitrogen

UNIROYAL CHEMICAL, Division of Uniroyal, Inc., Naugatuck, CT 06770

Uniroyal Antioxidants

Naugard 445

Chemical Type: Aromatic amine. White powder form.

Physical Properties: Specific Gravity, 1.126 and Melting Range, 96°–99°C.

Key Properties: An excellent processing and heat stabilizer for polyolefins, polyacetal, nylon 6, etc., EVA and polyamide hot melts. Unique for an amine in that it is completely nondiscoloring under heat aging conditions. It is mildly discoloring under severe light-aging conditions. FDA approved for adhesives.

Naugard PHR

Chemical Type: Naugard P + 1% triisopropanolamine. Straw colored liquid.

Physical Properties: Specific Gravity, 0.98.

Key Properties: More resistant to hydrolysis than Naugard P. Recommended for use where contact with moisture at elevated temperatures is possible. FDA approved for adhesives.

Naugard P

Chemical Type: Tri(mono- and dinonylphenyl)phosphite. Straw colored liquid.

Physical Properties: Specific Gravity, 0.98.

Key Properties: A versatile processing and color stabilizer for a variety of polymers: polyolefins, impact polystyrene, ABS, polycarbonate, PVC, SAN and EVA. Naugard P is FDA approved, nondiscoloring, stable at high temperatures and effective at low concentrations.

Naugawhite

Chemical Type: A general purpose antioxidant for thermoplastics. Nonstaining, resistant to volatility. Recommended for petroleum resins, styrene based polymers and EVA hot melt adhesives.

Physical Properties: Specific Gravity, 0.96.

Key Properties: A general purpose antioxidant for thermoplastics. Nonstaining, resistant to volatility. Recommended for petroleum resins, styrene based polymers and EVA hot melt adhesives. FDA approved for adhesives.

Naugard SP

Chemical Type: Phenolic.

Key Properties: An economical, nondiscoloring, general purpose stabilizer with low volatility compared to Naugard BHT.

Naugard BHT

Chemical Type: Di-tert-butyl-p-cresol. White crystals.

Physical Properties

	Technical	*Food Grade*
Specific Gravity	0.899	0.899
Melting Range (°C)	68–71	69.4–71

Key Properties: Food Grade—A widely used antioxidant and processing stabilizer in the plastics industry. Nondiscoloring, FDA approved and highly active, Naugard BHT is often used in combination with Naugard P for optimum stability.

Technical—Slightly lower in purity than Naugard BHT food grade. Used when FDA approval is not necessary.

VIRGINIA CHEMICALS, INC., 3340 W. Norfolk Rd., Portsmouth, VA 23703

Virset 656-4

Chemical Type: Improved, highly reactive melamine condensate insolubilizer, in clear, water-white form.

Chemical Analysis and Physical Properties

Nonvolatile Content (P_2O_5 Dehydration Method)	69–71%
Weight per Gallon @ 25°C	10 lb
Brookfield Viscosity @ 25°C	120–250 cp
Solubility in Water	Infinite
pH (Phthalein Red Indicator Method)	9–10

Key Properties: Used to impart maximum wet-rub and water resistance to coatings and adhesive systems containing starch, protein, polyvinyl alcohol and latex or combinations thereof. It is believed that Virset 656-4 provides more reactivity per unit of resin solids than any other commercially available insolubilizer. Another important property is its lack of viscosity build-up in sensitive adhesive systems.

FDA Status: Ingredients approved.

Section XLVI
Suppliers' Addresses

Abbott Laboratories
North Chicago, IL 60064
(312)-688-5160

Allied Colloids, Inc.
161 Dwight Place
Fairfield, NJ 07006
(201)-227-7750

American Bio-Synthetics Corp.
710 West National Ave.
Box 04275
Milwaukee, WI 53204
(414)-384-7017

American Cyanamid Co.
Wayne, NJ 07470
(201)-831-1234

The Ames Laboratories, Inc.
200 Rock Lane
Milford, CT 06460
(203)-874-2463

Amoco Chemicals Corp.
200 East Randolph Drive
Chicago, IL 60601
(312)-856-3414

Anglo-American Clays Corp.
14 Executive Park Drive
Atlanta, GA 30329
(404)-321-6304

ASARCO, Inc.
120 Broadway
New York, NY 10005
(212)-732-9500

The Asbury Graphite Mills, Inc.
Asbury, Warren County, NJ 08802
(201)-537-2155

Bareco Division
6910 East 14th St.
P.O. Drawer K
Tulsa, OK 74112
(918)-836-1601

BASF Corp.
491 Columbia Ave.
Holland, MI 49423
(616)-392-2391

Burgess Pigment Co.
P.O. Box 349
Sandersville, GA 31082
(912)-552-2544

Cabot Corp.
125 High St.
Boston, MA 02110
(617)-423-6000

Calcium Carbonate Co.
Division of J.M. Huber Corp.
P.O. Box 4005
3150 Gardner Expressway
Quincy, IL 62301
(217)-224-1100

Cargill, Inc.
Chemical Products Division
P.O. Box 9300
Minneapolis, MN 55440
(612)-475-7575

Chemical Components, Inc.
20 Deforest Ave.
P.O. Box 291
East Hanover, NJ 07936
(201)-887-7750

W.A. Cleary Chemical Corp.
P.O. Box 10
1049 Somerset St.
Somerset, NJ 08873
(201)-247-8000

Colloids, Inc.
394-8 Frelinghuysen Ave.
Newark, NJ 07114
(201)-243-8500

Commercial Minerals Co.
6899 Smith Ave.
P.O. Box 363
Newark, CA 94560
(415)-797-8080

Composition Materials Co., Inc.
26 Sixth St.
Stamford, CT 06905
(203)-359-0200

Cosan Chemical Corp.
400 Fourteenth St.
Carlstadt, NJ 07072
(201)-460-9300

Crosby Chemicals, Inc.
600 Whitney Bldg.
New Orleans, LA 70130
(504)-581-7047

Cyprus Industrial Minerals Co.
555 South Flower St.
Los Angeles, CA 90071
(213)-489-3700

Daniel Products Co.
400 Claremont Ave.
Jersey City, NJ 07304
(201)-432-0800

Davison Chemical Division
W.R. Grace & Co.
P.O. Box 2117
Baltimore, MD 21203
(301)-659-9010

Degussa Corp.
Pigments Division
Route 46 at Hollister Road
Teterboro, NJ 07608
(201)-288-6500

Diamond Shamrock Corp.
P.O. Box 2386R
Morristown, NJ 07960
(201)-267-1000

Dow Chemical U.S.A.
Midland, MI 48640
(517)-636-1000

Dow Corning Corp.
Midland, MI 48640
(517)-496-4000

Dover Chemical Corp.
Davis at West Fifteenth St.
P.O. Box 40
Dover, OH 44622
(216)-343-7711

Emery Industries, Inc.
1300 Carew Tower
Cincinnati, OH 45202
(513)-762-6200

Emkay Chemical Co.
319-325 Second St.
Elizabeth, NJ 07206
(201)-352-7695

The English Mica Co.
Ridgeway Center Bldg.
Stamford, CT 06905
(203)-324-9531

Essential Chemicals Corp.
Merton, WI 53056
(414)-691-3000

Ferro Chemical Division
7050 Krick Road
Bedford, OH 44146
(216)-641-8580

Filter-Media Co.
P.O. Box 19156
Houston, TX 77024
(713)-780-9000

Flintkote Stone Products Co.
Executive Plaza IV
Hunt Valley, MD 21031
(301)-628-4000

Franklin Mineral Products Co.
635 Main St.
P.O. Box O
Wilmington, MA 01887
(617)-658-2310

GAF Corp.
Chemical Division
140 West 51 St.
New York, NY 10020
(212)-582-7600

General Electric Co.
Silicone Products Division
Waterford, NY 12188
(518)-237-3330

Georgia Kaolin Co.
433 North Broad St.
Elizabeth, NJ 07207
(201)-352-9800

Georgia Marble Co.
2575 Cumberland Parkway, N.W.
Atlanta, GA 30339
(404)-432-0131

Glidden Pigments
Chemical/Metallurgical Division
SCM Corp.
2700 Hollins Ferry Road
Baltimore, MD 21230
(301)-633-6400

Gold Bond Building Products
2001 Rexford Road
Charlotte, NC 28211
(704)-365-0950

W.R. Grace & Co.
Organic Chemicals Division
55 Hayden Ave.
Lexington, MA 02173
(617)-861-6600

Great Lakes Chemical Corp.
P.O. Box 2200
Highway 52 N.N.
West Lafayette, IN 47906
(317)-463-2511

Grefco, Inc.
Minerals Division
3450 Wilshire Blvd.
Los Angeles, CA 90010
(213)-381-5081

Gross Minerals Corp.
P.O. Box 116
Aspers, PA 17304
(717)-677-7313

Halox Pigments
425 Manor Oak One
1910 Cochran Road
Pittsburgh, PA 15220
(412)-344-5811

Harris Mining Co.
P.O. Box 628
Spruce Point, NC 28777
(704)-765-4251

Henkel Corp.
425 Broad Hollow Road
Melville, Long Island, NY 11746
(516)-293-5215

Hercules, Inc.
910 Market St.
Wilmington, DE 19899
(302)-575-6500

J.M. Huber Corp.
P.O.Box 310
Havre de Grace, MD 21078
(301)-939-3500

Humphrey Chemical Corp.
P.O. Box 2
Edgewood Arsenal, MD 21010
(301)-676-1000

ICI Americas, Inc.
Concord Pike & New Murphy Rd.
Wilmington, DE 19897
(302)-575-3000

Illinois Minerals Co.
2035 Washington Ave.
Cairo, IL 62914
(618)-734-4172

Indusmin, Ltd.
365 Bloor St. East
Suite 200
Toronto, Ontario, Canada M4W 1H7
(416)-967-1900

O.G. Innes Corp.
10 East 40th St.
New York, NY 10016
(212)-679-6180

International Dioxcide, Inc.
136 Central Ave.
Clark, NJ 07066
(201)-499-9660

Interstab Chemicals, Inc.
A Part of Akzo
500 Jersey Ave.
P.O. Box 638
New Brunswick, NJ 08903
(201)-247-2202

Isochem
Cook St.
Lincoln, RI 02865
(401)-723-2100

ITT Rayonier, Inc.
1177 Summer St.
Stamford, CT 06904
(203)-348-7000

Johns-Manville
Ken Caryl Ranch
Denver, CO 80217
(303)-979-1000

Kay-Fries, Inc.
Member Dynamit Nobel Group
200 Summit Ave.
Montvale, NJ 07645
(201)-573-1050

Kelco Division
Merck & Co., Inc.
20 North Wacker Drive
Chicago, IL 60606
(312)-372-1352

LDL Technology, Inc.
137 Pennsylvania Ave.
Paterson, NJ 07503
(201)-345-9111

Mackenzie Chemical Works, Inc.
Rt. 2-Box 219-M
Bush, LA 70431
(504)-886-5848

Madison Industries, Inc.
P.O. Box 175
Old Bridge, NJ 08857
(201)-727-2225

Mayco Oil and Chemical Co., Inc.
Beaver & Canal Streets
Bristol, PA 19007
(215)-788-0862

McCloskey Varnish Co.
7600 State Road
Philadelphia, PA 19136
(215)-624-4400

McWhorter Resins
International Minerals & Chemical Corp.
P.O.Box 308
Cottage Place
Carpentersville, IL 60110
(312)-428-2657

The Meadowbrook Corp.
30 Rockefeller Plaza
New York, NY 10020
(212)-582-0420

M. Michel and Co., Inc.
90 Broad St.
New York, NY 10004
(212)-344-3878

Micro Powders, Inc.
1730 Central Park Ave.
Yonkers, NY 10710
(914)-793-4058

Minerals and Chemicals Division
Engelhard Minerals & Chemicals Corp.
Menlo Park
Edison, NJ 08817
(201)-321-5000

Mississippi Lime Co.
Alton, IL 62002
(618)-465-7741

Mobay Chemical Products
Plastics and Coatings Division
Pittsburgh, PA 15205
(412)-777-2000

Mobil Chemical Co.
P.O. Box 26683
Richmond, VA 23261
(804)-798-4291

Mona Industries
65 East 23 St.
Patterson, NJ 07524
(201)-274-8220

Mooney Chemicals, Inc.
2301 Scranton Road
Cleveland, OH 44113
(216)-781-8383

NL Chemicals
NL Industries, Inc.
P.O. Box 700
Hightstown, NJ 08520
(609)-443-2000

Nalco Chemical Co.
2901 Butterfield Road
Oak Brook, IL 60521
(312)-887-7500

Nyacol, Inc.
Megunco Road
Ashland, MA 01721
(617)-881-2220

Ottawa Chemical Division
Ferro Corp.
700 North Wheeling St.
Toledo, OH 43605
(419)-691-3507

Pacific Anchor Chemical Co.
1145 Harbour Way South
Richmond, CA 94804
(415)-233-7660

Pacific Smelting Co.
22219 South Western Ave.
Torrance, CA 90510
(213)-775-3421

Pennsylvania Glass Sand Corp.
Three Penn Center
Pittsburgh, PA 15235
(412)-243-7500

Pfizer Minerals, Pigments & Metals Division
Pfizer, Inc.
640 North 13th St.
Easton, PA 18042
(215)-253-6261

Pilot Chemical Co.
11756 Burke St.
Santa Fe Springs, CA 90670
(213)-723-0036

Pioneer Division
Witco Chemical Corp.
277 Park Ave.
New York, NY 10017
(212)-644-6364

Polymer Research Corp. of America
2186 Mill Ave.
Brooklyn, NY 11234
(212)-444-4300

PPG Industries, Inc.
One Gateway Center
Pittsburgh, PA 15222
(412)-434-3131

Procter and Gamble Distributing Co.
P.O. Box 599
Cincinnati, OH 45201
(513)-562-2655

PVO International, Inc.
World Trade Center
San Francisco, CA 94111
(800)-227-4652

Raybo Chemical Co.
Huntington, WV 25722
(304)-525-5171

Reheis Chemical Co.
Division of Armour Pharmaceutical Co.
235 Snyder Ave.
Berkeley Heights, NJ 07922
(201)-464-1500

Reichhold Chemicals, Inc.
RCI Bldg.
White Plains, NY 10603
(914)-682-5700

Rohm and Haas Co.
Independence Mall West
Philadelphia, PA 19105
(215)-592-3000

The Shepherd Chemical Co.
4900 Beech St.
Cincinnati, OH 45212
(513)-731-1110

Sherwin-Williams Co.
Chemicals Division
1500 Higgins Road
Park Ridge, IL 60068
(312)-823-1180

Stauffer Chemical Co.
Specialty Chemical Division
Westport, CT 06880
(203)-222-3000

Synthron, Inc.
44 East Ave.
Pawtucket, RI 02860
(401)-723-4567

Tammsco, Inc.
North Front St.
Tamms, IL 62988
(618)-747-2327

Chas. S. Tanner Co.
Box 1848
Greenville, SC 29602
(803)-277-7080

Texaco, Inc.
2000 Westchester Ave.
White Plains, NY 10650
(914)-253-4000

Thiokol/Ventron Division
150 Andover St.
Danvers, MA 01923
(617)-774-3100

Thompson-Hayward Chemical Co.
A North American Philips Co.
P.O. Box 2383
Kansas City, KS 66110
(913)-321-3131

Ultra Adhesives, Inc.
460 Straight St.
Paterson, NJ 07543
(201)-279-1306

Union Camp Corp.
1600 Valley Road
Wayne, NJ 07470
(201)-628-9000

Union Chemicals Division
Union Oil Co. of California
1345 N. Meacham Road
Schaumburg, IL 60196
(312)-885-5461

U.S. Mica Co., Inc.
26 Sixth St.
Stamford, CT 06905
(203)-324-9531

Uniroyal Chemical
Division of Uniroyal, Inc.
Naugatuck, CT 06770
(203)-723-3000

R. T. Vanderbilt Co., Inc.
30 Winfield St.
Norwalk, CT 06855
(203)-853-1400

Virginia Chemicals, Inc.
3340 W. Norfolk Rd.
Portsmouth, VA 23703
(804)-483-7000

Vikon Chemical Co., Inc.
P.O. Box 1520
Burlington, NC 27215
(919)-226-6331

S. Winterbourne & Co., Inc.
P.O. Box 316-E
Rahway, NJ 07065
(201)-352-6666

Witco Chemical
Organics Division
277 Park Ave.
New York, NY 10017
(212)-644-6454

Other Noyes Publications

HANDBOOK OF PAINT RAW MATERIALS

by Ernest W. Flick

This handbook contains descriptions of numerous raw materials currently available to the paint industry. It is the most complete and up-to-date listing of its type on the market today. It will be a valuable tool to technical and managerial personnel in the coatings field, useful to those with extensive experience as well as those new to the field.

The book consists of data on paint raw materials, obtained from the manufacturers but prepared with no influence from nor cost to the manufacturer or distributor. Only the most recent data are used and, basically, only trademarked raw materials are included.

Raw material descriptions include the following, as available: **chemical name and grade, chemical type, chemical analysis, physical properties, key properties**—those the supplier considers of special importance, **specifications and regulations** met by the raw material, and **toxicity.** Suppliers' addresses are included at the beginning of each section, and a complete alphabetical listing is given in the last chapter.

The table of contents is organized so as to serve as a basic raw material and company index. Each raw material subject category is a chapter which is further subdivided by company and then by specific product. The **condensed table of contents below lists the chapter titles.**

ISBN 0-8155-0881-6 (1982)

340 pages